Programming ASP.NET for ArcGIS Server

Vincent Zhuang
Dave Wrazien
Minhua Wang
Xiaowen Huang

THOMSON
DELMAR LEARNING

Australia Canada Mexico Singapore Spain United Kingdom United States

THOMSON

DELMAR LEARNING

OnWord Press

Programming ASP.NET for ArcGIS Server

Vincent Zhuang, Dave Wrazien, Minhua Wang, and Xiaowen Huang

Vice President, Technology and Trades SBU:
Alar Elken

Editorial Director:
Sandy Clark

Senior Acquisitions Editor:
James DeVoe

Senior Development Editor:
John Fisher

Marketing Director:
Dave Garza

Channel Manager:
Dennis Williams

Marketing Coordinator:
Stacey Wiktorek

Production Director:
Mary Ellen Black

Production Manager:
Andrew Crouth

Senior Production Editor:
Stacy Masucci

Technology Project Manager:
Kevin Smith

Technology Project Specialist:
Linda Verde

Printed in Canada
1 2 3 4 5 XXXX 08 07 06 05

For more information, contact:
Thomson Delmar Learning
Executive Woods,
5 Maxwell Drive, PO Box 8007
Clifton Park, NY 12065-8007.
Or find us on the World Wide Web at
http://www.delmarlearning.com

ISBN: 1-4180-1875-9

About the Authors

Vincent Zhuang has been working in web GIS as a senior software architect, GIS analyst, and GIS specialist for more than nine years. He is a GIS lead at the Science Application International Corporation (SAIC) in Falls Church, Virginia. He started EnviroMapper in 1997—the enterprise-wide web GIS tool for the Environmental Protection Agency (EPA)—and continuously led the development of the EnviroMapper suite of tools for the EPA from 1998 to 2004. Vincent has promoted the building and consuming of geospatial web services throughout his projects over the years and has helped facilitate e-Gov among agencies and with commercial vendors. He has also promoted the delivery of GIS content to mobile devices, including PDAs, cellular phones, and BlackBerry devices. Vincent holds a Master's degree and Ph.D. in Remote Sensing and GIS from Purdue University at West Lafayette, Indiana. Dr. Zhuang has served as a professional journal reviewer for the International Journal of Remote Sensing since 1993.

David R. Wrazien has more than 20 years of experience in the GIS industry and is currently employed as a Senior Solution Architect at Environmental Systems Research Institute (ESRI) of Redlands, California. As an accomplished programmer and technologist, Dave has spent eight years designing, building, and deploying web-based GIS applications for organizations in both the public and private sectors. Most recently, Dave has worked with industry leaders such as SAP, IBM, Microsoft, and Information Builders to deliver server-based enterprise GIS capabilities to core IT platforms and applications. Dave received his Bachelor's degree from the University of Delaware and his Master's degree from the University of Wyoming.

Minhua Wang is a system architect, data modeler, and GIS developer with more than 18 years of experience in the IT and GIS industries—in particular including extensive experience with ArcGIS. He is currently a technical architect for GeoDecisions, a GIS/IT consulting company. Minhua holds a Ph.D. in GIS from the University of Waterloo, Canada. Dr. Wang provides leadership in enterprise architecture design, data modeling, system development for client/server applications, desktop applications and web applications, and enterprise GIS system integration.

Xiaowen Huang has 10 years of experience in GIS and is currently employed as a Senior GIS Specialist at the Science Application International Corporation (SAIC) in Falls Church, Virginia. She holds a Master's degree in Geography from the University of North Carolina at Charlotte, North Carolina, and has developed numerous Internet mapping applications.

Acknowledgments

We would like to acknowledge the following people in a chronological order for their kindness and support in making this book a reality. We would first like to thank Clint Brown for writing such a friendly foreword for the first edition of this book. He spent his valuable time reviewing and conveying the history of GIS from the desktop-only environment of the past to the server-based enterprise GIS of today. We feel fortunate that we are "at the forefront of this GIS revolution" in bringing this book to publication.

Special thanks are due Dennis Smith for his support for Vincent Zhuang's initial idea of writing a book of web GIS development; Christian Harder, who helped facilitate the initial idea and found a match for it; John Sun for his valuable comments on the initial proposal; Roxanne Sandoval for getting us the evaluation copy of the software in a timely fashion; and Andy Potts for making a great effort to get our questions answered.

No book can be published without the help of a great technical editor. Extra special thanks to the technical editor Theresa Urban, who worked tirelessly and with impeccable professionalism. Also, her immense knowledge of general IT, the Web, and GIS played a great role in improving the clarity of the book.

We are grateful to the team at Thomson Delmar Learning for their help in publishing this book, including acquisitions editor James DeVoe, editorial assistant Tom Best, developmental editor John Fisher, production editor Stacy Masucci, marketing assistant Patti Garrison, designer and compositor Carol Leyba, and copy editor Daril Bentley.

To Jing, Janice, and Cynthia

To my father and
the memory of my mother and grandma

—Vincent Zhuang

In dedication to Alisa, Haley, and Brittney—May happiness follow you always and the pursuits of your dreams come true.

—Dave Wrazien

I am dedicating this writing to my wife, Mary. She has been always very supportive in my career development, including writing this book.

—Minhua Wang

Foreword

IN *THE EARLY DECADES OF GIS,* professionals concentrated primarily on data compilation and focused application projects, spending a majority of their time creating GIS databases and authoring geographic knowledge. Gradually, GIS professionals began to use and exploit these knowledge collections in numerous GIS applications and settings. Users applied comprehensive GIS workstations to compile geographic data sets, build workflows for data compilation and quality control, author maps and analytical models, and document their work and methods.

This reinforced the traditional view of a GIS user with a professional scientific workstation that connects to data sets and databases. The workstation had a comprehensive GIS application with advanced GIS logic and tools that were used to accomplish almost any GIS task.

This concept of a GIS software seat has proven invaluable and continues to grow. GIS workstations are widely adopted by GIS professionals in over 200,000 organizations worldwide. In fact, this client/server computing model has been so successful that many only think of GIS within this context. However, the GIS vision is expanding.

Recent developments in computing—the growth of the Internet; advances in DBMS technology, object-oriented programming, and mobile computing; and widespread GIS adoption—have led to an evolving vision and role for GIS. The vision of a GIS platform is also expanding.

In addition to GIS desktops, GIS software can be centralized in application servers and web servers to deliver GIS capabilities to any number of users over networks. Focused sets of GIS logic can be embedded and deployed in custom applications. And increasingly, GIS is deployed in mobile devices for field GIS. Today, in thousands of organizations and servers enterprise GIS users connect to central GIS servers using traditional advanced GIS desktops as well as web browsers, focused applications, mobile computing devices, and digital appliances.

Software developers are at the forefront of this GIS revolution. They have been the early adopters of server GIS and have developed a broad range of applications using both the Microsoft .NET and Java programming frameworks.

In this book, Vincent Zhuang, Dave Wrazien, Minhua Wang, and Xiaowen Huang have built a highly useful guide for web GIS developers using .NET. It provides useful, detailed application examples and code samples for developers who work with ESRI's ArcGIS Server. Any member of the ESRI Developer Network (EDN) program who uses Microsoft Developer tools will find that this book provides invaluable insights and programmer examples.

Clint Brown
ESRI Director of Software Products
May 2005

Contents

Code Listings

Preface

THIS BOOK IS ABOUT PROGRAMMING GIS ON THE WEB using ArcGIS Server and ASP.NET, which is a professional guide for developing ArcGIS Server web applications and web services using Visual Basic .NET. This preface briefly describes the following.

- Audience for this book
- Conventions used in this book
- Book content
- Code samples

Who Should Read This Book?

Programming ASP.NET for ArcGIS Server is written for developers with a basic understanding of GIS and the Web and who are now ready to learn how to write ASP.NET web pages or web services using ArcGIS Server and Visual Basic .NET. This book assumes that you have a basic understanding of web-based GIS and ASP.NET programming. The book will teach you how to program in ASP.NET, which is the latest Microsoft web programming solution. Of course, a background in GIS is required to use any ESRI development tool, but knowing early-generation ASP is not important. If you are familiar with previous generations of ESRI Internet mapping technology—including ArcIMS, MapObjects Internet Map Server, and ArcView Internet Map Server—you will immediately notice similarities and differences with ArcGIS Server. If you have never used any ESRI Internet GIS tool, you will become a pro after you complete Chapters 1 through 9.

We hope that you will find the materials included in this book interesting, well-rounded, concise, and useful. You can build your own ArcGIS Server web applications and web services in a few hours after you have gone through the entire book. There are two development frameworks that are shipped with ArcGIS Server: ASP.NET and J2EE. We have found ASP.NET to be the easiest and most immediately accessible solution for developing web-based applications in Microsoft Windows XP, Windows

2000 Server, and Windows Server 2003. ASP.NET is also the most popular industry standard for the Web. Examples are written in Visual Basic .NET. Whether you are a Visual Basic or Visual C++ programmer or a C# programmer, you should be able to walk through the examples without a problem.

Conventions Used in This Book

There are four types of code examples found in this book: Visual Basic .NET, XML, DHTML, and JavaScript. Code examples and fragments are set in a fixed-width font. The Letter Gothic font (an example of which follows) is used for Visual basic .NET code.

```
'Declare a string variable for the book title
Dim BookTitle As String = "Programming ASP.NET for ArcGIS Server"
```

The `Courier` font (an example of which follows) is used for XML, DHTML, and JavaScript code.

```
<?xml version="1.0" encoding="UTF-8"?>
```

The names of files are printed in *italic*. If a space character is required, it will be marked as the symbol " ". Sometimes we have a hint or alert to share with you. These notes are highlighted as follows.

✔ **NOTE:** *The sample code in this book has been tested with version 1.1 of the .NET Framework. Be sure to install Visual Studio .NET 2003 before you try to work with the coding samples.*

What's in This Book?

This book is organized into nine chapters, each of which begins with concepts, followed by methods, and ending with examples. Chapter 1 introduces you to what ArcGIS Server is and where it fits into the new ArcGIS 9 product line. This chapter compares ArcGIS Server and ArcIMS, and then describes the ArcGIS Server architecture. Information on installing and configuring ArcGIS Server is included in this chapter. At the end of the chapter, concepts for developing ArcGIS Server applications are discussed.

Chapter 2 introduces you to ASP.NET and describes the difference between classic ASP and ASP.NET. It then goes on to talk about

ASP.NET pages and component models. This chapter also lists all web controls, validation controls, and mobile controls and describes how to do debugging in ASP.NET (a new functionality of ASP.NET) and provides guidelines for developing ASP.NET applications.

Chapter 3 introduces you to Visual Basic .NET and walks through the core of Visual Basic .NET. It can be used as a crash course in Visual Basic .NET. This chapter talks about defining variables and constants, manipulating strings, using regular expressions, defining arrays and collections, writing functions and subroutines, handling errors and exceptions, and utilizing object-oriented programming in Visual Basic .NET.

Chapter 4 walks you through a simple ASP.NET application based on ArcGIS Server. It starts with creating ArcGIS Server project files—map document and address locator—and then talks about how to create a map server object and a geocode server object. The chapter also covers how to create a map server object based on the Geography Network or any ArcIMS service. It continues with writing an ASP.NET web application using Visual Basic .NET and the ArcGIS Server API.

Chapter 5 begins with an exploration of web forms, an inventive technology by Microsoft for web applications. It continues with web controls and validation controls used on web forms. This chapter then goes on to address template web controls and web templates. This chapter also briefly discusses the data access component of the Microsoft .NET Framework (called ADO.NET) by covering one of mostly used ADO.NET component, OLE DB.

Chapter 6 begins by examining the ArcGIS Server Application Developer Framework (ADF) for .NET, through which ArcObjects can be exposed on the Web. The ArcGIS Server ADF for .NET is built on top of Microsoft's .NET Framework and includes a set of ESRI web controls and templates used to build GIS web applications. This chapter discusses ArcGIS web controls and web templates and includes an application built based on an ArcGIS Server template and another application example built directly based on web controls.

Chapter 7 discusses how to extend existing web controls and how to create your own composite controls. The chapter then explores how to convert a page to a user control and compares composite controls and user controls. Finally, the chapter discusses how to customize ArcGIS web controls and web templates.

Chapter 8 introduces you to building GIS-centric web services for ArcGIS Server. It starts with the introduction to the infrastructure of web services, including the definition, the Web Services Description Language (WSDL), the Simple Object Access Protocol (SOAP), and Universal Description, Discovery, and Integration (UDDI). The chapter continues with discussion of how to build and consume a web service and goes on to discuss how to build and consume ArcGIS Server web services. The chapter also briefly talks about how to secure web services, manage the state of web services, and publish and deploy web services.

Chapter 9 introduces you to building mobile GIS applications for PDAs, cellular phones, and BlackBerry handheld devices. The chapter begins with an introduction to the mobile Internet and talks about mobile emulators, which help developers test their mobile applications. This chapter then discusses mobile web controls, which are customized web controls for mobile devices, and mobile web forms. ESRI's ArcWeb Services are also covered in terms of building wireless GIS web applications. This chapter ends with an exploration of how to build mobile GIS applications.

Code Samples

This book serves as a workbook, providing many examples in every chapter and section. The sample code used in this book is available on the book's web site at *http://www.onpress.com/*. From the web site, click on Online Companions and select the book title *Programming ASP.NET for ArcGIS Server*. The code is organized by chapter. For example, to find the source code in Chapter 6, which talks about ArcGIS Server web controls and web templates, you would look up the code in the project directory *Chapter6*. Some additional steps may be required to install virtual directories for some samples. These are documented in the README file available at the sample code web site.

✔ **NOTE:** *The sample code in this book has been tested with version 1.1 of the .NET Framework. Be sure to install Visual Studio .NET 2003 before using the examples.*

System Requirements

You will need the following software to work through the samples in this book.

- Visual Studio .NET 2003
- Microsoft Windows XP, Windows 2000 Professional, Windows 2000 Server, or Windows Server 2003
- Microsoft Internet Explorer 5.5 or later

The minimal hardware specifications follow those specified in the system requirements for ArcGIS Server from ESRI. We suggest that you have at least 512 MB of RAM. If you are using Windows 2000 Server or Windows Server 2003, these values are the bare minimum according to Microsoft. ArcGIS Server applications run much more smoothly on systems with ample RAM and plenty free disk space. We would suggest that you install 1 GB of RAM on your Windows 2000 Server or Windows Server 2003 production server. You can also develop your application on Windows XP.

CHAPTER 1
ArcGIS Server Overview

ARCGIS SERVER IS A PLATFORM FOR BUILDING and delivering enterprise GIS applications and web services. Those applications and web services are centrally managed on the server, support multiple users, and offer advanced GIS functionality. Built based on industry standards, ArcGIS Server provides the framework to build and deploy centralized GIS applications and web services to meet a variety of needs and to target different clients, including other enterprise servers, desktop workstations, and portable and mobile devices.

The ArcGIS Server platform consists of two primary components: a GIS server and an application developer's framework (ADF). The GIS server manages a set of server objects that currently include the map server object and the geocode server object. The ADF is cross platform and supports the development of web applications and web services in both .NET and J2EE.

What Is ArcGIS Server?

Over the years, an increasing number of industries and disciplines have embraced Geographic Information System (GIS) technology to view and analyze information from a spatial context. Retail businesses, for example, have been using GIS to research customer markets, evaluate site suitability, analyze service areas, manage delivery fleets, and visualize the supply chain. Many of these early GIS applications were implemented on a small scale by select professionals within dedicated departments of an organization. As GIS functionality became more widely used and adopted within organizations, it became necessary to support these processes on not only traditional desktop PCs but on centrally managed servers and on mobile devices. Server-based GIS computing has enabled organizations to bring spatial processing and visualization into mainstream IT computing environments and architectures such as enterprise application integration

(EAI) platforms, service-oriented architectures (SOAs), event-driven architectures (EDAs), and web services.

ESRI ArcGIS represents a comprehensive GIS platform for deploying geospatial functionality to mobile clients, desktop workstations, and enterprise servers. The platform is responsible for providing software functionality to support data management, editing, analysis, and visualization of geographical content. ArcGIS is founded upon a common component software library called ArcObjects. ArcObjects provides a wide variety of GIS components, from fine-grained objects representing individual features or functions to coarser-grained objects representing aggregated functions and capabilities. ArcObjects provides the building blocks upon which the ArcGIS family of products is developed (Figure 1–1). ESRI not only uses ArcObjects to build its product offerings but makes them available to developers for building and deploying applications to the desktop, embedded, mobile, and server frameworks.

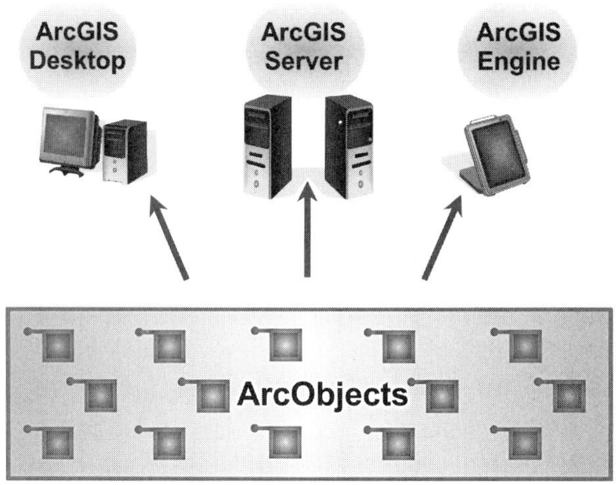

Figure 1–1. ArcObjects—the building blocks for ArcGIS.

ArcGIS Server is an enterprise application server that implements and exposes ArcObjects from a server context for the delivery of rich GIS functionality, including mapping, geocoding, complex spatial queries, transactional editing, feature-level geoprocessing, and complex analytics. As a platform, ArcGIS Server supports the construction and deployment of centrally managed web-based applications, web services, and other enterprise applications through standard computing environments such as J2EE and .NET. For example, web developers can develop applications or web services that provide tracing along linear networks such

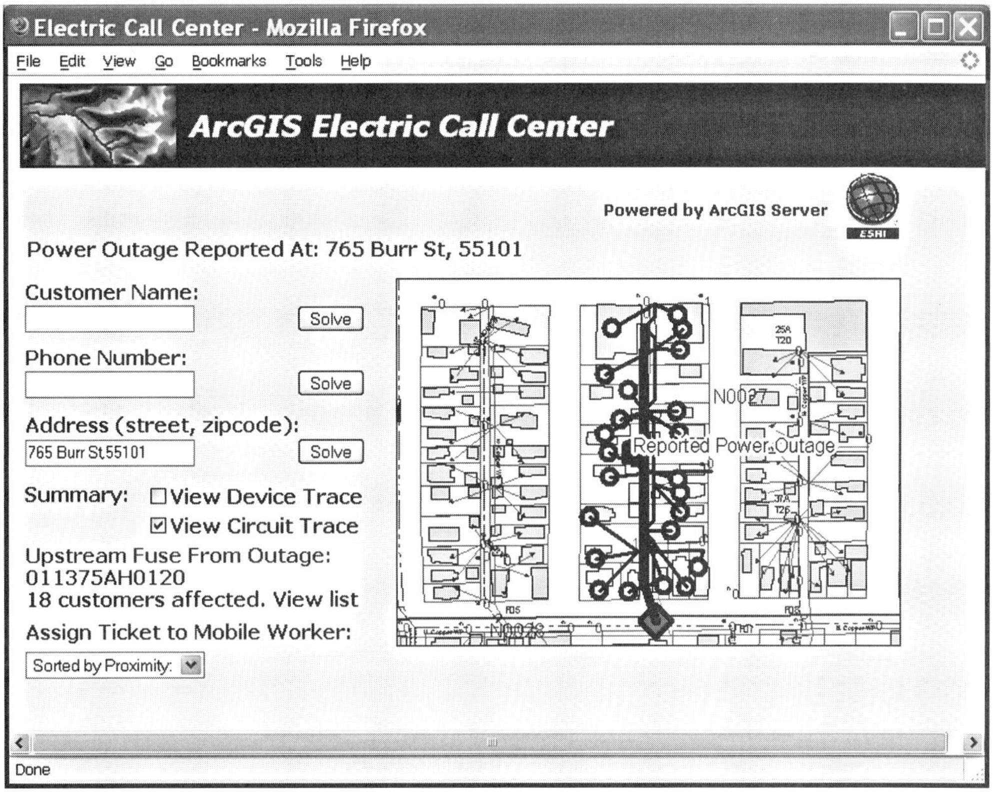

Figure 1–2. Electric power outage application leveraging ArcGIS Server's network tracing algorithms.

as an electric utility outage application (Figure 1–2) or geocoding of customer addresses as a web service in support of a customer relationship management (CRM) system.

As GIS becomes more pervasive within organizations, its overhead and the ability to integrate it with other core business systems (e.g., systems for asset management, financials, and supply chain management) become critically important. The centralization of GIS applications and services reduces overall costs by simplifying installation and administration. For example, the financial burdens associated with adopting a centralized GIS server approach can lessen expenses associated with numerous desktop deployments. In addition, the incorporation of GIS functionality and processes into other key business systems may extend the value of those systems by increasing accuracy, efficiency, and productivity. For example, a fleet management system that used GIS to dispatch and route delivery vehicles can reduce mileage and other vehicle/driver costs and can more accurately determine delivery time windows, which ultimately results in greater customer satisfaction. Successful long-term interoperability with other enterprise systems is contingent upon the adherence to industry standards and protocols. ArcGIS Server provides a highly cost effective, extremely interoperable, and IT-compliant platform for supporting organizations' comprehensive GIS needs.

ArcGIS Server and ArcIMS

At first glance, the distinction between ESRI's ArcIMS and ArcGIS Server may be confusing. To help understand the differences between these two GIS server products it is important to understand the purpose and functional capabilities of each.

ArcIMS is a highly scalable map server designed with the purpose of publishing GIS maps, data, and metadata to users of the Internet and private intranets. The ArcIMS server is responsible for the dynamic rendering and delivery of maps in web-friendly image formats such as GIF, JPEG, and PNG to desktop and browser-based clients. The server is also capable of delivering data streams that contain the actual GIS features/objects to capable clients, as opposed to just delivering a rendering of those features/objects in the form of an image map. Finally, for those organizations that have the need to publish metadata records and catalogs of their geographic data repositories, ArcIMS delivers this information through standard transmission protocols such as the ANSI/NISO Z39.50. In summary, ArcIMS is intended to be used as GIS publishing platform—delivering GIS content to many users across the Web.

ArcGIS Server, on the other hand, has been designed as an enterprise GIS computing platform. Although ArcGIS Server also has the ability to publish GIS maps and content such as ArcIMS, the similarity ends there. ArcGIS Server extends the capabilities of ArcIMS with a comprehensive set of functions for spatial data management, editing, and analysis. Organizations that embrace ArcGIS are often looking to provide distributed data management using versioned geodatabases; central, multi-user geodatabase editing capabilities; server-based geoprocessing operations; and advanced hosted GIS web services and browser-based applications.

From a developer's perspective, ArcIMS provides an ActiveX Connector, Java Server Page (JSP) Connector, and ColdFusion Connector (and upcoming .NET Connector) for programming web applications in ASP, JSP, and ColdFusion, respectively. In addition, ArcIMS offers the WMS and WFS connectors, which provide an Open GIS Consortium (OGC) standard web map service (WMS) and web feature service (WFS) interface into maps and data served by ArcIMS. ArcGIS Server ships with the ADF for both .NET and Java, which includes a collection of web controls, convenience classes, data objects, and application templates. The ArcGIS Server ADF for .NET, which is built on top of the Microsoft .NET framework, allows you to integrate GIS functionality into ASP.NET web applications. When working with Visual Studio .NET, you can drag and drop ArcGIS Server web controls onto a web form just like you do with other web controls.

Thus, although the initial perception is that these two products compete with each other the reality is that they do not due to their minor functional overlap. In fact, many organizations may embrace both technologies in an effort to handle all of their GIS publishing and application delivery needs.

ArcGIS Server Architecture

The ArcGIS Server platform consists of two primary components: a GIS object server (which manages a set of server objects) and the ADF, which supports the development of .NET- and J2EE-based web services and web-based applications. The GIS server's responsibility is to manage course-grained objects (e.g., MapServer or GeocodeServer) and fine-grained objects (e.g., LayerDescription or AddressLocator) within a server context (GIS server objects are essentially ArcObjects running on a server). Application developers can take advantage of the ADF to interact with both the coarse-grained and fine-grained server objects via a library of proxy objects, convenience classes, web controls, web application templates, and web services (Figure 1-3).

The primary component of the ArcGIS Server platform is the GIS server. The GIS server consists of a server object container (SOC)—which hosts the processes of the actual server objects—and a server object manager (SOM), which manages the various SOCs. The SOC is a process in which one or more server objects are running. For scalability purposes, multiple container processes can run on one or more individual

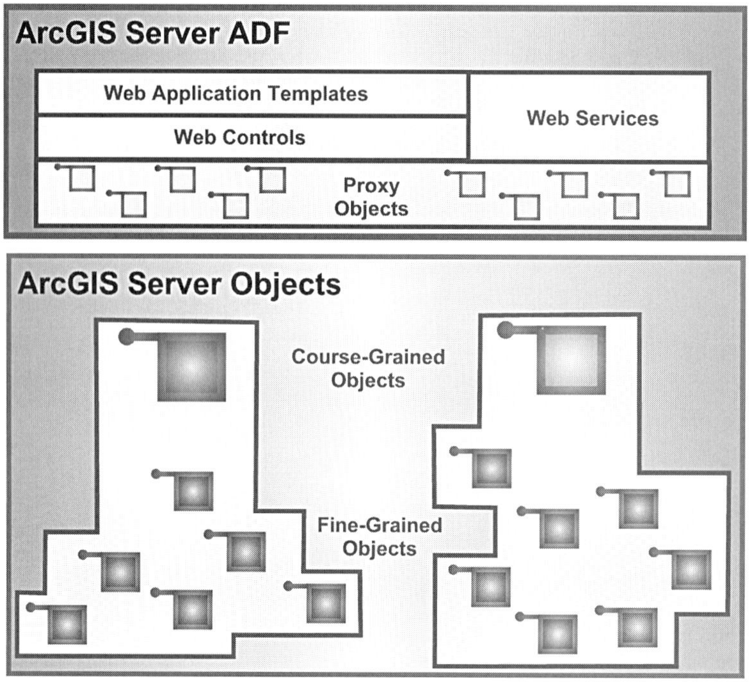

Figure 1–3. ArcGIS Server components and the ADF.

machines. However, it should be noted that the SOM regards all container process to be equally configured. When managing the various SOCs, it is important to consider that each container process requires that all necessary and applicable resources (objects, data, and file system directories) be available for execution. The SOM is a Windows service or a UNIX daemon that manages the activation, deactivation, and load balancing among the various SOCs. ArcGIS Server system administrators can manage the SOM and associative SOCs with the ArcGIS desktop application called ArcCatalog. The SOC and SOM are requisite components of the GIS server that support the distributed architectural options ArcGIS Server has to offer (Figure 1–4).

The second primary component of the ArcGIS Server platform is the ADF. The ADF contains both a software developer's kit (SDK) and a web runtime application that allows developers to build and deploy web service and web applications. The SDK contains software objects, web controls, web application templates, and numerous developer samples, whereas the web runtime application permits the deployment of web services and applications without having to install ArcObjects on a web application server.

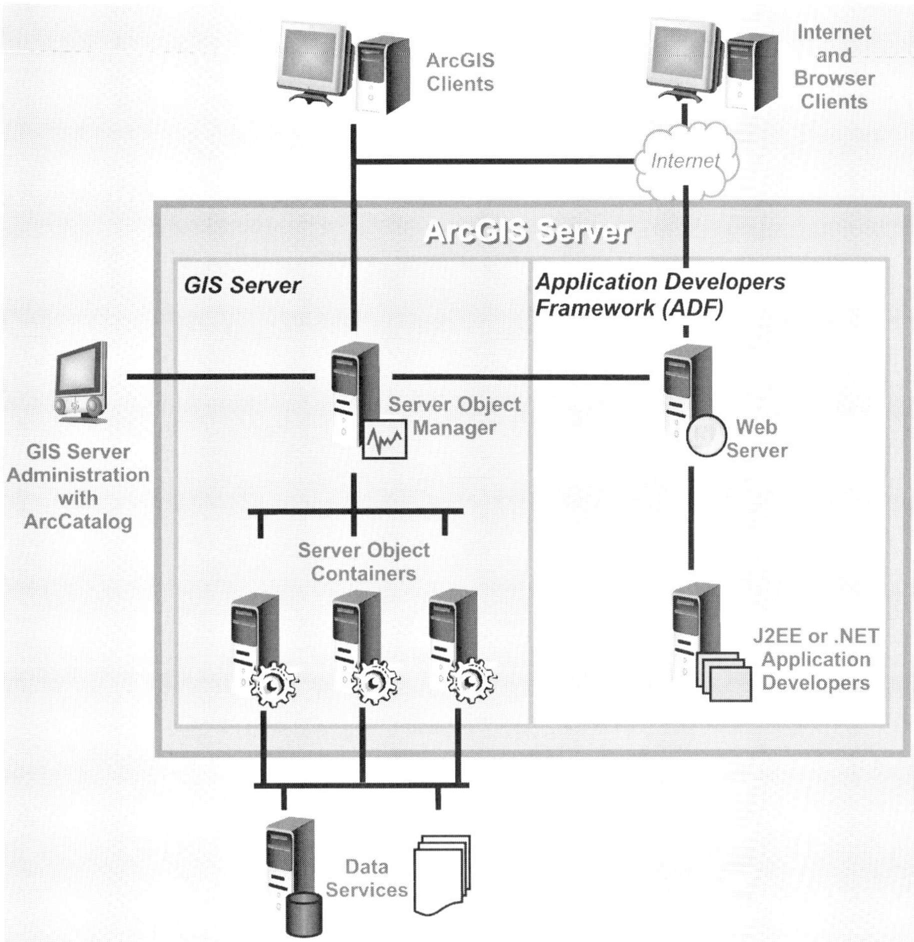

Figure 1–4. ArcGIS Server in a distributed computing environment.

The foundation of the SDK is a library of proxy objects—delivered with both the .NET and JAVA application programmer interface (API)—that represent remote GIS server objects (ArcObjects). These proxy objects have all of the supported interfaces of the remote objects and respond accordingly to both method and property calls. The difference between the proxy objects and the remote objects is that when a method or property call is placed on a proxy object the actual invocation of that process occurs on the corresponding remote server object hosted by the SOC. This remote invocation between the proxy and remote server objects is performed via binary communications over standard network protocols.

Built on top of the ArcGIS Server proxy object library is a set of web controls and convenience classes for web developers to leverage when building applications. ArcGIS server web controls represent the necessary user interface (UI) components—such as a map, north arrow, table of contents, and overview map—required for many of today's GIS applications. Developers can use ArcGIS Server web controls within integrated development environments (IDEs) by simply dragging and dropping these components onto a design "canvas" or "form" and setting their associative properties such as height, width, font, and color. Whereas the web controls enable developers to design the look and feel of their applications, it is the convenience classes behind the web controls that actually invoke the requisite GIS functions. Convenience classes simplify how developers can access the various ArcGIS Server functions by providing an entry point into the ArcGIS Server and ArcObjects API. Working with both web controls and convenience classes, developers can create GIS applications that exploit specific capabilities of ArcGIS Server in both .NET and JAVA development environments. Table 1–1 outlines the ArcGIS Server web controls and their associative convenience classes.

Table 1–1: ArcGIS Server Web Controls and Associative Convenience Classes

Control	Description	Convenience Class
Map	Displays a map for a specific data frame within a map service.	WebMap
Overview Map	Displays a map and reference extent for a particular map control.	WebMap
Page Layout	Displays a specific map layout, including all map elements (e.g., map titles, scale bars, north arrows).	WebPageLayout
Table of Contents (TOC)	Displays a list of map layers being displayed in a particular map or page layout control.	WebTOC
Toolbar	Displays a toolbar containing functions for working with other web controls.	None
North Arrow	Displays a north arrow for a particular map, overview map, or page layout control.	None
Scale Bar	Displays a scale bar for a particular map, overview map, or page layout control.	None

Control	Description	Convenience Class
GeocodeConnection	Provides access to geocoding server objects. There is no runtime visual display of this control.	WebGeocode
Impersonation	Provides security credentials for accessing and invoking ArcGIS Server objects.	None

The ArcGIS Server ADF provides a series of customizable templates based on the ADF web controls and convenience objects. These templates help in the construction of some more common GIS web applications such as map viewers, search applications, thematic mapping applications, geocoding applications, buffer/spatial query applications, web service catalogs, and page layout applications (Figure 1–5). These templates are functionally complete and can be used "as is" by simply making a few property settings. In addition, because all of the underlying application code is available these templates can be customized and extended to meet the needs of application developers.

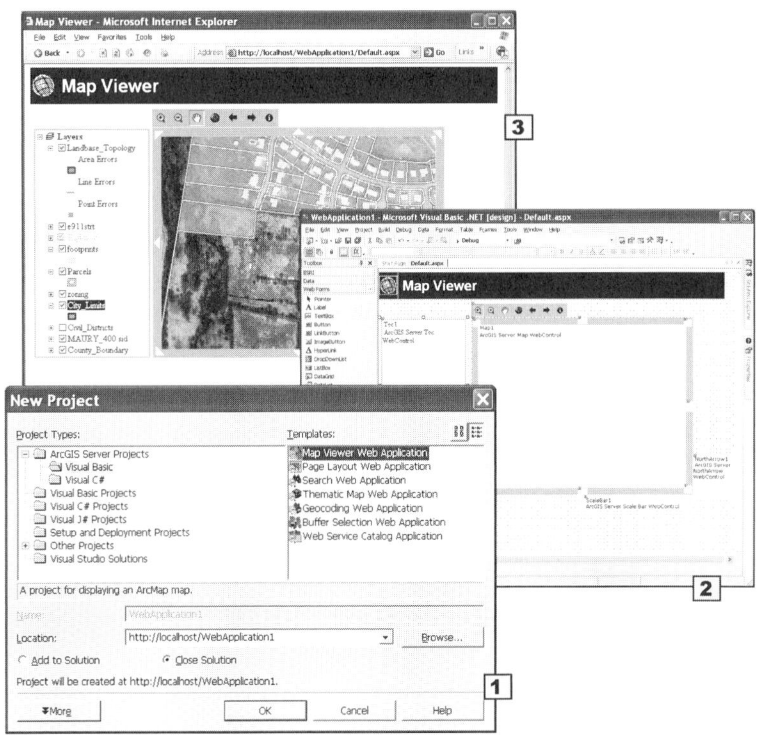

Figure 1–5. Accessing the ADF templates from the Visual Studio .NET New Project dialog: (1) Map Viewer template in the Visual Studio .NET design-time environment and (2) Map Viewer template running within a web browser (3).

ArcGIS Server Installation and Configuration

ArcGIS Server and its various components have been designed and constructed to run on one or more computers. The simplest installation involves installation and configuration of all three components (SOM, SOC, and ADF) on a single machine. However, it is more common that these components will be installed across multiple machines. For example, in a distributed installation the SOM may be installed on a single machine as a Windows service. This service will manage the various objects running on different SOC machines.

The actual number of SOC machines depends on many factors, including available hardware resources (e.g., disk space, memory, processors, and so on), software application requirements, and performance goals. Finally, the ADF may be installed on several machines. The ADF runtime application needs to be installed on the same machine running either the J2EE or ASP.NET application server, whereas the other ADF components (proxy objects, convenience classes, web controls, and application templates) might be installed on multiple application-developer machines. Once all software components are installed on the desired machines, the next step is to configure and manage these components.

The configuration of the GIS server components is dependent on various aspects of application design, including the type of server objects being hosted, the number of object instances, whether the objects are pooled or non-pooled, the object's isolation level, and the server's security model. As mentioned earlier in this chapter, the coarse-grained server objects have a specific type of MapServer or GeocodeServer. Each of these types has specific initialization parameters. For example, the MapServer type requires initialization parameters that include service name, map document name (*.mxd* or *.pmf*), image output type, and output directory location, whereas the GeocodeServer type requires parameters such as service name, Locator name (*.loc* or *.mxs*), and batch size.

Both MapServer and GeocodeServer type objects can be further configured to be either pooled or non-pooled instances (Figure 1-6). Pooled object instances can be shared among multiple application sessions at the request level. With pooled object instances, connections to the server last for only the duration of one request/response cycle. Because object instances are shared, more users can be supported with fewer resources being expended. To configure object pooling, the minimum and maximum number of instances needs to be set. When a server object is first started, the number of instances is based on the minimum number defined in the pooling configuration. If the number of concurrent requests exceeds the minimum number of instances, the SOM will automatically spawn another instance of that server object in order to handle the increased number of requests. The SOM will not spawn server object instances that exceed the maximum number defined in the pooling configuration. A pooled object configuration is best for supporting applications that do not require object state to persist on the server.

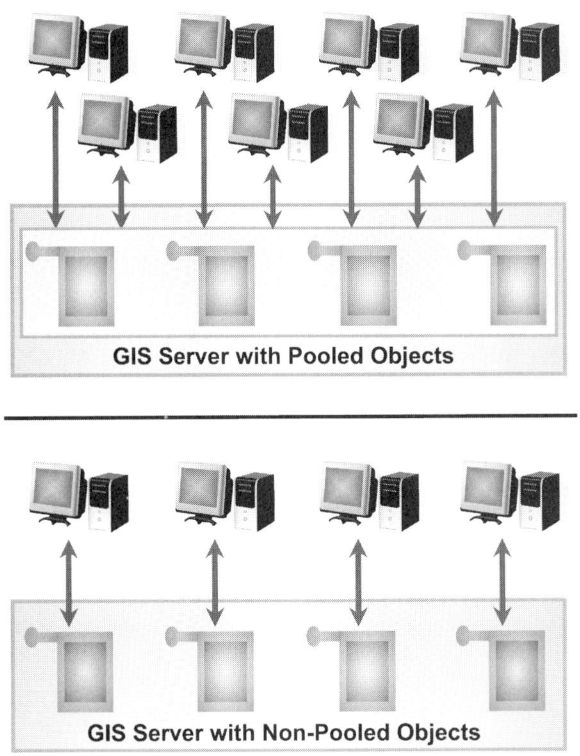

Figure 1–6. Pooled and non-pooled objects.

Non-pooled object instances are not shared among multiple applications but are created for each application session that uses them. Applications utilizing non-pooled objects will maintain reference to a particular server object for the duration of the application's use, and the server object's use will typically span multiple request/response application cycles. Therefore, the number of non-pooled instances is equivalent to the number of applications using that object. Non-pooled objects are configured by defining a maximum number of instances that can be spawned. When the maximum number of instances is reached, applications will be unable to gain access to those object resources and requests will be queued until the number of utilized instances drops below the maximum allowed value. A non-pooled object configuration is best for supporting applications that require object state to be persistent on the server. For example, an editing application may store multiple edits within the server object's state and only commit those edits to the database when all editing transactions of that session are complete.

Pooled and non-pooled objects run within processes on the SOC machine. These objects can be configured to run either in high isolation (dedicated processes) or low isolation (shared processes) (Figure 1–7.)

High-isolation object instances do not share processes with other object instances and therefore require more server resources to run. Low-isolation object instances share processes with other server objects and are more efficient in their utilization of server resources. When determining whether or not to have objects running in high or low isolation it is important to consider system resource use and to understand how each reacts with process failure. If an error occurs with an object running in high isolation and that process is forced to shut down or terminate, only that particular server object is affected. Conversely, if an object is running in low isolation and a process is shut down or terminated *all* objects sharing that process will be affected.

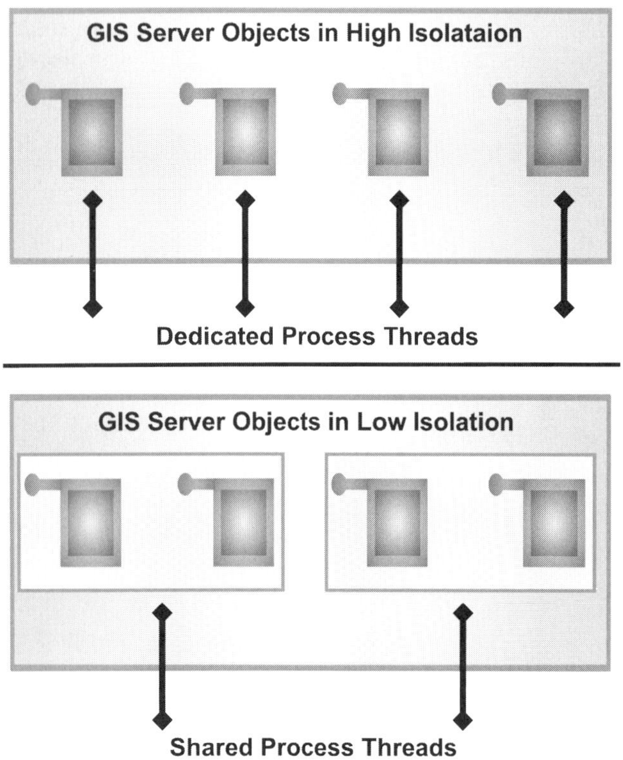

Figure 1–7. High and low GIS server object isolation.

In an effort to maintain performance and system stability, GIS server objects can be destroyed and recreated through a process called recycling. Over time, some GIS server objects may become stale or unusable by applications. A recycling process refreshes those objects and reclaims any system resources the unusable objects may have held. The frequency interval for recycling specific GIS server objects is completely configurable by GIS Server administrators. Regardless of whether or not the GIS server objects are pooled or non-pooled, or in high or low isolation, GIS server objects are recycled by a low-priority background process. This process recycles only those GIS server objects that have been released by applications (so as not to disrupt the users of these objects).

When addressing security issues regarding the ArcGIS Server, it is important to manage security issues on the GIS server itself and on the applications and web services accessing the GIS Server. Upon completing the installation of ArcGIS Server, a post-installation routine will automatically create two groups of users in the local operating system: a group for ArcGIS Server users called *agsusers* and a group for ArcGIS Server administrators called *agsadmin*. Because the post-installation routine does not add any users to either of these groups, it is important to add individuals or groups of users depending on their access requirements.

The ArcGIS Server user group is designed for "consumers" of the GIS server objects, whereas the ArcGIS Server administrator group is designed for server administrators that need to add/remove/modify SOC machines, server objects, and server directories, for example. Applications and web services that access the GIS server must do so through a valid member of either the ArcGIS Server administrator or ArcGIS Server user group. Because the ASP.NET and J2EE application servers have their own defined security model, it is often necessary to have applications running in these environments that impersonate a user in one of the ArcGIS Server groups (*agsuser* or *agsadmin*).

ArcGIS Server administration is primarily performed with the ArcGIS desktop application called ArcCatalog. ArcCatalog users who are members of the ArcGIS Server administrator (*agsadmin*) group have the ability to manage various aspects of the GIS server objects, including the following.

- Adding, deleting, and modifying server objects
- Setting any requisite initialization and start-up parameters
- Specifying response type and any necessary output directories
- Determining whether server objects are pooled or non-pooled
- Specifying the minimum and maximum number of server object instances
- Specifying the maximum usage and wait times
- Setting the isolation level
- Establishing a recycling interval

The overall performance, stability, and accessibility of the ArcGIS Server platform are critically dependent on the configuration of the platform's various components. ArcGIS Server administrators must give careful consideration to the initial configuration and must monitor the configuration as users start leveraging the platform. When applications start accessing the GIS server objects, statistics and log files are available for review to help identify and troubleshoot any issues that might arise. GIS

Server statistics are available to ArcGIS Server administrators through the server properties in ArcCatalog (log files are written to the < *install_directory* > *\log* directory on the SOM machine).

Building and Deploying Applications with ArcGIS Server

Application developers can build a wide variety of ArcGIS Sever applications that range from multiuser focused web applications (Figure 1–8) and web services to single-user desktop applications. The ArcGIS Server ADF contains the SDK and a web runtime application that allows developers to build and deploy their web services and web applications. Subsequent chapters supply ASP.NET developers with practical information and step-by-step instructions on how to write web-based GIS applications and services that leverage the capabilities of the ArcGIS Server platform.

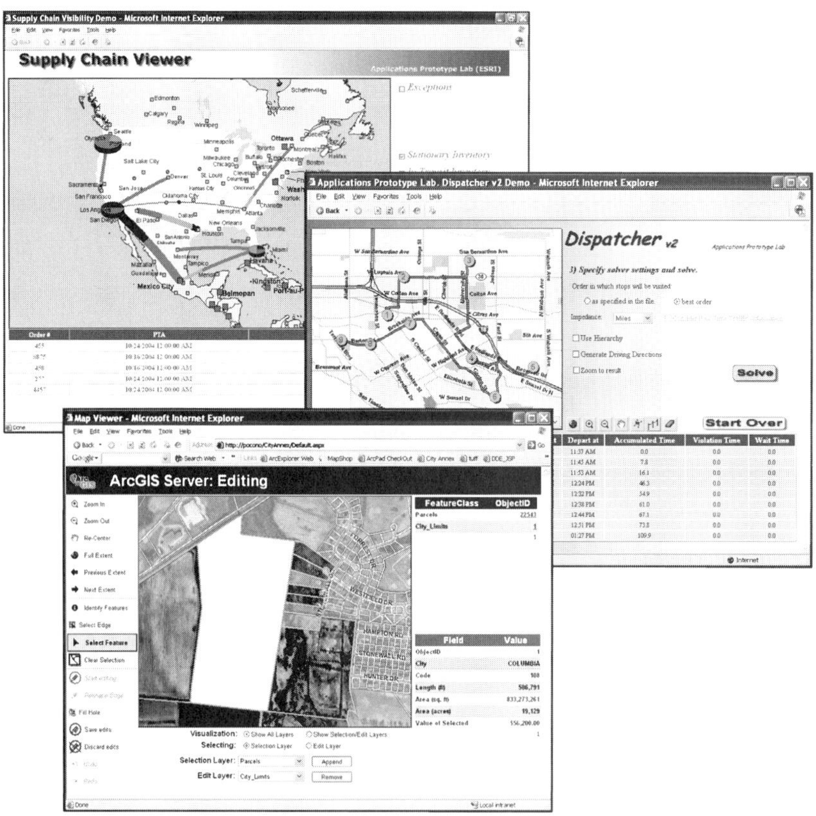

Figure 1–8. Focused web-based applications developed with ArcGIS Server.

Summary

In this chapter, we have introduced what ArcGIS Server is and where it fits in enterprise GIS. We discussed the difference and similarity between ArcIMS and ArcGIS Server. ArcIMS is a highly scalable map server targeted publishing maps, data, and metadata on the Internet and organizational intranets. ArcGIS Server, on the other hand, has been designed as an enterprise GIS computing platform. The ArcGIS Server architecture was illustrated, which consists of a GIS object server and the .NET and J2EE application development frameworks. The GIS server is used to manage course-grained objects and fine-grained objects within a server context. Application developers can take advantage of the ADF to interact with both the coarse-grained and fine-grained server objects via a library of proxy objects, convenience classes, web controls, web application templates, and web services.

The ArcGIS Server installation and configuration were also discussed. GIS server objects can be configured to be pooled and non-pooled. A pooled object configuration is best for supporting applications that do not require object state to persist on the server, whereas a non-pooled object configuration is best for supporting applications that require object state to be persistent on the server. Regarding the security issues, applications and web services that access the GIS server must run through a valid member of either the ArcGIS Server administrator or ArcGIS Server user group. It is necessary to impersonate a user in one of those two groups of users and have applications and web services running under that user.

ArcGIS Server is new but its potential is boundless. It offers an enterprise GIS solution. You can now build truly functional web-based GIS applications and web services based on Microsoft's ASP.NET. In the next chapter, we will discuss ASP.NET.

CHAPTER 2

ASP.NET

ASP.NET IS MICROSOFT'S LATEST TECHNOLOGY for building web applications, web services, and mobile web applications. As a successor to Microsoft's Active Server Pages (ASP), ASP.NET is a part of the Microsoft .NET framework, and is available through Visual Studio .NET, Microsoft's Integrated Development Environment (IDE) for visual programming.

ASP.NET used to be called ASP+, where the plus suffix emphasized that it is the next version of ASP and a compiled environment. By the time Microsoft released its .NET (pronounced "dotnet") Framework, ASP+ was changed to ASP.NET, which is part of .NET Framework but is dedicated to provide a web development model for developers to build enterprise-class web applications.

The .NET Framework provides a platform for simplified rapid development of both web-based and Windows-based applications. The Framework has two primary components: the Common Language Runtime (CLR) and the Framework Class Library (FCL). The .NET Framework is a common environment for building, deploying, and running web applications, web services, and mobile web applications. Microsoft's Visual Studio.NET is a common development environment for the .NET Framework.

✔ **NOTE:** *The CLR is the execution environment for code written for the .NET Framework. The CLR manages the execution of .NET code, including memory allocation and garbage collection (which helps avoid memory leaks), security (including applying differing trust levels to code from different sources), thread management, enforcement of type safety, and many other tasks. The FCL is a set of reusable object-oriented classes that provide access to system functionality. It also contains all classes that make up ASP.NET.*

Introduction to ASP.NET

ASP.NET is a compiled programming framework built on the CLR that can be used on a server to build powerful web applications. You can create web applications using any .NET-compatible language, including Visual Basic .NET, C# (pronounced "C sharp"), Visual C++ .NET, and JScript .NET (as well as others provided by third-party vendors). This increases flexibility in development because different languages can be used to develop the same web application. It also broadens the development environment under which a variety of programmers with different skills and experience can work. Components and modules developed using different languages communicate with each other in the CLR. In addition, the entire .NET Framework is available to any ASP.NET application. Developers can easily access the benefits of these technologies, which include the managed CLR environment, type safety, inheritance, and so on.

In ASP.NET, a web client communicates with ASP.NET applications or web services through Microsoft's Internet Information Services (IIS), which compose the Microsoft web server. IIS interprets and optionally authenticates the request. If the Anonymous access setting is turned on, no authentication is required. IIS also finds the requested resource, whether it is an ASP.NET web application or web service, and returns the applicable resource (Figure 2–1).

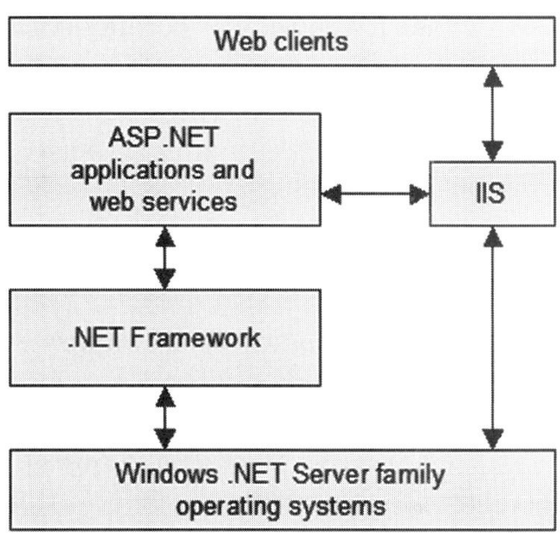

Figure 2–1. The ASP.NET architecture.

ASP.NET differs from classic ASP in that it separates presentation from the code, which can make the web application development more efficient. ASP.NET stores the code in *code-behind* files. Those files contain the code that is executed in response to the user interaction with the web form (web page). The presentation, also called the content, is stored in an *.aspx* file, whereas the code is stored in an *.aspx.vb* file. Listings 2–1 and 2–2 show a simple ASP.NET page written in Visual Basic .NET, where the content is stored in *WebForm1.aspx* and the code is stored in *WenForm1.aspx.vb*. The code is generated by Visual Studio .NET.

Listing 2–1. *WebForm1.aspx.*

```
<%@ Page Language="vb" AutoEventWireup="false"
Codebehind="WebForm1.aspx.vb" Inherits="helloArcGISServer.WebForm1"%>
<!DOCTYPE HTML PUBLIC "-//W3C//DTD HTML 4.0 Transitional//EN">
<HTML>
<HEAD>
<title>WebForm1</title>
<meta name="GENERATOR" content="Microsoft Visual Studio .NET 7.1">
<meta name="CODE_LANGUAGE" content="Visual Basic .NET 7.1">
<meta name="vs_defaultClientScript" content="JavaScript">
<meta name="vs_targetSchema"
content="http://schemas.microsoft.com/intellisense/ie5">
</HEAD>
<body MS_POSITIONING="GridLayout">
<form id="Form1" method="post" runat="server">
<asp:Label id="lblHellowArcGISServer"
style="Z-INDEX: 101; LEFT: 96px; POSITION: absolute; TOP: 64px"
runat="server">Hello ArcGIS Server</asp:Label>
</form>
</body>
</HTML>
```

Listing 2–2. *WebForm1.aspx.vb.*

```
Public Class WebForm1
Inherits System.Web.UI.Page
#Region " Web Form Designer Generated Code "
'This call is required by the Web Form Designer.
<System.Diagnostics.DebuggerStepThrough()>
Private Sub InitializeComponent()
End Sub
Protected WithEvents lblHellowArcGISServer As
System.Web.UI.WebControls.Label
'NOTE: The following placeholder declaration is required by
the Web Form Designer.
'Do not delete or move it.
Private designerPlaceholderDeclaration As System.Object
Private Sub Page_Init(ByVal sender As System.Object,
```

```
    ByVal e As System.EventArgs) Handles MyBase.Init
12: 'CODEGEN: This method call is required by the Web Form Designer
    'Do not modify it using the code editor.
13: InitializeComponent()
14: End Sub
15: #End Region
16: Private Sub Page_Load(ByVal sender As System.Object,
    ByVal e As System.EventArgs) Handles MyBase.Load
17: 'Put user code to initialize the page here
18: End Sub
19: End Class
```

As you can see, *WebForm1.aspx* is basically an HTML file except for the first line. The first line of code is a page directive generated automatically by Visual Studio .NET. The page directive is defined by special sequences of characters beginning with <%@ and ending with %>. The directive contains four attributes, including *Language, AutoEventWireup, CodeBehind*, and *Inherits*. The *Language* attribute specifies that the code is Visual Basic .NET, the *AutoEventWireup* attribute is set to *false* by Visual Studio .NET, the *CodeBehind* attribute specifies the name of the file (*WebForm1.aspx.vb*) that contains the Visual basic .NET code, and the *Inherits* attribute specifies the name of the class the web page inherits. These four attributes and other commonly used attributes are discussed further later in the chapter. Line 13 in Listing 2–1 shows that the web page uses a label control. You will learn more about ASP.NET pages and web controls later in this chapter. The code in *WebForm1.aspc.vb* is generated automatically by Visual Studio .NET.

An ASP.NET web page is essentially a web form. Web forms refer to dynamic web pages in ASP.NET. A web form can have objects such as buttons, controls, and text, each of which has behavior and appearance. Like Visual Basic forms, web forms offer an event-driven programming model to web programming so that individual objects can communicate with the server. With Visual Studio .NET, users can also enjoy rapid application development (RAD) by combining event-driven programming with the easy-to-use Web Form design tool. It is easier than ever to develop ASP.NET web pages with Visual Studio .NET.

ASP.NET ships with many new features and enhances many of the existing capabilities of ASP. It provides increased performance by running compiled code. ASP.NET is a completely new product that incorporates an environment in which web developers can take advantage of their development experience with ASP, but ASP.NET code is not fully backward compatible with ASP. The following are important features of ASP.NET.

- *Visual Studio .NET:* Visual Studio .NET is a brand new development environment that provides RAD to the server.
- *Compiled languages:* ASP.NET provides increased performance by running compiled code. Compiled languages allow you to verify that code is at least syntactically correct.
- *Web forms:* Web forms allow you to build powerful form-based web pages in Visual Basic's drag-and-drop style with server and HTML controls.
- *XML web services:* XML web services enable you to create services and make them available via HTTP.
- *ADO.NET:* ADO.NET is ADO for the .NET Framework that allows ASP.NET applications to more conveniently connect to databases.
- *ASP.NET Mobile:* ASP.NET Mobile is an extension to Microsoft ASP.NET, which targets mobile devices such as cellular phones and PDAs.
- *Better user authentication:* ASP.NET offers tighter integration with Windows-based authentication, as well as two new authentication modes: form-based authentication and passport authentication.

The Difference Between ASP.NET and ASP

As a developer, the major advantage of ASP.NET over ASP is that you can program and debug code using IDE. Moreover, ASP.NET is language independent. Many programming savvy people are developing ASP.NET using C#. Novices lean toward Visual Basic .NET. It is said that Visual Studio .NET was written in C#. With Visual Studio .NET, you do not necessarily have to know how to write the code in HTML. Instead, you can simply drag and drop web controls onto the web form and treat the coding as if you were doing it for Windows. Proper client-side code such as HTML is automatically generated for you. In ASP, however, you have to blend your VB Scripts and HTML codes. ASP.NET encompasses a completely new programming model, which is more consistent and easier to use. It provides a robust set of classes and has complete access to .NET Framework. Table 2–1 compares ASP.NET and ASP.

Table 2–1: Comparison of ASP.NET and ASP

ASP.NET	ASP
Visual Basic .NET, C#, Visual C++ .NET	VB Script
Control/event based programming–similar to VB6	Top-down programming–like most scripting languages
Separation between logic and presentation	No separation between logic and presentation
One web form per web page	N/A
Able to import	Only able to include
All data types	Only variants

Developing ASP.NET Applications Using Visual Studio .NET

Visual Studio .NET is Microsoft's IDE for visual programming based on the .NET Framework. It is much better than its predecessor, Visual InterDev, which is an IDE for developing ASP pages. Visual Studio .NET lets you drag and drop any server or HTML control onto a web form. It also allows you to create ASP.NET pages in multiple languages. You do not have to use Visual Studio .NET to create ASP.NET pages, but your life as a web developer will be improved if you do. If you happen to develop web services, you will feel the superiority of RAD when using Visual Studio .NET.

Starting the Internet Information Services (IIS)

Before you launch Visual Studio .NET, you need to make sure your Microsoft web server IIS is started. To check the status of IIS, open the Control Panel and click on Administration Tools (see Figure 2–2). Internet Information Services is one of the administration tools. Double click on Internet Information Services to open the IIS console (shown in Figure 2–3). By default, your machine name is listed under Internet Information Services (*ZHUANGV* is my machine name and local computer). Expand your machine folder and then the *Web Sites* folder and locate

Default Web Site. If IIS is started, a Running status will be shown under State on the right-hand side. If a Stopped status is shown under State, highlight Default Web Site and then click on the Start button (a button with a right arrow ▶) to start the web site.

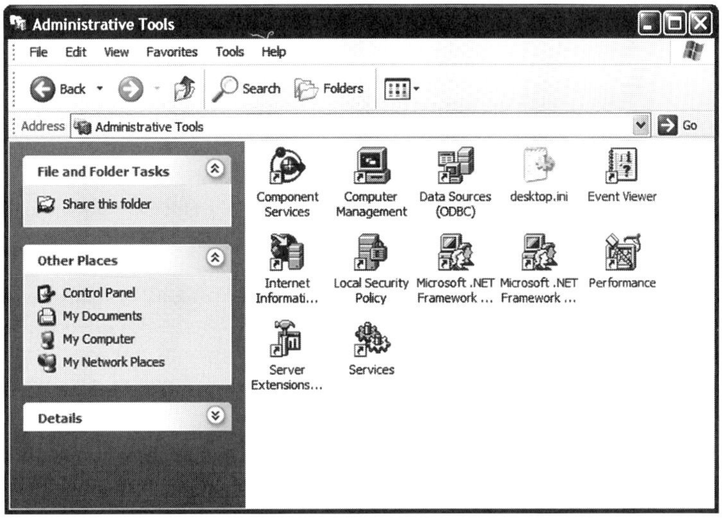

Figure 2–2. Administration tools.

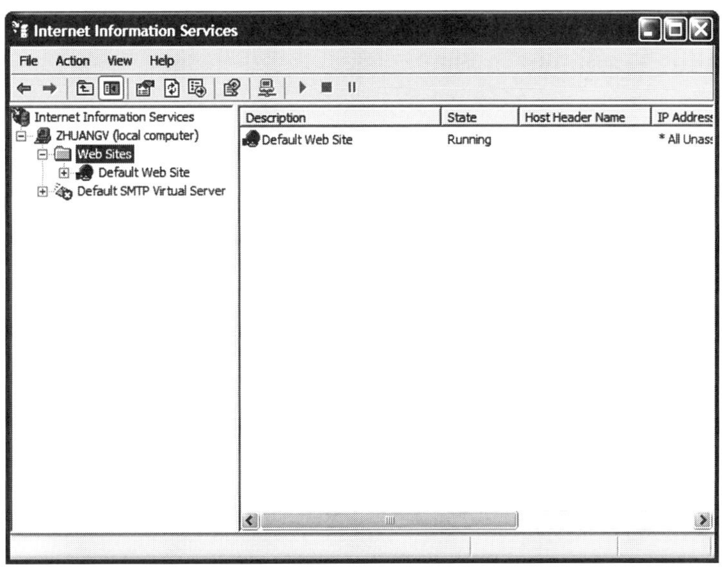

Figure 2–3. The Microsoft Internet Information Services console.

Creating a Visual Studio .NET Project

You build an ASP.NET page by creating a Visual Studio .NET project. First launch your Visual Studio .NET (see Figure 2-4) and then click on the New Project button to access the New Project dialog box, shown in Figure 2-5. When you create a new Visual Studio .NET project, you need to specify a virtual path instead of physical path. The Visual Studio .NET project will then be stored in that visual directory. Visual Studio .NET lets you create ASP.NET applications in Visual Basic .NET, Visual C#, Visual C++ .NET, and Visual J#. As mention in the Introduction, this book focuses on Visual Basic .NET. In Figure 2-5, the project type was chosen as Visual Basic Projects. There are more than a dozen templates that ship with Visual Studio .NET. Those most commonly used for developing web applications are ASP.NET Web Application, ASP.NET Web Service, ASP.NET Mobile Web Application, Web Control Library, and Class Library. If you want to build a web application, select the ASP.NET Web Application template (shown in Figure 2-5).

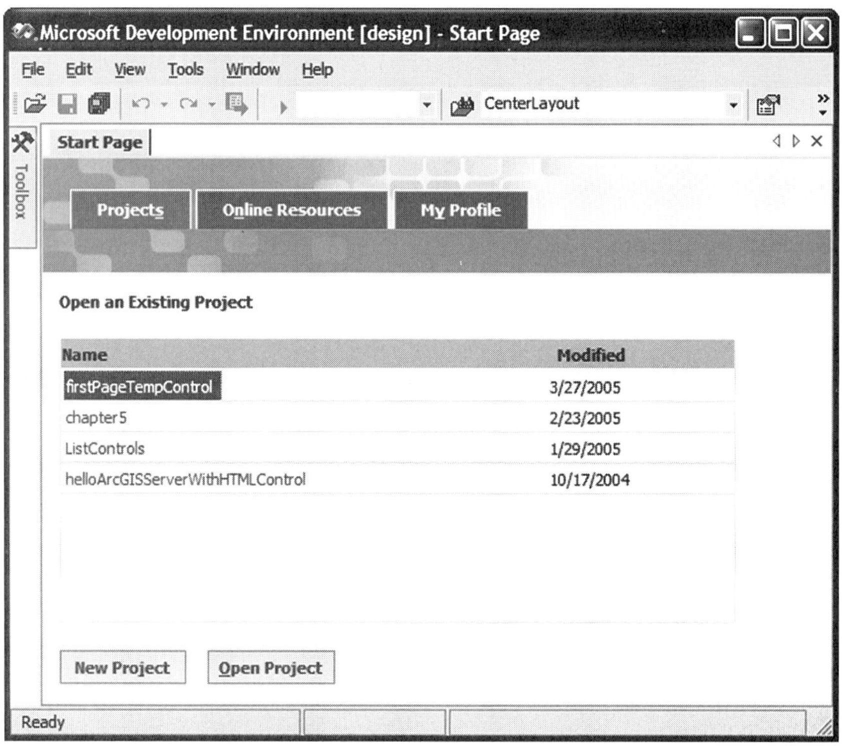

Figure 2-4. Start page of Visual Studio .NET 2003.

✔ **NOTE:** *The start page of Visual Studio .NET, which is also called Visual Studio .NET 2002, is different from what is shown in Figure 2–4. As mentioned in the Introduction, you need to use Visual Studio .NET 2003 to run all examples in this book. So if you don't see a similar Start Page, please refer to your version of Visual Studio .NET.*

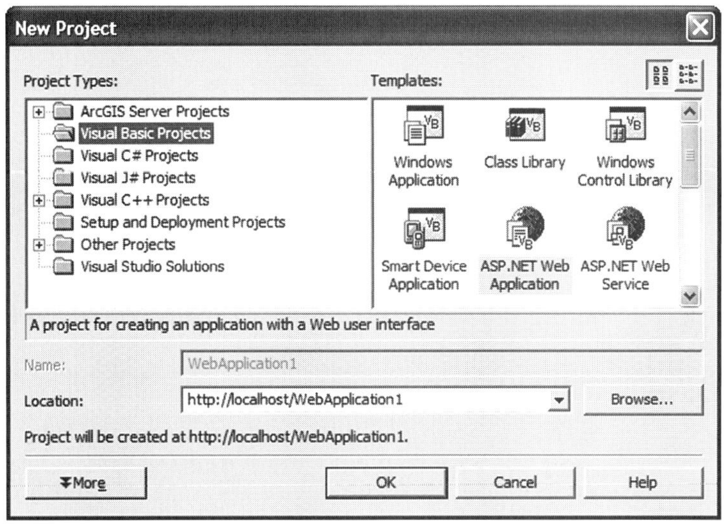

Figure 2–5. Visual Studio .NET New Project dialog box.

Visual Studio .NET generates classes for a project and assigns them a namespace named after the project. The classes will be compiled into an assembly with the same name as the project. The assembly is created when you build the solution and place it in the *bin* subdirectory of the virtual directory. The namespace name, as well as the name of the project's assembly, can be changed by right-clicking on the project folder in Solution Explorer and choosing Properties, or by clicking on the Project menu and selecting Properties. Figure 2–6 shows the project Property Pages dialog box for a project in Visual Studio .NET.

Visual Studio .NET assigns a default project name of *WebApplication1* (or *WebApplication2* if *WebApplication1* is taken) if you do not specify one. A virtual directory named *WebApplication1* is created after you click on the OK button on the New Project dialog box. Upon the creation of the project, an empty ASP.NET web page named *WebForm1.aspx* is generated in the project virtual directory (Figure 2–7). If your web server is not started, no project will be created because Visual Studio .NET cannot create a virtual directory under the web root (most commonly *C:\inetpub\wwwroot*).

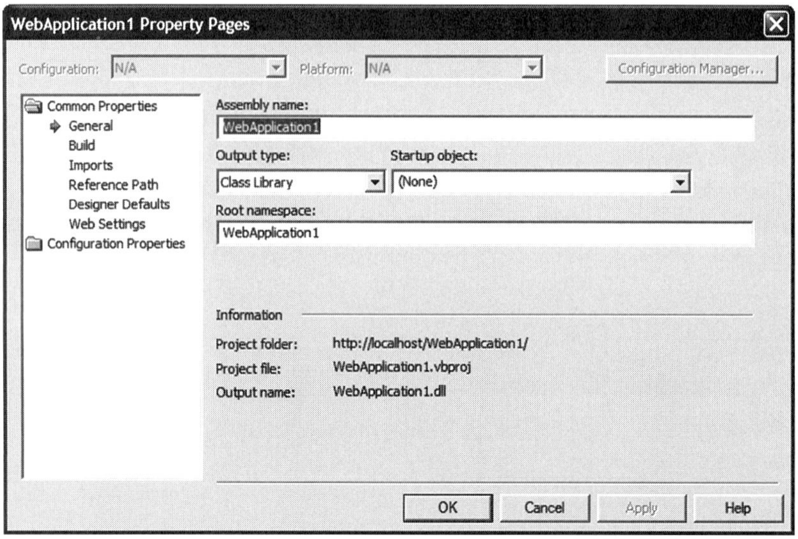

Figure 2–6. Project and assembly Property Pages dialog box.

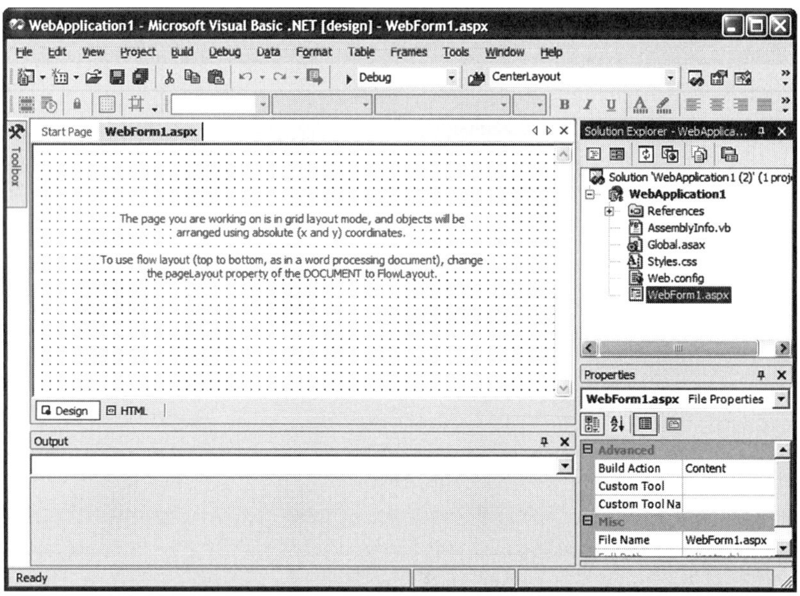

Figure 2–7. The WebApplication1 *project shown through Visual Studio .NET.*

In general, a Visual Studio .NET project upon opening displays a web form in a grid-layout area, a Solution Explorer window on the right-hand side, and a Properties window beneath the Solution Explorer window. If you do not see the Solution Explorer or the Properties Window, go to the Project menu and select Solution Explorer and Properties Window to display them.

Create an ASP.NET Page of "Hello ArcGIS Server"

A great tradition in any introductory programming book is to write the "Hello World" program in that language. We will change that tradition a little bit and derive a "Hello ArcGIS Server" program instead of the traditional "Hello World" program. Hello ArcGIS Server is the name given to a short program that demonstrates how the simplest web page can be built in Visual Basic .NET and run. You can write the web page by creating one object, setting one property, and entering one line of code. The following are the steps.

1 Click on Label Control on the Web Forms tab of the Visual Studio .NET toolbox and drag it onto the Web Forms Designer surface (the grid-layout area of the web form). (See Figure 2–8.)

Label Control is used to display any HTML text within a predefined area. Labels are static text, meaning that users cannot edit the text once the HTML page is rendered. There is no handler event associated with the Label Control option. After you have placed the label control on the web form, you can resize it to your requirements.

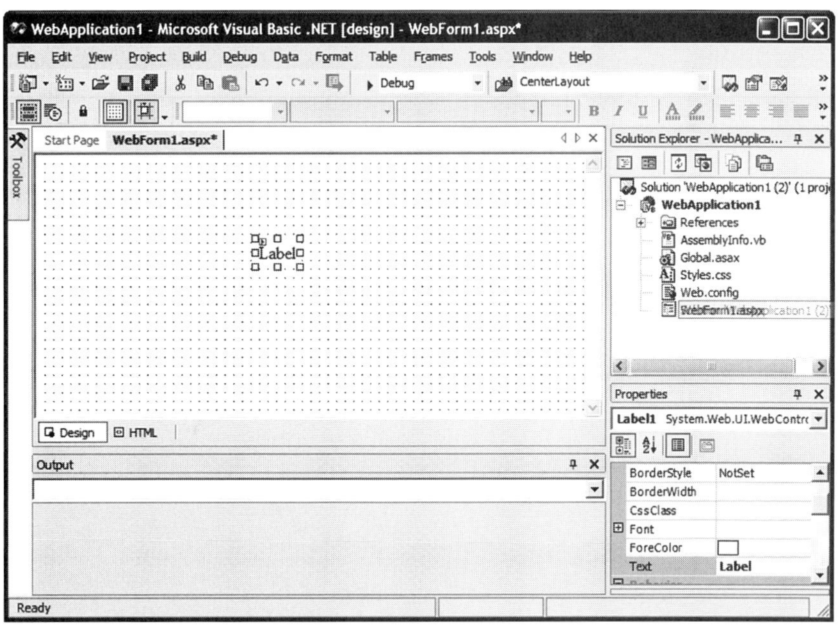

Figure 2–8. Web form with a label control.

2 Set the Text property for the label control to *Hello ArcGIS Server.*

By default, the Text property should show up in the Properties window. If not, scroll down in the Properties window until you see the Text property. The default Text property value is *Label*. To change the value, highlight *Label*, type *Hello ArcGIS Server*, and press Enter. The greeting message *Hello ArcGIS Server* will display on the web form (Figure 2–9).

3 Select the Debug menu and then Start, or click on the Start button on the Standard toolbar (or simply press F5).

After a couple of seconds a web page containing the HTML text *Hello ArcGIS Server* will display (Figure 2–10).

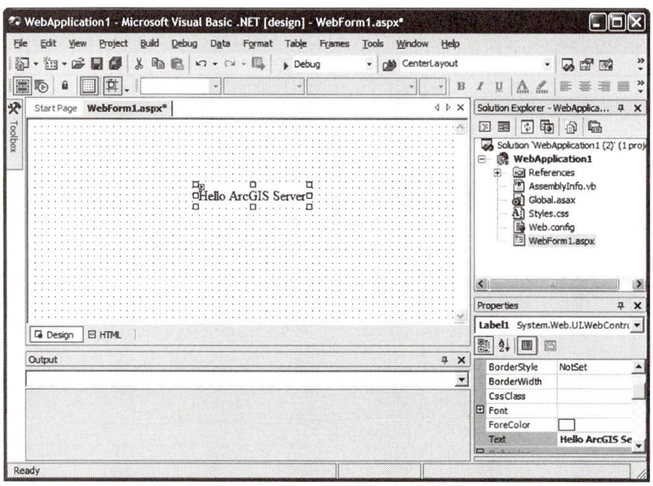

Figure 2–9. Label control with the Text property set to Hello ArcGIS Server.

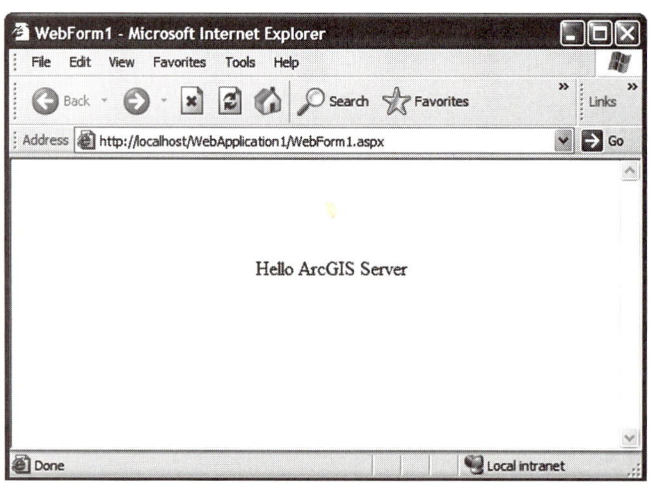

Figure 2–10. An ASP.NET web page showing Hello ArcGIS Server.

Congratulations—you have just successfully created an ASP.NET web page and have joined the club of programmers who have written a Hello ArcGIS Server program. As you have experienced, it is very easy to create an ASP.NET web page using Visual Studio .NET.

Visual Studio .NET Project Files

Aside from the pages, classes, and resources that actually comprise the application, a number of additional files and an application folder called *bin* form a Visual Studio .NET project. In a Visual Basic .NET project named *WebApplication1*, all of these files are stored at the project web folder *C:\Inetpub\wwwroot\WebApplication1* (Figure 2-11).

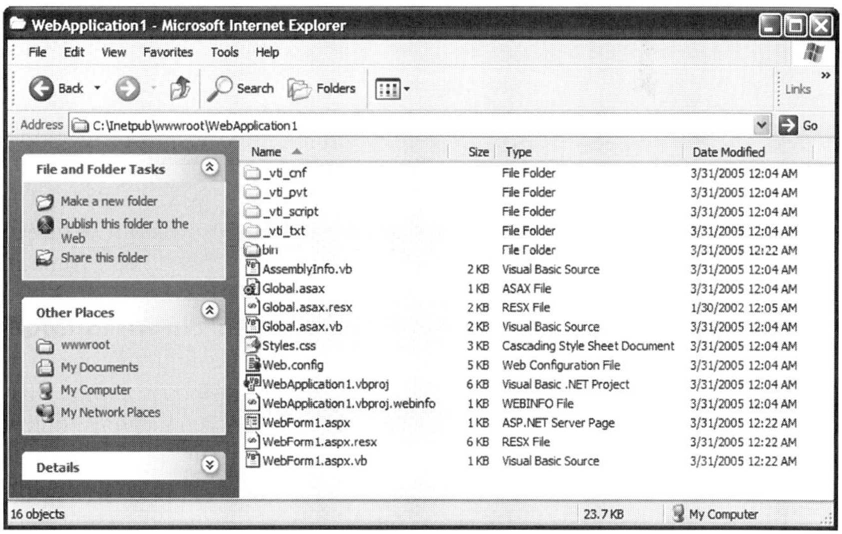

Figure 2-11. Project files for the WebApplication1 *ASP.NET project.*

- *Application* \bin *folder:* This is located immediately under the application root, as the local assembly cache. The *\bin* directory is also configured to deny any browser access so that a remote client cannot download and steal the code. For the *WebApplication1* project, *WebApplication1.dll* and *WebApplication1.pdb* are stored under the *\bin* folder, where the DLL file is the assembly for the project and the *.pdb* file is the Program Debug Database file. The benefit is that no registration is required to make components available to the ASP.NET Framework application. Components can be deployed by simply copying to the *\bin* directory or performing an FTP file transfer.

- *Global.asax:* This is a collection of event handlers you can use to change and set settings in your web site. High-level application events such as *Application_Start*, *Application_End*, *Session_Start*, *Session_End*, and so on are handled in this file. In *Global.asax* there are many more events to work with (and you can create your own) than there are in *global.asa* (used in classical ASP), which gives you a lot more customization when it comes to your application.
- *Global.asax.resx:* This is an XML schema file that contains the resources for *Global.asax*.
- *Global.asax.vb:* This is the code-behind class for *Global.asax*. The class name *Global* by default inherits from *HttpApplication*.
- *Styles.css:* This is a Cascading Style Sheets (CSS) file for the project.
- *Web.config:* This is an XML file that contains configuration information for the web application.
- *WebApplication1.vbproj:* This is an XML file that contains build and file information for the project. The `<Build>` section contains the application settings and the assemblies to link to, whereas the `<Files>` section contains the list of the files in the project.
- *WebApplication1.vbproj.webinfo:* This is an XML file that contains the *.vbproj* file with the absolute virtual path (i.e., URL). The information in this file complements the content of the previous file in that *WebApplication1.vbproj* does not contain absolute path information.
- *WebForm1.aspx:* This is an ASP.NET page that renders in HTML and contains the processing directive information for the page.
- *WebForm1.aspx.resx:* This is an XML schema file that contains the resources for the web form.
- *WebForm1.aspx.vb:* This is a Visual Basic .Net file that contains the Visual Basic .NET code that is code-behind.

ASP.NET Page Object Model

An ASP.NET web page consists of web controls that generate a rich HTML-based user interface and interact with the user through events. When a web browser issues an HTTP request, the HTTP runtime in ASP.NET handles the request, routes it through the applicable components for processing, generates a response, and then sends the request back to the web client.

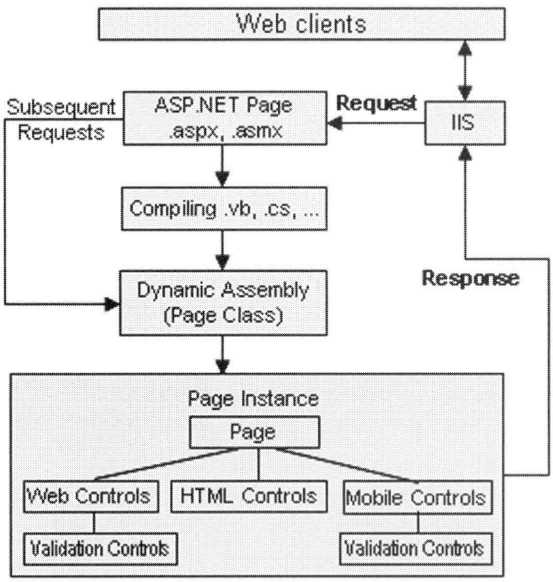

Figure 2–12. ASP.NET page object model.

As shown in Figure 2–12, an incoming request is handled by IIS, which hands the request to the HTTP runtime in ASP.NET and routes it to the page handler for processing. If it is the first request, ASP.NET parses and compiles the page into an assembly that includes the page class and other classes that support the page and stores the assembly on the disk. ASP.NET then executes the binary code to create an instance of the page class. The instantiated page object is used to process the applicable events and produces the HTML output for the page. The resulting response is then sent back to the web client that issued the request. After the first request, the assembly is retrieved ASP.NET pages are simply text files with the *.aspx* file name extension that can be placed on the web server with ASP.NET installed. When a browser requests an ASP.NET page, the ASP.NET runtime parses and compiles the requested *.aspx* page into an instance of a .NET Framework class that inherits from the base *Page* class. An ASP.NET page consists of the following elements.

- Directive
- ASP.NET server controls
- Code-behind

These elements are brand new and innovative, which again shows how advanced ASP.NET technology is over classic ASP.

Page Directive

An ASP.NET web page is represented as a Page class. The Page class serves as a “container” for all elements on the page and is globally available throughout the page. Thus, you do not have to explicitly reference it when accessing its properties and methods.

An ASP.NET web page starts with the *Page* directive defined by special sequences of characters beginning with < %@ and ending with % >. The *Page* directive defines page-specific attributes used by the ASP.NET page parser and compiler and can only be included in *.aspx* files. This directive is where you specify the scripting language, code-behind, inherits, and so on. You can also set a proper attribute to see detailed error messages while debugging the page. Fortunately, Visual Studio .NET automatically generates four attributes of *Language*, *AutoEventWireup*, *CodeBehind*, and *Inherits*. Table 2–2 outlines those attributes commonly used with the *Page* directive. The general format for this directive follows (text between [...] is optional).

```
<%@ Page attribute="value" [attribute="value"...]%>
```

The *Page* directive can be used only in web form pages. You can include only one *Page* directive per *.aspx* file. To define multiple attributes for the directive, use a space-separated list (do not include a space on either side of the equal sign of a specific attribute, as in *trace*= "*true*").

Table 2–2: Commonly Used Page Directive Attributes

Attribute	Description
AutoEventWireup	Indicates whether the page's events are autowired. *True* if event autowiring is enabled; otherwise, *false*. The default is *true*. *AutoEventWireup* requires that the page event handlers have specific, predictable names. This limits your flexibility in how you name your event handlers. Therefore, in Visual Studio .NET, the *AutoEventWireup* attribute is set to *false* by default and the designer generates explicit code to bind page events to methods.
CodeBehind	Specifies the name of the compiled file that contains the class associated with the page. This attribute is used by the Visual Studio .NET Web Forms designer. It tells the designer where to find the page class so that the designer can create an instance of it for you to work with at design time. For example, if you create a web form page in Visual Studio .NET called *WebForm1*, the designer will assign the *CodeBehind* attribute the value *WebForm1.aspx.vb* for Visual Basic .NET. This attribute is not used at runtime.
Debug	Indicates whether the page should be compiled with debug symbols. *True* if the page should be compiled with debug symbols; otherwise, *false*.
EnableSessionState	Defines session-state requirements for the page. *True* if session state is enabled; *ReadOnly* if session state can be read but not changed; otherwise, *false*. The default is *true*. These values are case insensitive.

Attribute	Description
EnableViewState	Indicates whether view state is maintained across page requests. *True* if view state is maintained; otherwise, *false*. The default is *true*.
Inherits	Defines a code-behind class for the page to inherit. This can be any class derived from the *Page* class.
Language	Specifies the language used.
Trace	Indicates whether tracing is enabled. *True* if tracing is enabled; otherwise, *false*. The default is *false*.
TraceMode	Indicates how trace messages are to be displayed for the page when tracing is enabled. Possible values are *SortByTime* and *SortByCategory*. The default, when tracing is enabled, is *SortByTime*.
ValidateRequest	Indicates whether request validation should occur. If *true*, request validation checks all input data against a hard-coded list of potentially dangerous values. If a match occurs, an *HttpRequestValidationException* class is thrown. The default is *true*.

ASP.NET Server Controls

The server controls are the core of ASP.NET. They represent dynamic elements with which a web page can interact. ASP.NET server controls make a web page truly dynamic, although the page is still an HTML page. Unlike a web page written in ASP, Visual Basic, Perl, or PL/SQL (if you are developing DHTML web pages in Oracle), ASP.NET allows you to individually interact with each server control without the need of refreshing the entire web page. A complete page refresh is required to publish a DHTML web page written in the languages previously cited. There are five basic types of server controls—including HTML controls, web controls, validation controls, and mobile controls—as well as user controls you can design yourself.

✔ **NOTE:** *According to Microsoft, ASP.NET includes four types of server controls: HTML controls, web controls, validation controls, and user controls. Because mobile controls become part of ASP.NET, there are five types of server controls now. However, mobile controls are based on user controls. There are also custom controls, which are controls derived from the base web control class.*

All ASP.NET server controls reside within a *<form runat="server">* tag in order to function properly. The only two exceptions are the *Html-*

GenericControl and the label control, which do not have any method associated with them. To a web client/browser, the resultant presentation of an ASP.NET web page is still in HTML, except that most of the HTML form elements are "live."

Because ASP.NET is all about server controls, we will be examining them in great details as we move through this book. For example, each family of server controls is discussed in the "ASP.NET Components" section. Chapter 5 deals with web controls and validation controls. Custom controls are explored in Chapter 7, and ASP.NET mobile controls are examined in Chapter 9.

Code-behind

Code-behind is a new feature designed to clearly separate code and presentation in an ASP.NET page. This feature reduces labor and time among a team of developers and designers and increases the readability and maintainability of code and content. With code-behind, developers can work on code while designers work on page layout. When you build a Visual Studio .NET project, the page class is compiled into the project's assembly and the runtime link between the *.aspx* file and its base class is defined by the *Inherits* attribute (*helloArcGISServer.WebForm1* in Listing 2-1, for example) in the *Page* directive. The design-time link between the *.aspx* file and its code-behind class is defined by the *CodeBehind* (*WebForm1.aspx.vb* in Listing 2-1, for example) attribute in the *Page* directive. The separation is achieved by using the *.aspx* file and the *.aspx.vb* file for the layout and the code, respectively.

ASP.NET also allows you to simply embed code in a *runat* = "*server*" script block within the *.aspx* file itself, which is sometimes referred to as inline code. It looks like the classic ASP coding style, although it specifies the script to be run at the server (*runat* = "*server*"). If you do not have Visual Studio .NET, the form of inline code may work well with any text editor (such as Notepad).

ASP.NET Components

ASP.NET pages are web forms. Each page has one web form. Four native types of web controls can be used on web forms: HTML controls, web controls, validation controls, and mobile controls. Mobile controls are discussed in detail in Chapter 9. All of these controls are server controls, meaning that they are rendered by the web server. You can also build your own custom controls or modify existing controls.

HTML Controls

HTML controls correspond to standard HTML 3.2 elements, such as buttons, input boxes, tables, and so on. However, standard HTML elements are interpreted by a web browser. To make those elements programmable, you can simply add a *runat* = "*server*" attribute to each HTML element. This attribute indicates that the element should be treated as a server control. The advantage over standard HTML elements is that the properties of HTML controls can be referenced programmatically. Before HTML controls existed, we could write code to generate HTML dynamically and add some JavaScript code to communicate among standard HTML elements. With HTML controls, you can simply write server-side code to deal with that specific control without refreshing the web page. You can also change the properties or behaviors through programming. HTML controls are defined in the namespace of *System.Web.UI.HtmlControls*. Table 2–3 outlines all HTML controls available in ASP.NET. An HTML control can be selected from the Visual Studio .NET toolbox under the HTML tab (shown in Figure 2–13).

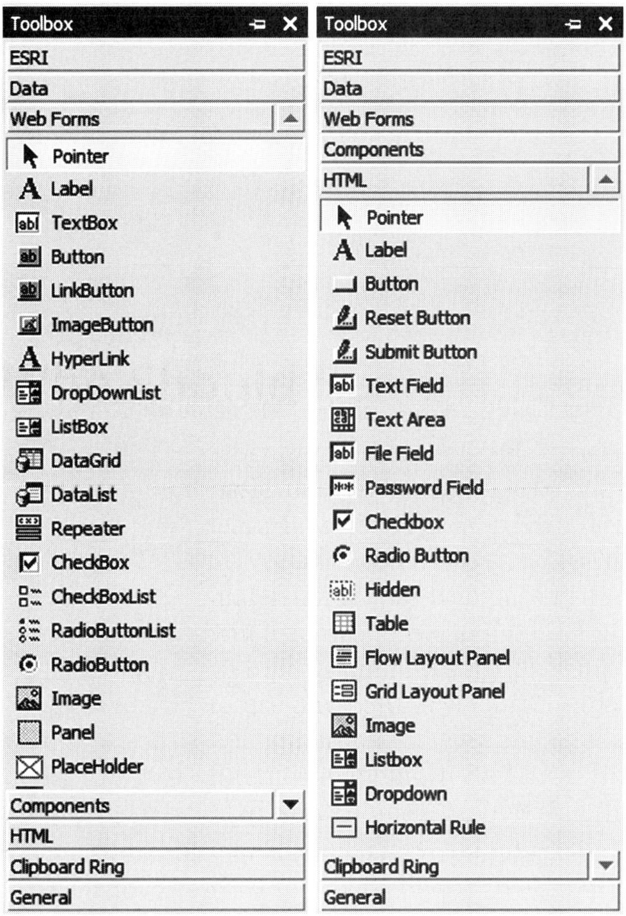

Figure 2–13. HTML controls and web controls in the Visual Studio .NET toolbox.

Table 2–3: HTML Controls in ASP.NET

HTML Control Class	HTML Control	HTML Element Description
HTMLContainer-Control	HtmlAnchor	Controls HTML hyperlink *<a>*
	HtmlButton	Controls HTML push button *<button>* (only available in HTML 4.0)
	HtmlForm	Controls an HTML form element *<form>*
	HtmlGenericControl	Controls other HTML elements not specified by a specific HTML server control, such as *<body>*, *<div>*, and *<span>*
	HtmlSelect	Controls HTML list of options *<select>*
	HtmlTable	Controls an HTML table *<table>*
	HtmlTableCell	Controls HTML table cell elements *<td>* and *<th>*
	HtmlTableRow	Controls HTML table row *<tr>*
	HtmlTextArea	Controls an HTML text area *<textarea>*
HTMLImage	HtmlImage	Controls HTML image *<img>*
HTMLInputControl	HtmlInputButton	Controls HTML form buttons, including *<input type="button">*, *<input type="submit">*, and *<input type="reset">*
	HtmlInputCheckBox	Controls HTML checkbox *<input type="check-box">*
	HtmlInputFile	Controls HTML input file *<input type="file">*
	HtmlInputImage	Controls HTML input image *<input type="image">*
	HtmlInputHidden	Controls HTML input hidden *<input type=" hidden">*
	HtmlInputRadioBut-ton	Controls HTML radiobutton element *<input type="radio">*
	HtmlInputText	Controls HTML input text or input password *<input type="text">* and *<input type= "password">*

Web Controls

Like HTML controls, web controls are also created on the server and require a *runat* = "*server*" attribute to work. They provide richer functionality than HTML controls. Unlike HTML controls, web controls are newly designed by Microsoft for ASP.NET without needing to be compliant with the HTML syntax but generating valid HTML. Web controls are characterized by the *asp* prefix. For example, the web control namespace defines the Button control and makes it available through the *<asp:button>* tag, whereas the HTML control namespace provides the *HtmlInputText* control and declares it using the *<input>* tag. The output that HTML controls and web controls produce is almost the same. Choosing one over the other is largely a matter of preference. Using web controls provides you a far richer development model and exception management. Web controls are defined in the namespace of *System.Web.UI.WebControls*. Table 2–4 outlines all web controls available in ASP.NET. A web control can be selected from the Visual Studio .NET toolbox under the Web Forms tab (see Figure 2–13).

Table 2–4: Web Controls in ASP.NET

Web Control	Description
AdRotator	Displays a sequence of images
Button	Displays a push button
Calendar	Displays a calendar
CheckBox	Displays a check box
CheckBoxList	Creates a multiselection check box group
DataGrid	Displays fields of a data source in a grid
DataList	Displays items from a data source by using templates
DropDownList	Creates a drop-down list
HyperLink	Creates a hyperlink
Image	Displays an image
ImageButton	Displays a clickable image
Label	Displays programmable static content (lets you apply styles to its content)
LinkButton	Creates a hyperlink button

Web Control	Description
ListBox	Creates a single-selection or multiselection drop-down list
Literal	Displays programmable static content (does not let you apply styles to its content)
Panel	Provides a container for other controls
PlaceHolder	Reserves space for controls added by code
RadioButton	Creates a radio button
RadioButtonList	Creates a group of radio buttons
Repeater	Displays a repeated list of items bound to the control
Table	Creates a table
TableCell	Creates a table cell
TableRow	Creates a table row
TextBox	Creates a text box
Xml	Displays an XML file or the results of an XSL transform

Validation Controls

With classic ASP, we often write extensive JavaScript code to check the validity of an HTML form. Things such as data, time, phone number, e-mail address, and latitude and longitude coordinates require validation. Those JavaScript codes are browser specific and are not very easy to maintain. They are not part of the ASP scripts and may change from application to application. On the other hand, a web page needs validation. The stability of a web site relies on the acceptance of valid data from the user. Validation at the server level also reduces the complexity of coding on the server. Fortunately, in ASP.NET a series of server controls called validation controls are provided for validating data. Validation controls can be used to validate the data of the form fields within a web page. If the data do not pass validation, the control will display an error message to the user. Validation controls inherit from the *BaseValidator* class, which is a child of the *Label* class. Table 2–5 lists all validation controls in ASP.NET. A validation control can be selected from the Visual Studio.NET toolbox under the Web Forms tab. Validation controls are grouped with web controls in the toolbox. You need to scroll down the Web Forms tab to see all validation controls (Figure 2–14).

Table 2–5: ASP.NET Validation Controls

Validation Control	Description
CompareValidator	Compares the value of one input control to the value of another input control or to a fixed value using an operator such as *LessThan*, *Equal*, or *GreaterThan.*
CustomValidator	Allows you to write a custom code to handle the validation of the value entered.
RangeValidator	Checks that a value entered by the user falls between two values.
RegularExpressionValidator	Validates that the value of an input control matches a pattern defined by a regular expression.
RequiredFieldValidator	Makes an input control a required field.
ValidationSummary	Displays all validation error messages found on a web page.

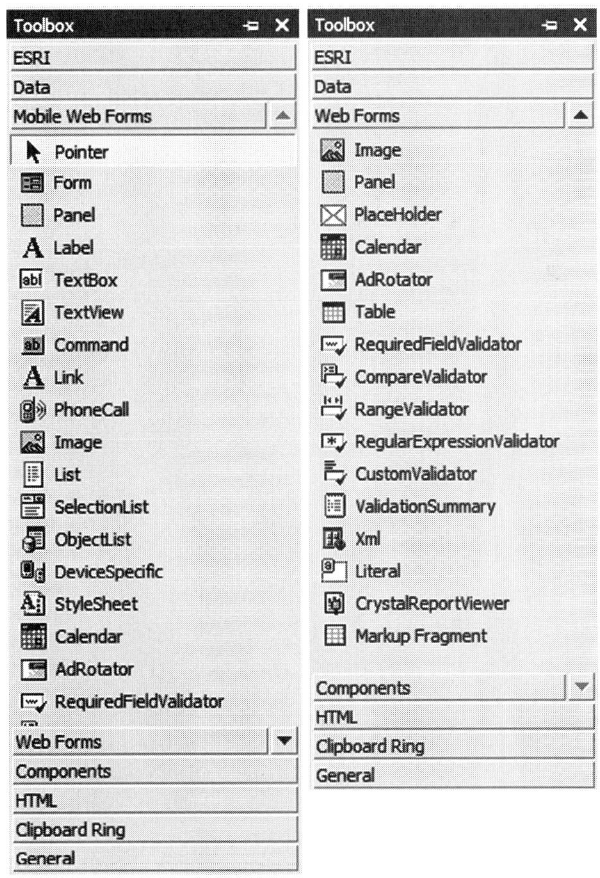

Figure 2–14. Validation controls and mobile controls in the Visual Studio .NET toolbox.

Mobile Controls

The ASP.NET mobile controls are formerly known as the Microsoft Mobile Internet Toolkit (MMIT) and are now an extension for ASP.NET version 1.1. They provides the tools you need to quickly develop mobile web applications for a wide variety of wireless devices, including cell phones, pagers, Pocket PCs, Palm, Blackberry, and other PDAs. Each individual device uses its own web browser. Most of those web browsers are based on Compact HTML (cHTML) or Wireless Markup Language (WML). A few of them are based on HTML version 3.2. In addition, wireless devices come with different screen sizes, orientations, and device capabilities. The ASP.NET mobile controls make it possible to develop a single application that can target all of the devices. They render the appropriate markup language (cHTML, WML, HTML 3.2) while dealing with different screen sizes, orientations, and device capabilities. Because ASP.NET mobile controls are part of ASP.NET, they extend the power of the Microsoft .NET Framework and Visual Studio .NET to build mobile web applications by enabling ASP.NET to deliver markup to a wide variety of mobile devices. ASP.NET mobile controls are defined in the namespace of *System.Web.UI.MobileControls*. A mobile control can be selected from the Visual Studio .NET toolbox under the Mobile Web Forms tab. The mobile controls are grouped into three categories.

- User Interface Controls
- Validation Controls
- Utility Controls

Table 2–6 lists all ASP.NET mobile controls by these three categories.

✔ **NOTE:** *Mobile controls can only be placed on mobile web forms. When you create your project, make sure to choose the ASP.NET Mobile Web Application template to begin with. Otherwise, all mobile controls will be dimmed out in the Visual Studio .NET toolbox.*

Table 2–6: ASP.NET Mobile Controls

Control Type	Mobile Control	Description
UI Controls	Command	Performs an action
	Form	Defines a container for mobile controls
	Image	Defines an image

Control Type	Mobile Control	Description
	Label	Defines a text
	Link	Defines a hyperlink
	List	Defines a list
	MobilePage	Defines a base class for all mobile pages
	ObjectList	Defines a list of data objects
	Panel	Defines a container for other controls
	SelectionList	Defines a list to select from
	StyleSheet	Defines styles to apply to other controls
	TextBox	Defines a single line input box
	TextView	Defines a multiline input box
Validation Control	CompareValidator	Compares the value of one input control to the value of another input control or to a fixed value
	CustomValidator	Allows you to write a method to handle the validation of the value entered
	RangeValidator	Checks that the user enters a value that falls between two values
	RegularExpressionValidator	Ensures that the value of an input control matches a specified pattern
	RequiredFieldValidator	Makes an input control a required field
	ValidationSummary	Displays a report of all validation errors in a page
Utility Control	AdRotator	Displays advertisements
	Calendar	Displays a calendar
	PhoneCall	Calls a telephone number

The Use of View State for State Management

In classic ASP, the *Session* object on the server and cookies on the client were used to control the state of a web page. However, *Session* state is not scalable, and cookies cannot be depended on for all applications. ASP.NET provides a new mechanism called *View State*. View State maintains the state of a web page as it travels back and forth between the server and the client. ASP.NET server controls automatically retain their data when a user sends a page to the server. In the past, we typically used hidden variables or text inputs of an HTML form to keep track of the status of HTML form elements. We no longer need to do that. ASP.NET pages automatically retain all information contained within HTML form elements.

View State is enabled for every web page by default. If you do not intend to use View State, you can turn it off, which will result in a slight performance gain in your web pages. To do this, set the *EnableViewState* property of the *Page* directive to *false* as follows.

```
<%@ Page EnableViewState="False" %>
```

Debugging ASP.NET

No matter how skilled a programmer you are, you are bound to make mistakes once in a while. Tracking down bugs in your code can be baffling without the appropriate tool. Fortunately, the compiled nature of ASP.NET means that debugging web applications is no different from debugging any other managed applications. Visual Studio .NET, like any IDE, provides a powerful debugging tool.

Debugging a Page

In Visual Studio .NET, there several commands under the Debug menu (Figure 2-15) to assist you with debugging. (Note that the Debug menu changes when you switch to debugging mode.)

- *Windows:* This command allows you to open the Breakpoints window or Immediate Command window.
- *Start:* This command initiates the debugging process.
- *Start Without Debugging:* This command runs the program without debugging.
- *Processes:* This command accesses the Processes dialog box to let you view and manipulate programs.

- *Exceptions:* This command accesses the Exceptions dialog box to let you change the way the debugger handles exceptions or categories of exceptions.
- *Step Into:* This command steps into a statement or a function, and then halts at the first line of code if it is a function.
- *Step Over:* This command executes the entire function, and then halts at the first line of code immediately after the function.
- *New Breakpoints:* This command accesses the New Breakpoints dialog box to allow you to create a new breakpoint.
- *Clear All Breakpoints:* This command clears all break points that have been set.

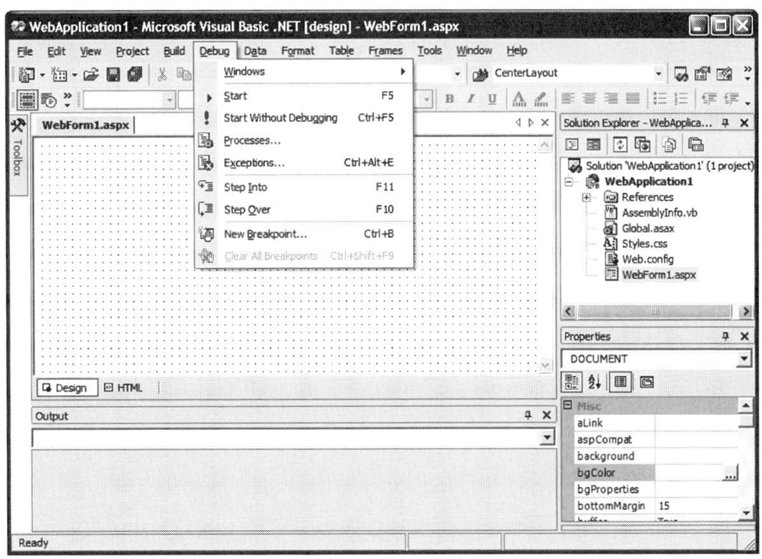

Figure 2–15. Debug menu of Visual Studio .NET.

Let's create a new ASP.NET project and learn how to use the debugging tool. The Visual Studio .NET. debugging tool let you debug the code-behind Visual Basic .NET code. We are going to modify the Hello ArcGIS Server project by adding a Button control to it. Instead of showing *Hello ArcGIS Server* at the startup of the web page, the greeting message will display when the button is clicked. The following are the steps for creating a new project.

To create a web form:

1 Launch Visual Studio .NET and then click on the New Project button. The New Project dialog box appears.

2 In the Project Type pane, click on Visual Basic Projects and in the Templates pane select ASP.NET Web Application.

3 Name your application *WebApplication2* in the Location box (*http://localhost/WebApplication2*).

4 Click on OK.

To add the web controls:

5 Click on the Label control on the Web Forms tab of the toolbox and then drag it onto the Web Forms Designer surface. Change the text property setting for the label control from Label to blank. The text property of the Label control will be set at runtime.

6 Click on the Button control on the Web Forms tab of the toolbox and then drag it onto the Web Forms Designer surface. Align the button with the label (Figure 2–16).

7 Double click on the Button control and add the following Visual Basic .NET code to the *Button1_Click* method (subroutine).

```
'Declare a string variable for the greetings
Dim myGreetings As String = "Hello ArcGIS Server"
Label1.Text = myGreetings
```

8 Click on the Start button or press F5 to run the application. A web page with the button displays (Figure 2–17).

9 Click on the button on the web page. A web page with the button and a text of *"Hello ArcGIS Server"* displays (Figure 2–18).

10 Close the web browser.

To debug the application:

11 Locate the Dim line in *Button1_Click* and then click on the outline area in front of the line. A solid dot is placed on the outline area and the entire Dim line is highlighted in the same color (brown by default) as the solid dot (Figure 2–19).

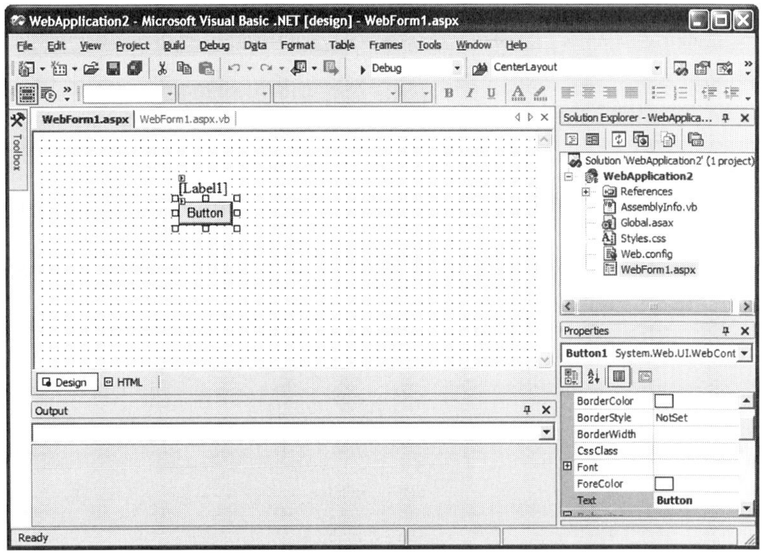

Figure 2–16. Web form with label control and button control.

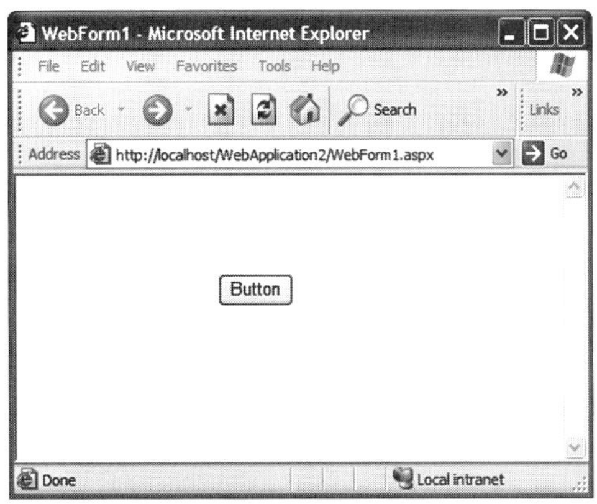

Figure 2–17. A web page before showing Hello ArcGIS Server.

Figure 2–18. A web page showing the button and the text "Hello ArcGIS Server".

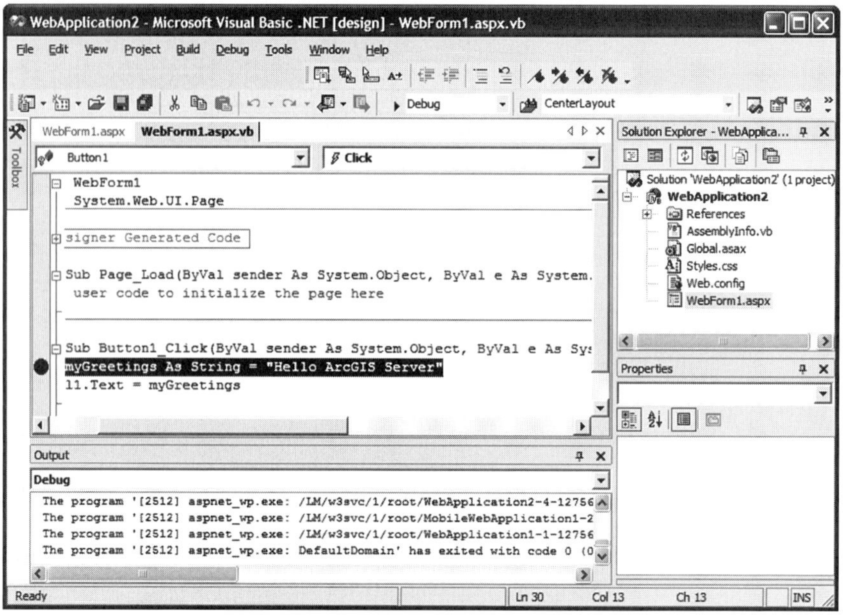

Figure 2–19. An example of setting a break point for debugging.

12 Click on the Start button or press F5 to start running the application in debug mode. A web page similar to that shown in Figure 2–17 displays.

13 Click on the button on the web page. The application stops at the line where the break point is set. A right arrow (yellow by default) displays inside the solid dot, and the line is highlighted in yellow by default (Figure 2–20). Note also that the Autos window appears in the same area where the Output window was, showing *Label1.Text* with a blank string and *myGreetings* with nothing because both declaration and assignment are not executed yet.

14 Select the Debug menu and then the Step Into command (or simply press F11). Now the yellow arrow moves to the next line and that line is highlighted in yellow by default, with the cursor blinking at the beginning of the line (Figure 2–21). Note also that the Autos window still shows *Label1.Text* empty because it has not been assigned yet, and shows *myGreetings* with a value of *Hello ArcGIS Server* because *myGreetings* has been just declared and assigned. You can also move your mouse cursor onto *Label1.Text* or *myGreetings* to see a display of the variable value at the mouse cursor (Figure 2–22).

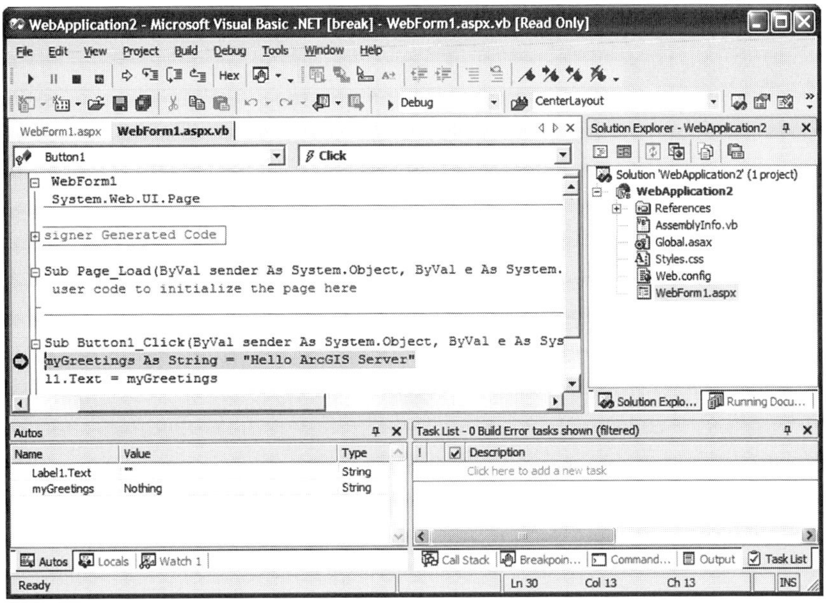

Figure 2–20. An example of showing the break point at runtime.

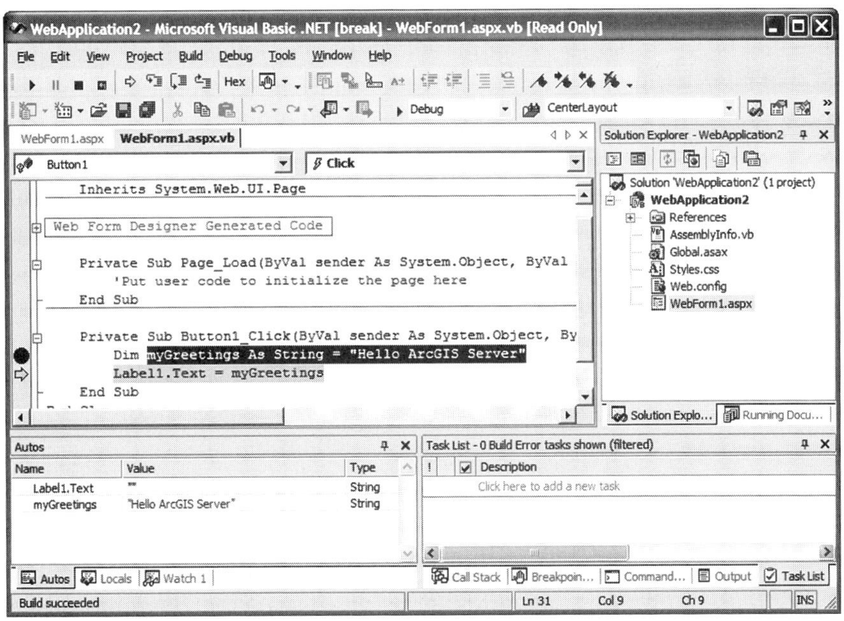

Figure 2–21. An example of executing the Step Into command.

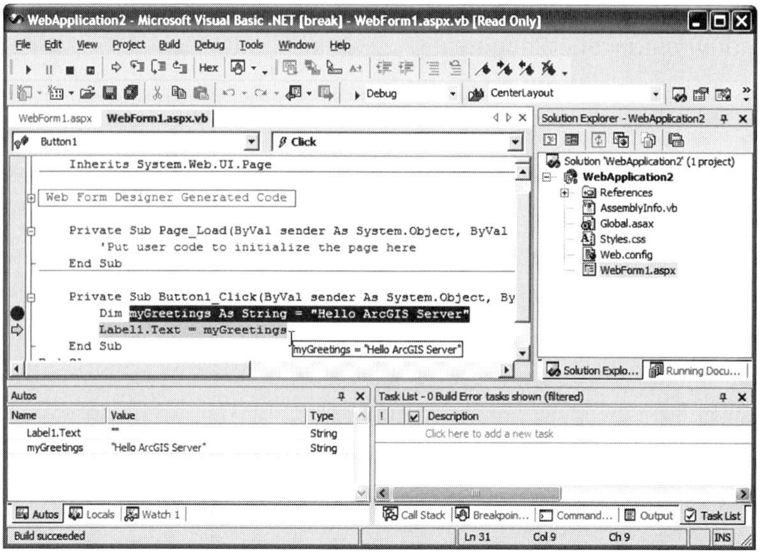

Figure 2–22. An example of instantly showing a variable value at the mouse cursor.

15 Highlight the *myGreetings* variable and click the right mouse button to bring up a menu. Select QuickWatch from the menu. The QuickWatch dialog box displays and shows the current value of the *myGreetings* variable (Figure 2–23).

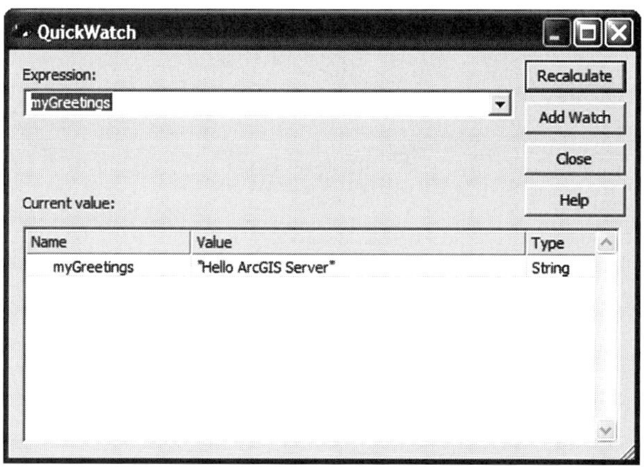

Figure 2–23. An example of using the QuickWatch dialog box to view a variable value.

16 Press F11 again (you can also press F10 and select Step Over). Now the yellow arrow moves to the End Sub line and that line is highlighted.

17 Select the Debug menu and then Continue, or click on the Continue button to finish the debugging. A web page similar to that shown in Figure 2–18 displays.

18 Close the web browser.

19 Select the File menu and then Close Solution to close the *WebApplication2* project.

We have just walked through how to debug using Visual Studio .NET. Those commands such as setting up break points, Step Into, Step Over, using QuickWatch, and instantly viewing a variable are used daily by any .NET programmer. More sophisticated debugging techniques can be explored at the Visual Studio .NET on-line help or via the software manual.

Tracing a Page

ASP.NET includes a *Trace* object that allows you to write debug statements that appear when you enable tracing for a page or the entire application. Debug statements can be directly written in your code without having to remove them from your application when it is deployed to production servers. You can write variables or structures in a page, assert whether a condition is met, or simply trace through the execution path of your page or application.

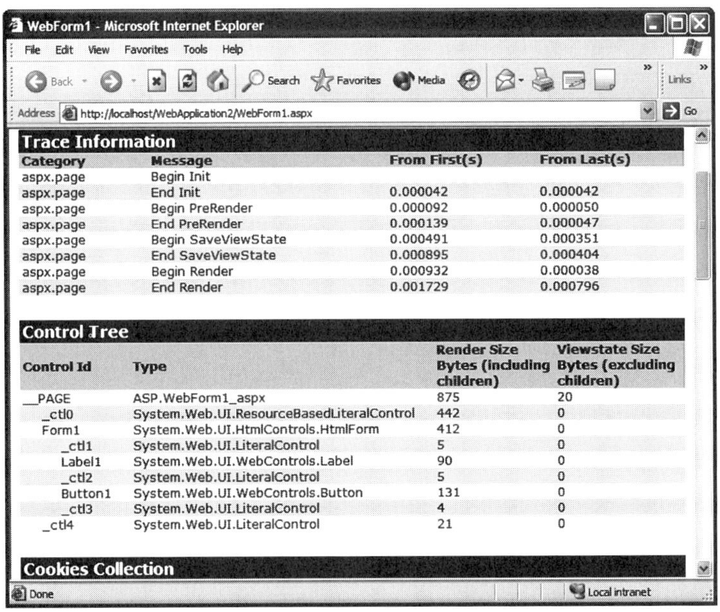

Figure 2–24. The Trace information shown for the WebApplication2 *project.*

For these debug messages and other tracing information to be gathered and displayed, you must enable tracing for the page or application. You can control whether tracing is enabled or disabled for a page with the *Trace* attribute of the *Page* directive. Regardless of whether you write messages to the trace log, when you enable tracing and the page is requested ASP.NET appends a series of tables containing performance information about the page request (Figure 2–24).

Tracing is disabled by default. If you do not include a *Trace* attribute in your page, trace information is not gathered, and trace statements you write do not appear on the page. The following steps show you how to enable tracing for a web page.

To enable tracing for a page:

1 As follows, add the *Trace* attribute to the *Page* directive and set its value to *true*.

```
<%@ Page Trace="True" %>
```

2 If you would like to see better-organized trace messages, add the *TraceMode* attribute to the *Page* directive to specify the order in which you want the trace messages to appear. Set the attribute value to *SortByTime* to sort the messages in the order in which they are processed. Set the attribute value to *SortByCategory* to sort them by category. You can specify the categories in your code and call them in the trace-rendering methods of *Trace.Warn* and *Trace.Write*.

```
<%@ Page Trace="True" TraceMode="SortByCategory" %>
```

As you have seen, it is very easy to set up a tracing. The greatest advantage of the *Trace* object is that you do not have to remove it from the application for the deployment. You can consider it part of the code or design.

Summary

In this chapter, we have introduced what ASP.NET is and what ASP.NET web pages are. We have also discussed the differences between classic ASP and ASP.NET. ASP.NET is part of Microsoft's .NET Framework. An ASP.NET web page is essentially a web form. You can place HTML controls on the web form, which look like HTML form elements. However, these are really server controls, meaning that you can set their properties and change their behaviors at the server. You can also place web controls, validation control, and user controls. If you build ASP.NET mobile applications, you can use mobile controls. Visual Studio .NET is an ideal IDE for visual programming. It is the only tool you need to build ASP.NET applications, web services, and mobile web applications. With Visual Studio .NET, you can also perform sophisticated debugging, which is critical to RAD. You can develop ASP.NET web pages in multiple languages, but Visual Basic .NET works just as well as any other language for this purpose. In the next chapter, we will examine Visual Basic .NET in detail.

CHAPTER 3
Programming in Visual Basic .NET

VISUAL BASIC IS ONE OF MOST POPULAR LANGUAGES since the early 1990s. It has been used for both Windows and web development for the last decade. Visual Basic has two children: Microsoft Visual Basic for Applications (VBA) and Visual Basic Script Edition (VBScript). VBA is a powerful development technology for rapidly customizing desktop applications and integrating them with existing data and systems. It makes Windows software such as ArcMap customizable. VBScript, like JavaScript, provides client-side scripting for web applications specifically designed for Internet Explorer. With the birth of Microsoft .NET Framework, so goes Visual Basic .NET.

With Microsoft .NET Framework, Visual Basic .NET can be used to build a Windows application, a web application, and an XML web service. It can use robust multithreading, manipulate relational data and XML, harness COM+ services, and more. Visual Basic .NET can access the entire .NET class library. With the .NET Framework, Visual Basic .NET can access other .NET components built with the .NET family of languages (such as C#) and provide true object-oriented programming.

Visual Basic .NET comes with Visual Studio .NET. Visual Studio .NET provides IDE for the .NET family of programming languages, including Visual Basic .NET. It also has templates for developing Windows applications, web applications (ASP.NET), and web services in Visual Basic .NET. Visual Basic .NET has become a true web development tool. You can now add graphical functionality using GDI+ (incorporated in the *System.Drawing* namespace of the .NET library) and easily build wireless web applications using Visual Basic .NET.

This chapter is intended for those who have little experience with Visual Basic .NET. It talks about defining variables and constants, manipulating strings, using regular expressions, defining arrays and collections, writing functions and subroutines, handling errors and exceptions, and discussing object-oriented programming in Visual Basic .NET. If you know Visual Basic .NET already, you can skip this chapter. However, if you only know Visual Basic, we recommend that you go through this chapter before proceeding. Even if you know Visual Basic .NET, this chapter would still serve as a good review and as a VB.NET reference source.

Variables, Constants, and Operators

All programming starts with declaring variables and manipulating those variables with operators. Variables are memory locations. Their values can be changed in code. If those values cannot be changed, those memory locations refer to constants. Anytime you put a variable or constant to work, you need operators.

Variables

Variables are temporary storage places. These memory locations can be present at runtime. Like most programming languages, Visual Basic .NET uses variables for storing values. A variable has a name (the word you use to refer to the value the variable contains) and a data type (which determines the type of data the variable can store). A variable can represent an array if it needs to store an indexed set of closely related data items.

You declare a variable to specify its name and characteristics. The declaration statement for variables is a `Dim` statement. Its location and content determine the variable's characteristics.

```
Dim i, j as Integer' two integer variables
Dim a, b as Double' two real variables with double precision
Dim mapTitle as String' one string variable
Dim aTag as Boolean' one Boolean variable
```

In previous examples, the `As` clause in the declaration statement allows you to specify the data type or object type of the variable you are declaring. You can define any of the following types for a variable.

- An elementary data type, such as *Integer*, *Double*, or *Boolean*

- A composite data type, such as an array or structure
- An object type, or class, from Visual Basic .NET, such as *Label* or *TextBox*

✔ **NOTE:** *There used to be an undocumented convention for naming variables—Hungarian notation. You start with the data type and then concatenate with the name qualifier. For example,* strMapTitle *can be used for the variable of a map title, where its prefix* str *stands for string and the name qualifier* MapTitle *is in camel case. You do not have do use the Hungarian notation anymore with .NET because data types no longer pose the same problems. In addition, Visual Studio .NET will alert you to invalid conversions and refuse to compile.*

However, we suggest that you keep using the Hungarian notation with web controls in order to distinguish different types of controls. For example, *txtMapTitle* is for a *TextBox* control of map title and *btnSubmit* is for a *Button* control of submission.

You can declare several variables in one statement without having to repeat the data type. Table 3–1 outlines the Visual Basic .NET data types, their supporting common language runtime types, their nominal storage allocation, and their value ranges.

Table 3–1: Visual Basic Data Types

Type	Runtime Type	Storage (bytes)	Description
Boolean	System.Boolean	2	*True* or *False* value
Byte	System.Byte	1	0 to 255, binary data
Char	System.Char	2	Single unicode character
Date	System.DateTime	8	1/1/0001 through 12/31/9999
Decimal	System.Decimal	16	Decimal fractions
Double (double-precision floating point)	System.Double	8	Double-precision floating-point numbers with 6 digits of accuracy
Integer	System.Int32	4	Whole number in the range -2,147,483,648 through 2,147,483,647

Type	Runtime Type	Storage (bytes)	Description
Long (long integer)	System.Int64	8	Larger whole numbers
Object	System.Object (class)	4	Any type of data
Short	System.Int16	2	Small integers in the range -32,768 through 32,767
Single (single-precision floating-point)	System.Single	4	Single-precision floating-point numbers with 14 digits of accuracy
String (variable-length)	*System.String* (class)	Varies	Alphanumeric data, including letters, digits, and other characters
User-Defined Type (structure)	(inherits from *System.ValueType*)	Varies	User-defined type

Constants

Memory locations that hold data that cannot change at runtime are called *constants*. For example, *Pi*, which stands for the ratio of the circumference of a circle to its diameter, is a numerical constant. However, it is irrational, and we often use an approximation value of 3.14159. You will find two important advantages to using named constants rather than the actual values in code. The code is easier to read; for example, seeing the identifier *PI* is more meaningful than seeing a number, such as 3.14 or 3.14159. In addition, if you need to change the value with different decimal places, you only need to change the constant declaration; you do not have to change every reference to it throughout the code. The following is a definition for named constants in Visual Basic .NET:

```
' declare a constant for Pi
Const PI_Double = 3.14149 As Double
' declare a constant for Earth's equatorial circumference in mile
Const EARTH_CIRCUMFERENCE_MILES_AT_EQUATOR_Double = 24902R As Double
' declare a string constant
Const ESRI_STREET_ADDRESS = "380 New York Street" As String
```

As you can see from this example, you declare a named constant using the keyword `Const`, followed by a name, a value assignment, and an `As` clause that actually declares the data type. We suggest that you use all uppercase letters for the name (with individual words separated by underscores) and pad with an extra underscore plus the data type at the end of the name (such as *EARTH_CIRCUMFERENCE_MILES_AT_EQUATOR_Double*).

The value assignment for constants follows certain rules. Text values must be enclosed in quotation marks. Numeric values may contain only the digits (0 through 9), a decimal point, and a sign (+ or –) at the left side. You cannot include a comma, dollar sign, any other special characters, or a sign at the right side for numeric values. You can declare the data type of numeric constants by appending one of the following type-declaration characters.

- Decimal: *D*
- Double: *R*
- Integer: *I*
- Long: *L*
- Short: *S*
- Single: *F*

If you do not append a type-declaration character to a numeric constant, any whole number is assumed to be *Integer* and any fractional value is assumed to be *Double.*

Text values (also called string constants) may contain letters, digits, and special characters, such as $#@%&*. To include a double-quote character for a string constant, Visual basic .NET requires the use of two double-quote characters as an escape sequence representing a double quote in a string. You can define a date and time constant by enclosing formatted date and time within pound signs (#), as in the following.

```
' declare a data and time constant
Const FIREWORKS_TIME As Date= #07/04/2005 21:30:00#
```

✔ **NOTE:** *There are many system-defined constants that can be found in system class libraries in .NET. They are called* intrinsic *constants. Visual Basic .NET also offers a convenient way of working with sets of related constants via* enumerations. *For example, you can declare an enumeration called* Days *for a set of integer constants associated* *with the seven days of the week, and then use the names of the days rather than their integer values in your code.*

Operators

Wherever you need to add, multiply, concatenate, or compare, you need an operator. An operator is a code unit that performs an operation whether it is arithmetic, concatenation, comparison, or logical. There are four types of operators in Visual Basic .NET.

- Arithmetic operators
- Concatenation operators
- Comparison operators
- Logical operators

Arithmetic operators include addition, subtraction, multiplication, division, exponentiation, and modulus, as outlined in Table 3–2. Visual basic .NET also added assignment operators (outlined in Table 3–3) that have existed in other programming languages. Now you can write a statement such as `i += 1` (equivalent to `i = i + 1`).

Table 3–2: Arithmetic Operators

Operator	Use
^	Exponentiation
*	Multiplication
/	Division
\	Integer division
Mod	Modulus arithmetic
+	Addition
-	Subtraction or indication of negative value
<<	Left shift
>>	Right shift

Table 3–3: Assignment Operators

Operator	Use
=	Assignment
^=	Exponential assignment
*=	Multiplication assignment
/=	Division assignment
\=	Integer division assignment
+=	Addition assignment
-=	Subtraction assignment
<<=	Left shift assignment
>>=	Right shift assignment
&=	String concatenation

Concatenation operators (outlined in Table 3-4) join multiple strings into a single string. There are two concatenation operators: + and &. Both carry out the basic concatenation operation. However, you must set `Option Strict` to `On` if you use + to concatenate strings. If `Option Strict` is set to `Off`, Visual Basic .NET will convert *String* to *Double* and then add.

```
Dim myNumber As Integer
Dim myStr1 As String
Dim myInt1 As Integer
Option Strict On
' Initialize mixed variables.
myStr1 = "16"
myInt1 = 4
myNumber = myStr1 + myInt1  ' Generates a compile-time error.
Option Strict Off
myStr1 = "16"
myInt1 = 4
myNumber = myStr1 + myInt1
' Returns 20 (addition) after the string in myStr1 is
' converted to a numeric value. Use of Option Strict Off
' for these operations is not recommended.
```

Table 3–4: Concatenation Operators

Operator	Use
+	String concatenation
&	String concatenation

Comparison operators (outlined in Table 3–5) compare two expressions and return a Boolean value that represents the result of the comparison. Those expressions can be numerical, string, or objects. Strings can also be compared using the *Like* operator. The *Like* operator allows you to specify a pattern. The string is then compared against the pattern, and if it matches the result is *True*. Otherwise, the result is *False*.

Table 3–5: Comparison Operators

Operator	Use
=	Equality
<>	Inequality
<	Less than
>	Greater than
>=	Greater than or equal to
<=	Less than or equal to

```
myLike As Boolean
myLike = "A" Like "A" ' Does "A" match "A"? Returns True.
myLike = "A" Like "a" ' Does "A" match "a"? Returns False.
myLike = "A" Like "AAA" ' Does "A" match "AAA"? Returns False.
myLike = "aBBBa" Like "a*a"
    ' "aBBBa" starts and ends with "a”, and
    ' three other letters in between? Returns True.
myLike = "M" Like "[A-Z]"
    ' Is "M" a letter between A to Z? Returns True.
myLike = "M" Like "[!A-Z]"
    ' Is "M" not a letter between A to Z? Returns False.
```

As you can see from these examples, there are different types of patterns. You match strings using wildcard characters, character lists, or character ranges, in any combination (outlined in Table 3–6). The *Like* operator is similar to *regular expressions*, discussed in the next section.

Table 3–6: Patterns for the *Like* Operator

Characters in *pattern*	Matches in *string*
?	Any single character
*	Zero or more characters
#	Any single digit (0–9)
[A-Z] [a-z] [0-9]	Any single character or digit in the list
[!A-Z] [!a-z] [!0-9]	Any single character or digit not in the list

If you want to compare two objects, you need to use the *Is* operator. For example, you can use the *Is* operator to make a comparison between the new map extent object and the old map extent object, as shown in the following.

```
Dim newExt As New MapExtent
Dim oldExt As New MapExtent
If newExt Is oldExt then
' Insert code to continue program execution.
```

You can test whether an object is of a particular type with the *TypeOf...Is* operator, as follows.

```
Dim newExt As New MapExtent
If TypeOf(newExt) Is MapExtent then
' Insert code to continue program execution.
```

The logical operators compare Boolean expressions and return a *True* or *False* result. The *And*, *Or*, *AndAlso*, *OrElse*, and *Xor* operators take two operands, and the *Not* operator takes a single operand (see Table 3–7). The *And* operator performs a logical conjunction on two Boolean expressions and returns *True* if both expressions evaluate to *True* (and *False* otherwise). The *Or* operator performs logical disjunction on two Boolean expressions and returns *True* if either expression evaluates to *True* (and *False* otherwise). The *Xor* operator performs logical exclusion on two expressions and returns *True* if either expression evaluates to *True* (and *False* otherwise). The *Not* operator performs logical negation on a single Boolean expression and returns *False* if the expression evaluates to *True* (and *True* otherwise).

Table 3–7: Logical Operators

Operator	Use	Example
Not	Negation	Dim x As Boolean x = Not 23 > 12 ' x equals False x = Not 23 > 67 ' x equals True
And	Conjunction	x = 23 > 12 And 12 > 4 ' x = True x = 12 > 23 And 12 > 4 ' x = False
AndAlso	Conjunction	12 > 45 And MyFunction(4) ' MyFunction() is called 12 > 45 AndAlso MyFunction(4) ' MyFunction() is not called
Or	Disjunction	x = 23 > 12 Or 4 > 12 ' x = True x = 23 > 45 Or 4 > 12 ' x = False
OrElse	Disjunction	45 > 12 Or MyFunction(4) ' MyFunction is called 45 > 12 OrElse MyFunction(4) ' MyFunction is not called
Xor	Disjunction	x = 23 > 45 Xor 12 > 4 ' x = True x = 23 > 12 Xor 12 > 4 ' x = False x = 12 > 23 Xor 4 > 12 ' x = False

The *AndAlso* operator is very similar to the *And* operator but performs short-circuiting logical conjunction on two Boolean expressions. If the first expression in an *AndAlso* expression evaluates to *False*, the second expression is not evaluated and *False* is returned for the *AndAlso* expression. Similarly, the *OrElse* operator also performs short-circuiting logical disjunction on two Boolean expressions. If the first expression in an *OrElse* expression evaluates to *True*, the second expression is not evaluated and *True* is returned for the *OrElse* expression.

Strings and Regular Expressions

The *String* class, part of the System namespace, provides the data types for all strings. The *String* class offers a variety of methods for string manipulations, comparison, formatting, and so on. There are two types of methods in *String*: static methods and instance methods. A static method is a method that stems from the *String* class itself and does not require an instance of that class to work. These methods can be qualified with the name of the *String* class rather than with an instance of the *String* class, as shown in the following example.

```
Dim aString As String
bString = String.Copy("The Map of U.S.")
```

In this example, the *Copy* method is a static method but acts as a function, which takes the given string value and assigns it to a resultant string *bString*. An instance method, by contrast, stems from a particular instance of *String* and must be qualified with the instance name.

```
Dim aString As String = " The Map of U.S."
Dim bString As String = aString.Substring(4, 6)
```

In this example, the *SubString* method is a method of the instance of *String* (i.e., *aString*). It performs an operation on *aString* and assigns that value to *bString*. Table 3–8 outlines the most commonly used *String* methods.

Table 3–8: Commonly Used Methods of the *String* Class

Type	Name	Description
Static	Compare	Compares two specified ***String*** objects
	CompareOrdinal	Compares two ***String*** objects, without considering the local national language or culture
	Concat	Concatenates one or more strings
	Copy	Creates a new instance of ***String*** with the same value as a specified ***String***
	Format	Replaces each format specification in a specified ***String*** with the textual equivalent of a corresponding object's value
	Intern	Returns the system's reference to the specified ***String***
	IsInterned	Returns a reference to a specified ***String***
	Join	Concatenates a specified separator ***String*** between each element of a specified ***String*** array, yielding a single concatenated string
Instance	Clone	Returns a reference to this instance of ***String***
	CompareTo	Compares a string with another
	CopyTo	Copies a specified number of characters from a specified position in this instance to a specified position in an array of characters
	EndsWith	Determines whether the end of this instance matches a string

Type	Name	Description
	IndexOf	Reports the index of the first occurrence of a string, or one or more characters, within this instance
	IndexOfAny	Reports the index of the first occurrence in this instance of any character in a specified array of Unicode characters
	Insert	Inserts a string at a specified index position
	LastIndexOf	Returns the index position of the last occurrence of a specified character
	LastIndexOfAny	Reports the index position of the last occurrence of one or more characters in a string
	PadLeft	Right-aligns the characters in a string, padding on the left with spaces or a specified character for a specified total length
	PadRight	Left-aligns the characters in a string, padding on the right with spaces or a specified character for a specified total length
	Remove	Deletes a specified number of characters from a string beginning at a specified position
	Replace	Replaces all occurrences of a specified character or string with another specified character or string
	Split	Identifies the substrings in this instance that are delimited by one or more characters specified in an array, and then places the substrings into a string array
	StartsWith	Determines whether the beginning of a string matches a string
	Substring	Returns a substring from a string
	ToCharArray	Copies the characters in this instance to a Unicode character array
	ToLower	Returns a copy of a string in lowercase
	ToUpper	Returns a copy of a string in uppercase
	Trim	Deletes all occurrences of a set of specified characters from the beginning and end of a string
	TrimEnd	Deletes all occurrences of a set of characters specified in a character array from the end of a string

String manipulation is commonly employed but also very challenging. In this section, we will break those string methods into four groups.

- Creating and forming strings
- Manipulating string
- Comparing strings
- Formatting strings

Declare and Assign a String Variable

To use a string variable, you must explicitly define storage for it in your code. A string variable can be defined as follows.

```
Dim myString as String
```

Then you can assign a value for it, as follows.

```
myString = "This is my first string variable."
```

This is a legacy way of defining and initializing a variable. Visual Basic .NET offers a more economical way of combining two steps of a process into one. To do that, you declare a variable with the *Dim* statement and then add an assignment operator (=) along with the value you wish to assign to the variable, as follows.

```
Dim myString as String = "This is my first string variable"
```

Concatenate Strings

There is often a need to combine two or more strings. Some people call this string addition. However, it is confusing to refer to this as "adding strings." It is basically a concatenation. For example, we want to create a map for each individual state to show population distribution. We would need a map title saying something like "The Ohio Population." We could define the map title as follows.

```
' concatenating strings for Map Title
Dim myString1 as String = "The "
Dim myStringState as String
Dim myString2 as String = " Population Distribution"
Dim myMapTitle as String
myMapTitle = myString1 & myStringState & myString2
```

Here, *myMapTitlebyState* is assigned dynamically by the program. You can also use the *Concat* method of the *String* class to achieve the same goal, as follows.

```
' concatenating strings for Map Title
Dim myString1 as String = "The "
Dim myStringState as String
Dim myString2 as String = " Population Distribution"
Dim myMapTitle as String
myMapTitle = String.Concat(myString1, myStringState, myString2)
```

Copy a String

If you want a new string with the same value as the old one, you can use the *Copy* method of the *String* class to create a new instance of *String* with the same value as a specified *String*.

```
myNewMapTitle = String.Copy(myMapTitle)
```

Insert a String

You can also use the *Insert* method of the *String* class to build a dynamic map title. The *Insert* method allows you to place a substring in the middle of another string, where the position on which the new string is placed is specified by an integer.

```
Dim myMapTitle as String = "The Population Distribution"
Dim myMapTitleByState as String

' concatenating map title with Insert
myMapTitle = myMapTitle.Insert(4, myMapTitleByState)
```

Make sure to leave an extra blank space where you want to insert. Otherwise, the inserted word would "stick" with *Population*. For example, if you assign *Virginia* to *myMapTitleByState*, you would get *The Virginia Population Distribution* for *myMapTitle*. You need an extra blank space between *Virginia* and *Population*.

To make some HTML look nicer, we often need to pad "*
 & vbCrLf*" in classic ASP, where `vbCrLf` is a constant for a new line. In Visual basic .NET, you can use a `NewLine` property out of the *System.Environment* class.

```
Dim myText as
MyText &= Environment.NewLine
```

There are a certain number of characters (such as copyright and trademark characters) that cannot be entered using the keyboard. You have to enter them through code. You can enter a special character by using the *ToChar* method of the *System.Convert* class.

```
' insert a 'copyright' special character
Dim theCode as Integer = 169
Dim copyrightChar as Char = Convert.ToChar(theCode)
```

Remove Number of Characters from a String

If you happen to insert wrong text to a string, you can remove it. You can do that using the *Remove* method of the *String* class.

```
Dim myMapTitle as String = "The Virginia Population Distribution"
myMapTitle.Remove(16,1)
```

Trim a String

We often need to trim unintended white space at the beginning and end of a text input from the user. You can do this using the *Trim* method of the *String* class. You can alternatively trim leading spaces or tailing spaces using *TrimStart* or *TrimEnd*, respectively.

```
' trim leading and tailing spaces
Dim myString as String = " Map Title  "

myString = myString.Trim()
myString = myStringTrimStart()
myString = myStringTrimEnd()
```

Pad a String

Sometimes you may want to pad a number of characters to a string. You can do this using the *PadLeft* and *PadRight* methods of the *String* class. If you want to pad to the beginning of the string, use *PadLeft*; otherwise, use *PadRight* to the end of the string.

Subset a String

You can extract a portion of a string based on its position and length using the *Substring* method of the *String* class. You need to provide a start index and specify the length of the string in characters to be extracted.

```
' extract a 4-charater substring starting at 0
Dim fullName as String = "John Public"
Dim firstName as String = fullName.Substring(0, 4)
```

The *Substring* method is similar to *Mid* in classic ASP and Visual Basic 6. In addition, the old *Left* and *Right* functions are included in Visual Basic .NET but are not recommended. You can achieve *Left* and *Right* using the following code.

```
' The equivalent to Left(MyString, 4)
NewString =MyString.Substring(0, 4)
' The equivalent to Right(MyString, 4)
NewString = MyString.Substring(MyString.Length - 4, 4)
```

Find First Occurrence of Specific Text in a String

To extract a portion of a string, you need to locate where that portion is first. The *IndexOf* method of the *String* class allows you to do this. The *IndexOf* method returns the index of the first match in a string, or –1 if no match is found.

```
Dim myString As String = "The Virginia Population Distribution"
Dim myIndex As Integer
myIndex = myString.IndexOf("Virginia")
```

Replace All Occurrences of Specific Text in a String

Replacement is one of most commonly used string methods. In this case, you use the *Replace* method of the *String* class. When you apply the method to a string, all occurrences of the text pattern inside the string will be replaced.

```
' replace a portion of string
Dim myMapTitle as String = "The Virginia Population Distribution"
myMapTitle.Replace("Virginia", "Ohio")
```

Split a String into Segments

If you receive vertices in free text format, they often come in as a long and comma-delimited string. The easiest way to deal with them is to parse them out and place them in an array. The *Split* method of the *String* class allows you to accomplish this.

```
' split a String into segments
Dim myVertices as String = "-120.1256, 38.5623, -120.1212, 38.5689"
Dim delimiter as Char = (","c)
Dim myCoords() as String
' split the vertices into coords
myCoords = myVertices.Split(delimiter)
```

Join an Array of Strings

There is also a reverse method of *Spilt* called *Join*. You can concatenate an array of strings with a separator string using the *Join* method of the *String* class.

```
' join a string segments in to a String
Dim myCoords(4) as String
Dim myVertices as String
myCoords(0) = "-120.1256"
myCoords(1) = "38.5623"
myCoords(2) = "-120.1212"
myCoords(3) = "38.5689"
myVertices = String.Join(",", myCoords)
```

Note that if the array to be joined has empty members the *Join* method still adds a separator string between all empty instances in the array.

Convert the Case of All Characters in a String

It often happens that you need to convert a string into uppercase using the *ToUppper* method of the *String* class, or you need to convert a string to lowercase using the *ToLower* method of the *String* class. This happens frequently if you try to compare a string input with a predefined string constant. To make sure you compare apples with apples, you would use the following.

```
' convert an input string to upper case and then compare
if txtState.Text.ToUpper() = "OHIO" then
' add what you want to do here…
End If
```

Here, *txtState* is a text control.

Compare Strings

After you make two strings in either lowercase or uppercase, you can compare them using the *Compare* method of the *String* class. The *Compare* method compares two strings alphabetically and returns 0 if they are equal, –1 if the first string is less than the second, or 1 if the first string is greater than the second. With the *Compare* method, the previous example can be written as follows.

```
' compare two strings
if String.Compare(txtState.Text.ToUpper(), "OHIO", true) = 0 then
' add what you want to do here…
End If
```

As you can see, there is a third (Boolean) parameter called *ignoreCase*. The *ignoreCase* parameter is optional. If it is set to *True*, the *Compare* method performs a case-insensitive comparison.

Format a String

The *Format* method of the *String* class and the Visual Basic .NET Format command can generate a new string by applying formatting to a given string.

```
'format a string
myStateLine = String.Format("{0,2}, {1,2}, {2,15}", "VA", "51", "Virginia")
```

Regular Expressions

Regular expressions used to be the pride of UNIX and Perl. Not anymore! They were offered in Visual Basic 6.0 and classic ASP. Web programmers have dealt with them a lot in JavaScript. Regular expressions provide a concise and efficient way—far beyond the capabilities most conventional programming languages offer—of processing and manipulating text. Some people say that regular expressions are easy to use. However, most conventional programmers consider regular expressions a language different from other languages. The closest cousin of regular expressions would be UNIX shell scripts. Regular expressions are very powerful, flexible, and efficient. For example, it is very common to incorporate the ability to zoom in by ZIP code in a web-based GIS application. We all know that the text field needs to be in digit format, with a combination of 0 through 9. Although simple, it is not that easy to write a validation for it. However, validation can be done in one line of a regular-expression pattern as follows.

```
' 5-Digit ZIP code
Dim myRegex As New Regex ("^\d{5}$")
If myRegex.IsMatch("22042") then
   Response.write("valid ZIP code")
```

```
Else
   Response.write("invalid ZIP code")
```

Visual Basic .NET offers a complete namespace devoted to regular expressions called Regex at System.Text.RegularExpressions. In the previous example, the pattern "^\d{5}$" is the key. It says that "a match starts from the beginning of the string with 5 repetitions of any digit character (meaning 0 through 9) until it reaches the end of the string." As you can see, it is not even close to plain English. Without explanation, you are not likely to figure out what the expression means. Table 3–9 outlines common meta-characters of regular expressions.

Table 3–9: Common Meta-characters of Regular Expressions

Pattern	Inputs (Matches)
^	Matches the beginning of a string. ^A matches first A in Alabama, Alaska, Arkansas, Arizona.
$	Matches the ending of a string. a$ matches the last a in North Carolina.
+	Matches one or more occurrences. Equivalent to {1,}.
*	Matches zero or more occurrences. Equivalent to {0,}.
.	Matches any character except newline.
?	Matches zero or one occurrences. Equivalent to {0,1}.
()	Grouping a clause to create a clause. May be nested. (abc)+(def) matches one or more occurrences of abc followed by one occurrence of def.
|	Either of two matches (ab)|(cd)|(ef) matches ab or cd or ef.
[ ]	Matches any single character enclosed in the character set. [a-zA-Z] matches any letter including both lowercase and uppercase.
[^]	Matches any single character not enclosed in the character set. The caret indicates that none of the characters [^a-z] excludes all lowercase letters.
{m}	Matches exactly m occurrences of a regular expression. A{A} matches AAA.
{m,n}	Matches x to y number of occurrences of a regular expression. 5{2,3} matches 55 and 555 but not 5555.
\d	Matches any single digit. Equivalent to [0-9].
\D	Matches any nondecimal digit.

Pattern	Inputs (Matches)
\s	Matches any whitespace character.
\S	Matches any non-whitespace character.
\W	The negation of \w.
w	Matches any word character.

Now we can modify the previous ZIP code example that handles both 5-digit and 9-digit ZIP codes.

```
' 5-Digit and 9-digit ZIP code
Dim myRegex As New Regex ("^\d{5}$|^\d{5}-\d{4}$")
If myRegex.IsMatch("22042") or myRegex.IsMatch("22042-2342") then
   Response.write("valid ZIP code")
Else
   Response.write("invalid ZIP code")
```

You can now see the power of using regular expressions. It would be very obvious that you want to prebuild a certain number of often-used regular expressions, as outlined in Table 3–10.

Table 3–10: Commonly Used Regular Expressions

Subject	Expression	Description
Phone number	\(\d{3}\)\s?\d{3}\-\d{4}	(703) 555-1212
E-mail	/^[a-z0-9]([a-z0-9_\-\.]*)@([a-z0-9_\-\.]*)(\.[a-z]{2,3}(\.[a-z]{2}){0,2})$/	abc@aol.com
IP address	\b\d{1,3}\.\d{1,3}\.\d{1,3}\.\d{1,3}\b	192.1.20.200
ZIP code	(^\d{5}$)\|(^\d{5}-\d{4}$)	22042 or 22042-2342
State abbreviation	\w{2}	two characters
Date	(0[1-9]|1[012])[- /.](0[1-9]|[12][0-9]|3[01])[- /.](19|20)\d\d	mm/dd/yyyy
Time	([0|1|2]{1}\d):([0|1|2|3|4|5]{1}\d):([0|1|2|3|4|5]{1}\d)	hh:mm:ss
Latitude	/^-?\d{1,2}(\.?\d+$\|$)/	40.3421
Longitude	/^-?\d{1,3}(\.?\d+$\|$)/	–115.7834

Arrays and Collections

Visual Basic .NET still comes with arrays and collections. They are commonly used in classic ASP programming. Arrays are offered through the *System.Object* namespace, whereas collections are offered via the *System.Collections* namespace. There is also an *ArrayList* class, which works as an array and a collection at the same time but with a size dynamically increased as required.

Using Arrays

When you need to group the same category of variables, you use arrays. Arrays allow you to refer to a series of variables by the same name and to use a number, called an index or subscript, to tell them apart. With the use of indices of arrays, you can set up loops that deal efficiently with any number of elements within an array by using the index number. This could make your code shorter and simpler.

Arrays have dimensionality, which refers to the number of subscripts used to access an individual element. One, two, or three dimension arrays are most commonly used. Anything more than three dimensions is rarely used. Each dimension of an array has its own size, which is a nonzero length. The indices of an array start from 0. This is thus *zero* based, although much in Microsoft used to be *one* based. You define an array as follows.

```
' create an array from index 0 to 2 with three elements
Dim arrayColors() as String = {"red", "green", "blue"}

' or you can declare it in the following way
Dim arrayColors(2) As String
arrayColors(0) = "red"
arrayColors(1) = "green"
arrayColors(2) = "blue"
```

As you can see, there is no keyword *Array* here, as we used to declare with classic ASP. A pair of parentheses after the variable `arrayColors` identifies the variable as an array instead of as a string. Arrays are objects in Visual Basic .NET, and thus every array type is an individual reference type. An array has its own dimension but does not have to be fixed in size. You can change the size of an array using the `ReDim` statement at runtime (i.e., you can increase or decrease the size). However, all data in the array will be lost once the `ReDim` statement is used.

✔ **NOTE:** *The number you specify for each dimension in the declaration is the upper bound* (Ubound), *and the initial element count is equal to the upper bound plus one. The declaration in the preceding example reserves three elements for colors, with subscripts 0 through 2.*

You can also specify a zero-length array, which does not contain any elements. To declare a zero-length array, you declare its dimension to be –1. The array is then empty, but it still exists. Therefore, a variable pointing to the array is not equal to *Nothing*. You would declare such an as follows.

```
Dim myArrays(-1) As Integer
```

Technically, the `ReDim` statement releases the current array and creates a new one. To keep all content of an array when resizing it, you need to use the keyword `Preserve` (as shown in the following).

```
' redim an array
ReDim myArray(2) As String

' add three more colors to arrayColors and first three colors of Red,
' Green and Blue are kept unchanged.
'
ReDim Preserve arrayColors(5) As String
arrayColors(3) = "yellow"
arrayColors(4) = "cyan"
arrayColors(5) = "magenta"
```

When the `Preserve` keyword is present, it copies the data to the new array. Note that you can only resize the last dimension of an array. Thus, if you are resizing a 3D array you can only change the third dimension. If you are resizing a 2D array you can only alter the second dimension. If you are resizing a 1D array you can modify the first dimension.

Using Collections

The indices of an array are numerical. We often want more meaningful indices, such as using the state FIPS codes as the indices to retrieve state names and abbreviations. In this case, you should use *collections*. Collections are similar to the associated arrays in Perl. They have keys and values. In Visual Basic .NET, collections are objects. A collection is a related group of items. The items, or members, in a collection need only be related by the fact that they exist in the collection. Members of a collection do not have to share the same data type. The following example defines a collection of *States* for all middle Atlantic states.

```
' define a States collection
Dim States As New Collection

' populate the collection
States.Add("Delaware", "10")
States.Add("Maryland", "24")
States.Add("Pennsylvania", "42")
States.Add("Virginia", "51")
States.Add("West Virginia", "54")

' retrieve each state by its FIPS code
Response.Write(States("10") & "<BR>")
Response.Write(States("24") & "<BR>")
Response.Write(States("42") & "<BR>")
Response.Write(States("51") & "<BR>")
Response.Write(States("54") & "<BR>")
' and will produce the following
'    Delaware
'    Maryland
'    Pennsylvania
'    Virginia
'    West Virginia
```

As you can see, you can now retrieve each individual state using the FIPS code. This is much more intuitive than using arrays. We can further modify this example so that each collection value is a true object and can be retrieved by a FIPS code. We will define a *State* class first, which consists of three properties: name, abbreviation, and FIPS code. We will call these properties *Name*, *Abbr*, and *FIPS*, respectively.

```
Public Class State
  Public Name As String
  Public Abbr As String
  Public FIPS As String
End Class
```

Defining classes is further explored later in the chapter. For now, we can use the `State` class in our code. We define a new object called `myState`. Because `myState` is a `State` object, we can then assign values to its three properties: *Name*, *Abbr*, and *FIPS*. Our `States` is still a collection, but instead of adding a string *"Virginia"* we are now adding a `State` object `myState` with a reference key *"51"*. After the assignment, we can retrieve *Virginia* by the reference key or the collection index. Collections are 1 based. Because there is only one item in the collection, the index for *Virginia* is 1. As you can see, you do not have to know the index for *Virginia* in the collection and can use the key, which is an FIPS code in this case. Retrieving collection values with keys is much more convenient and sometimes more efficient than using arrays.

```
Dim myState As New State
myState.Name = "Virginia"
myState.Abbr = "VA"
myState.FIPS = "51"

Dim States As New Collection
States.Add(myState, "51")
Response.Write(States.Item("51").Name & "<BR>")
Response.Write(States.Item(1).Abbr & "<BR>")
Response.Write(States.Item(1).FIPS & "<BR>")
```

Functions and Subroutines

Functions and subroutines were invented to eliminate "spaghetti code." Spaghetti code arises when code execution flows from one area to another without organization. Functions and subroutines allow you to organize code and make your code more modular. They should be used to accomplish a single task instead of a lengthy list of tasks. The rule of thumb is that you choose a function if you want the procedure to return a value or object, and select a subroutine if you do not want the procedure to return anything. In Visual Basic .NET, functions and subroutines are often used to define methods for a class.

Using Functions

A function is a procedure that consists of a series of code statements and returns a result value to its calling program. A function is enclosed by the `Function` and `End Function` statements, as follows.

```
Function ComputeDistance() As Double
' your code to computer the distance between two points
'
End Function
```

Each function is referred to by name, as shown in the previous example. The function is called `ComputeDistance`. Because a function sends back a return value to wherever it is called, Visual Basic .NET requires that you designate the data type for the result, meaning that the function needs to have a data type as defined by the `As` clause. In the previous example, the function `ComputeDistance` is defined as *Double* because the return value is *Double*.

Some functions need input parameters and others do not. When parameters are needed, you can define those parameters by placing their names and types in parentheses following the function name. For example, *ComputeDistance* needs two pairs of points to be passed in. Therefore, that function would be upgraded as follows.

```
Function ComputeDistance(ByVal longMin as Double, ByVal latMin as Double,
   ByVal longMax as Double, ByVal LatMat as Double) As Double
'
ComputeDistance = Sqrt((longMax-longMin)**2+(latMax-latMin)**2)
'
End Function
```

Thus, the *x* and *y* values of a pair of diagonal vertices are passed in as *Double* because they are the real-world coordinates of latitudes and longitudes. Before you finish your function, you need to do one of the following two things to return a value.

- Assign an expression or a computed value in the current function to the function name.
- Place an expression or a computed value after the `Return` statement.

Thus, the previous example can also be written as follows.

```
Function ComputeDistance(ByVal longMin as Double, ByVal latMin as
   Double, ByVal longMax as Double, ByVal LatMat as Double) As Double
'
Return Sqrt((longMax-longMin)**2+(latMax-latMin)**2)
'
End Function
```

The first approach of assigning the return value to the function name has an advantage over the second. The control in the first approach does not return from the function until the program encounters an `Exit Function` or `End Function` statement. This allows you to assign a preliminary value and adjust it later, if necessary. In contrast, the second approach of placing the return value after the Return statement returns the control immediately. A function can also return an array, a collection, or an object. To call the function you have just defined, you can write as follows.

```
Dim myDistance As Double
myDistance = ComputeDistance(-77.123456,38.453456,-77.127890,38.459099)
```

Thus, the function needs to be used in an assignment statement when it is invoked. In the previous example, you likely noticed the keyword `ByVal`. Visual basic .NET allows you to pass an argument to a procedure by value or by reference by specifying the `ByVal` or `ByRef` keyword, respectively. Passing a parameter by value means that the content of the variable or object remains unchanged after the execution of the function, whereas passing by reference means that the content of the variable or object stays updated after the execution of the function. Passing a parameter by reference is referred to as *call by reference*. Call by reference passes the memory location of the variable used for the parameter.

Using Subroutines

In cases where you need to pass in multiple parameters but require no value to be returned, a subroutine is an optimal choice. A subroutine is also a procedure that consists of a series of code statements but does not return anything to its calling program. A subroutine is enclosed by the `Sub` and `End Sub` statements, as follows.

```
Sub drawCircle(ByVal pnt as Point, ByVal r as Single)
   ...
End Sub
```

Each time the procedure is called, its code statements are executed, starting with the first executable statement after the `Sub` statement and ending with the `End Sub`, `Exit Sub`, or `Return` statement encountered. Because a subroutine does not return any value to its calling program, you need to use the `Call` statement to invoke a subroutine. You cannot call a subroutine by using its name within an expression (as you do with invoking functions).

```
Call drawCircle(myPnt, 2.5)
```

You can pass parameters to a subroutine by value or by reference. You can also specify that a parameter is optional and does not have to be supplied when the subroutine is called. Optional parameters are indicated by the `Optional` keyword in the `Sub` definition. The following rules apply for specifying optional parameters.

- Every optional parameter must specify a default value.
- The default value for an optional parameter must be a constant expression.

- Every parameter following an optional parameter must also be optional.

The following syntax shows a subroutine declaration with an optional parameter.

```
Sub subname(ByVal arg1 As type1, Optional ByVal arg2 As type2 = default)
```

You can choose to supply one, some, all, or none optional parameters when you call the subroutine. If you do not supply any optional parameters, the subroutine uses the default value declared for that parameter. When you omit one or more optional parameters, you use successive commas to mark their positions. The following example call supplies the first and fourth arguments but not the second or third.

```
Call subname(arg1, , , arg4)  ' Leaves out arg2 and arg3.
```

A subroutine cannot detect at runtime whether a given parameter has been omitted or the calling program has explicitly supplied the default value. If you need to make this distinction, you can set an unlikely value as the default. The following procedure includes the optional argument *County*, and tests for its default value, *QJZ*, to see if it has been omitted.

```
Sub Notify(ByVal State As String, Optional ByVal County As String = "QJZ")
  If State = "QJZ" Then
   Debug.WriteLine("County not supplied -- notifying Richmond")
   County = "Richmond"
  End If
  ' Code to notify richmond or specified county.
End Sub
```

If the optional parameter is a reference type such as a *String*, you can use `Nothing` as the default value, provided that `Nothing` is not an expected value for the parameter.

Another way of defining a subroutine with optional parameters is to use *overloading*. For example, if you have one optional parameter you can define two overloaded versions of the subroutine with the same subroutine name: one with the parameter and one without it. This approach becomes more complicated as the number of optional parameter increases. However, the advantage of overloading is that you can determine whether the calling program supplied each optional parameter. The compiler uses the subroutine name and the parameter list to check that the call correctly matches its definition. Overload can also be used to change the data type for one or more parameters in the parameter list.

Error Handling and Exceptions

One of the main advantages of ASP.NET over ASP is its new error-handling features. The only way to capture errors in classical ASP was to use the "*On Error Resume Next*" statement and check each line for an error with "*If Err.Number < > 0 Then* " statements. Java programmers are familiar with *Try* and *Catch* for garbage collection—which is a much better method. Fortunately, all .NET languages (including Visual basic .NET) adopted *Try* and *Catch*.

Visual Basic .NET supports both unstructured and structured exception (error) handling. By placing specific code in your application, you can handle most of the errors users may encounter and enable the application to continue running. Unstructured and structured error handling allows you to plan for potential errors and prevents them from interfering with the intended purpose of the application.

Unstructured Exception Handling

Previous versions of Visual Basic had only unstructured error handling. This is the method of using a single error handler within a method that catches all exceptions. In unstructured exception handling, you place an `On Error` statement at the beginning of a block of code, and it handles any errors occurring within that block. When an exception is raised in a procedure after the `On Error` statement executes, the program branches to the line argument specified in the `On Error` statement. The line argument, which is a number or label, indicates the exception handler location. Visual Basic .NET still supports the following unstructured error handlers.

- *On Error GoTo ErrorHandler*
- *On Error Resume Next*
- *On Error GoTo 0*

The `On Error GoTo ErrorHandler` statement assumes that error-handling code starts at the line labeled `ErrorHandler`. *ErrorHandler* serves as a line label and can be anything. If a runtime error occurs, code jumps to the *ErrorHandler* line and activates the error handler portion of the code. The labeled *ErrorHandler* line must be in the same procedure as the `On Error GoTo ErrorHandler` statement. Otherwise, Visual

Basic .NET generates a compiler error. The following example illustrates the use of an error handler with a line label.

```
Sub TestSub
  On Error GoTo ErrorHandler
    ' Code that may or may not contain errors.
  Exit Sub

  ErrorHandler:
    ' Code that handles errors.
    Resume
End Sub
```

This example contains an error handler started at the line labeled `ErrorHandler`. If any code in the `TestSub` subroutine generates an error, Visual Basic .NET immediately executes the code that follows the `ErrorHandler` label. At the end of the error-handling block, the `Resume` statement passes control back to the line of code where the error first occurred. The rest of the subroutine then continues executing as if the error did not occur. Thus, you generally use `Resume` after an error handler corrects the error. Another variation on the `Resume` statement is `Resume Line`. `Resume Line` passes control to a line labeled `Line`. You can use `Resume Line` only within an error handler.

✔ **NOTE:** *You must place an* `Exit Sub` *statement immediately before the error-handling block. Otherwise, Visual Basic .NET continues to run the error-handling code, which is not designated for normal execution, until it reaches the end of the subroutine.*

The `On Error Resume Next` statement specifies that in the event of a runtime error code continues to the statement immediately following the one in which the error occurred. At that point, execution continues. `On Error Resume Next` enables you to include error-handling routines right where errors occur rather than jump to another location to execute the error handler in the procedure.

✔ **NOTE:** *If your procedure calls another procedure, the* `On Error Resume Next` *statement becomes inactive during the execution of the called procedure. Therefore, you should place* an `On Error Resume Next` *statement in each called procedure that needs one. This is necessary because the* `Resume Next` *behavior applies only to the procedure containing the* `On Error Resume Next` *statement. If an unhandled error occurs in a called procedure, the exception propagates back to the calling procedure, and execution resumes on the statement following* *the call. In such cases, the error is not handled. When debugging your code, you must disable the* `On Error Resume Next` *statement.*

The `Resume Next` statement directs control to the line immediately following the line of code that caused the error. You might use `Resume Next` for cases in which an error will not cause your application to stop working. You might also use it if an error will not change the expected results of your subroutine.

The `On Error GoTo 0` statement disables any error handler in the current procedure. If you do not include an `On Error GoTo 0` statement, the error handler is still disabled when the procedure containing the exception handler ends.

✔ **NOTE:** *The* `On Error GoTo 0` *statement is not meant to specify line 0 as the start of the error-handling code, even if the procedure contains a line numbered 0.*

Structured Exception Handling

The major disadvantage of unstructured exception handling is lack of the ability to distinguish the types of exceptions that occur. You need to be able to tell whether an exception is from your own code or is a system error, and where the exception occurs if it is a bug. Fortunately, Visual Basic .NET offers structured exception handling. Structured exception handling allows you to create and maintain programs with robust, comprehensive error handlers. Structured exception handling is encapsulated in blocks of code. Each block has one or more handlers designed to detect and respond to errors during execution by combining a control structure with exceptions, protected blocks of code, and filters. Each handler specifies some form of filter condition on the type of exception it handles. When an exception is raised by code in a protected block, the set of corresponding handlers is searched in order, and the first one with a matching filter condition is executed. A single method can have multiple structured exception-handling blocks. The blocks can also be nested within each other.

The *Try...Catch...Finally* statement is specifically designed for structured exception handling. Using the *Try...Catch...Finally* statement, you can protect blocks of code that are vulnerable to exceptions that may occur. You can nest exception handlers. The variables declared in each block will have local scope. The following code shows the structure of a *Try...Catch...Finally* statement.

```
Try
  ' Starts a structured exception handler.
  ' Place executable statements that may generate
  ' an exception in this block.
Catch [optional filters]
  ' This code runs if the statements listed in
  ' the Try block fail and the filter on the Catch statement is true.
[Additional Catch blocks]
Finally
  ' This code always runs immediately before
  ' the Try statement exits.
End Try
  ' Ends a structured exception handler.
```

The `Try` block of a `Try...Catch...Finally` exception handler contains the section of code you want your error handler to monitor. If an error occurs during execution of any of the code in this section, Visual Basic examines each `Try` statement within the `Try...End Try` until it finds one whose condition matches that error, where `Try` is the beginning of the section of code to protect and `End Try` is the end the block of code. The code in the `Catch` section deals with any exception that satisfies the `Catch` statement. There may be one or more `Catch` blocks. This process continues through the entire stack until a matching `Catch` block is found in the current procedure.

The code in the `Finally` section always executes last, just before the error-handling block loses scope, regardless of whether the code in the `Catch` blocks has executed. Place cleanup code, such as that for closing files and releasing objects, in the `Finally` section.

Object-oriented Programming

With Visual Basic 6.0, we were able to do component-based development and build three-tier application for the web. Visual Basic .NET provides another leap in Visual Basic development capabilities and features and provides for true object-oriented programming. It meets all four criteria a true object-oriented language requires: abstraction, encapsulation, polymorphism, and inheritance.

Defining a Class

Previously, we defined a *State* class consisting of three properties: name, abbreviation, and FIPS code. We called these *Name*, *Abbr*, and *FIPS*, respectively.

```
Public Class State
  Public Name As String
  Public Abbr As String
  Public FIPS As String
End Class
```

As you can see, a class is enclosed by the `Public Class` and `End Class` statements. In this example, three properties are defined as public property procedures. We can define them as private property procedures. Let's change the FIPS code property into a private one as follows.

```
Public Class State
  Public Name As String
  Public Abbr As String

   Private m_FIPS As String
   Property FIPS() As String
    Get
      Return m_FIPS
    End Get
    Set(ByVal Value As String)
      m_FIPS = Value
    End Set
   End Property

End Class
```

In Visual Basic .NET, there are only two types of property procedures: `Get` and `Set`. The `Get` procedure retrieves the property value from the class and the `Set` procedure assigns the property.

Creating Constructors and Destructors

Once you define a class, you can use it in your program. Using a class requires creating an instance of the class and then destroying it when you no longer need the class. The method for creating an instance of a class is called a constructor. In Visual Basic .NET, the constructor method is always named `Sub New`. If you do not define a constructor, the default is used when an instance of a class is created.

```
' define a new state object as myState
Dim myState as New State

' destroy the object myState afterwards
myState.Dispose
' or do the following
myState = Nothing
```

When you no longer need `myState`, you need to explicitly call the destructor `Dispose` to clean up the object. You can also set `myState` to `Nothing`.

✔ **NOTE:** *There is another destructor called* Finalize. *However, there may be a delay between the time an object is terminated and the time the garbage collector actually destroys the object. Microsoft recommends using* Dispose *instead.*

Understanding Inheritance

One of the most important new object-oriented features in Visual Basic .NET is *inheritance*. Inheritance allows one class to take on the characteristics of another class. The new class is then called a *derived class*, which inherits all behavior of the original class (called a *base class*) but can override base class functionality to perform specialized processing. Derived classes can also contain new class members, which act as a superset of the base class. To try this, let's add a method to the base class `State` to calculate the time zone for a given state.

```
Public Overridable Function TimeZone(ByVal aTime As Decimal) As Decimal
   ' Eastern Standard Time Zone, no calculation
   ' Return the passed in time
   Return aTime
End Function
```

This method uses the `Overridable` keyword. This keyword means that any derived classes can override this method and define their own implementation.

✔ **NOTE:** *If you build a class that may be used as a base class, be sure to use the* Overridable *keyword on any function a derived class may wish to override (unless you do not want the base class to be overridable).*

There may be a problem with the newly added method if a state is not in the Eastern Standard time zone. One solution is to modify the existing

State class to contain all time zones for all of the state types. You can use *Select/Case* statements or *If* statements to manage the logic to handle the different types. However, this can be tedious and complex, especially if there are extra rules or processing associated with the different types.

A second option is to create a separate class for each state type by specifying state (eastern), central state, mountain state, and west state. Each class can then contain its own logic for handling that particular state type. However, if there are similarities in some of the logic you may have duplication in the separate classes. Now we need to define three more new classes that are derived from the base class `State`. We call them `centralState`, `mountainState`, and `westState`.

```
' the centralState class
Public Class centralState: Inherits State
  Public Sub New()
   MyBase.New()
  End Sub
  Public Overrides Function TimeZone(ByVal aTime As Decimal) As Decimal
    Dim newTime As Decimal
    newTime = aTime + 1
    Return newTime
  End Function
End Class
' the mountainState class
Public Class mountainState: Inherits State
  Public Sub New()
   MyBase.New()
  End Sub
  Public Overrides Function TimeZone(ByVal aTime As Decimal) As Decimal
    Dim newTime As Decimal
    newTime = aTime + 2
    Return newTime
  End Function
End Class
' the westState class
Public Class westState: Inherits State
  Public Sub New()
   MyBase.New()
  End Sub
  Public Overrides Function TimeZone(ByVal aTime As Decimal) As Decimal
    Dim newTime As Decimal
    newTime = aTime + 3
    Return newTime
  End Function
End Class
```

The `Inherits` keyword defines the parent or base class. All of the public properties and methods of the base class are accessible to this derived class.

The first method in this derived class is the constructor. This constructor uses the *MyBase* keyword to call the constructor in the base class. The default constructor for the parent class is automatically called when the child is constructed, but you can manually specify the parent's constructor.

The `Overrides` keyword on the second function denotes that this is overriding the function defined within the base class. This allows a derived class to provide its own implementation of a particular function, which computes the correct time based on the time zone that state is in.

✔ **NOTE:** *If you do not put the* Overrides *keyword on the function in the derived class, the function is assumed to shadow the original function. A shadowed function is a function in the derived class that has the same name as a function in the base class but is not intended to override the base class function.*

There are two ways of creating an instance of the derived class.

- Declare an object variable as the derived class and then create the instance of the derived class, as follows.

```
Private m_ctrlState As centralState
m_ctrlState = New centralState()
```

- Declare an object variable as the base class and then create the instance of the derived class, as follows.

```
Private m_ctrlState As State
m_ctrlState = New centralState()
```

The first technique directly declares the object to be of type `centralState`. The derived object has access to any public properties and methods of the `State` base class, any properties and methods overridden in the `centralState` class, any properties and methods shadowed in the `centralState` class, and any public properties and methods of the `centralState` class.

The second technique declares the object to be of type `State`, allowing any type of `State` object to be defined and to leverage polymorphism. This allows reuse of the object variable for any type of `State` object.

However, by declaring the derived object using the parent class the derived object only has access to any public properties and methods of the `State` base class and any properties and methods overridden in the `centralState` class. It does not have access to any public properties and methods of the `centralState` class, nor to any shadowed properties and methods. There are two primary uses for inheritance in the applications you develop.

- You have objects of different types that have similar functionality. For example, the `centralState` class and the `westState` class inherit from the `State` class.
- You have general processing required for sets of objects. For example, every type of business object inherits from a *Business Object* (BO) class.

You should not use inheritance when:

- You only need one function from the base class. If this is the case, you should delegate to the class instead of inheriting from it.
- If you would need to override all of the functions. If this is the case, you should use an interface instead of implementation inheritance.
- The semantics of the hierarchy are not clear. If there is no clear *is a* type of relationship, such as a central *is a* state, delegation or an interface may be a better solution.

Summary

In this chapter, we have introduced what Visual Basic .NET is and survival skills needed for building ASP.NET applications. We discussed in detail variables, constants, and operators. There are three types of variables in Visual Basic .NET.

- An elementary data type, such as *Integer*, *Double*, or *Boolean*
- A composite data type, such as an array or structure
- An object type, or class, from Visual Basic .NET, such as *Label* or *TextBox*

Visual Basic .NET allows you to declare and initialize a variable in a single step. Strings are extremely important to developing web-based applications. It is one of most difficult subjects to deal with in Visual Basic programming. It might take you days to figure out manipulation of a single string. Thanks to regular expressions, such string processing suddenly seems trivial. The most difficult aspect is mastering the non English-like patterns of regular expressions. The good news is that you could prebuild commonly used patterns, such as those for e-mail, URL, and latitude and longitude.

Arrays, collections, functions, and subroutines can now be used in an object-oriented programming environment. They make the object-oriented programming less challenging, especially inside a web form. Garbage collection is no longer limited to Java. The enviable *Try...Catch...Finally* approach is now offered in all .NET families of languages, including Visual Basic .NET. Error and exception handing is easier than ever in Visual Basic .NET.

You have just gone through the crash course of Visual Basic .NET required for building ASP.NET web applications and web services. You had a glimpse at what an ASP.NET web form page was (and in Chapter 2 what web controls are). You may be eager to know how to build an ASP.NET web application. In Chapter 4 we explore the prerequisites of building a GIS web application and how to create a web application for ArcGIS Server.

CHAPTER 4
Building a Simple ArcGIS Server Application

ARCGIS SERVER REPRESENTS A NEW GENERATION OF GIS server products that allow you to run ArcObjects remotely on a GIS server. Working in a similar way to its predecessor ArcIMS, you need only author and publish a collection of ArcObjects bundled with .NET components. The collection can be a map document of ArcMap or an address locator. A map document published through a GIS server is called a map server object. An address locator served via the GIS server is called a geocode server object. The ArcGIS Server provides two application-programming interfaces for ASP.NET: the Server API and the .NET web controls. This chapter focuses on the server API. We will cover how to connect to a GIS server and access it. Then, we will build a simple application based on the server API. The ArcGIS Server web controls are discussed in Chapter 6.

Creating Server Project Files

As described in Chapter 1, an ArcGIS Server object is a coarse-grained ArcObjects component that runs in a process on the server object container machine. ArcGIS Server consists of two types of server objects:

- Map server object: *esriCarto.MapServer*
- Geocode server object: *esriLocation.GeocodeServer*

A map server object is a map document published on an ArcGIS server, whereas a geocode server object is an address locator served on an ArcGIS server. A map document is an ArcMap project file that can contain layers, rendering of the layers, labels of layers, legend, North arrow, scale bar, logo, text, and page layout. For a map document, there can be two types of views: data view and layout view. The data view is used for mapping and querying, whereas the layout view is used primarily for printout purposes. For example, if you plan to use a map document for a page layout control (discussed in Chapter 6), you will need to create a layout view in addition to the data view. To create a map document, you need to use ArcMap.

An address locator is a geocoding component that allows you to translate a street address into a pair of coordinates. According to ESRI, an address locator defines the path to the spatial reference data and the nonspatial address file, algorithms for standardizing addresses and matching them to the reference data, and parameters for reading address data, matching address data to the reference data, and creating output. To create an address locator, you need to use ArcCatalog.

Creating Map Documents

Creating a server project is the same as creating an ArcMap project file (which has the *.mxd* file extension). In the following, we will create three map documents:

- U.S. View
- U.S. Locator
- Arlington Roads
- GNET

The *U.S. View* map document contains state boundaries, county boundaries, and highways. This map document will be primarily used for the main map and page layout. The U.S. Locator map document has just the state and county boundaries. This map document will be used for the overview map. The *Arlington Roads* map document includes the 2000 Census TIGER line work for the streets in Arlington County, Virginia. The *GENT* map document includes an ArcIMS image service from the Geography Network web site and contains major 2000 Census TIGER line features for the entire United States.

To create map documents in ArcMap:

1 Launch ArcMap.

2 Click on the Add Data button (shows a + sign) from the toolbar.

3 Navigate to the folder *USA* of ESRI data such as *ESRI/USA* and select three shapefiles from that directory: *States*, *Counties*, and *Highways*. Load these layers in the order *Counties*, *States*, and *Highways*.

4 Set the *Counties* layer using a light yellow color, and the *Highways* layer using dark brown, but do not fill the *States* polygons. Set the map frame name to *US View*. You can do more with the shapefile-based project, such as setting scale dependency on layers and labels. Figure 4–1 shows the data view of the *US View* map document.

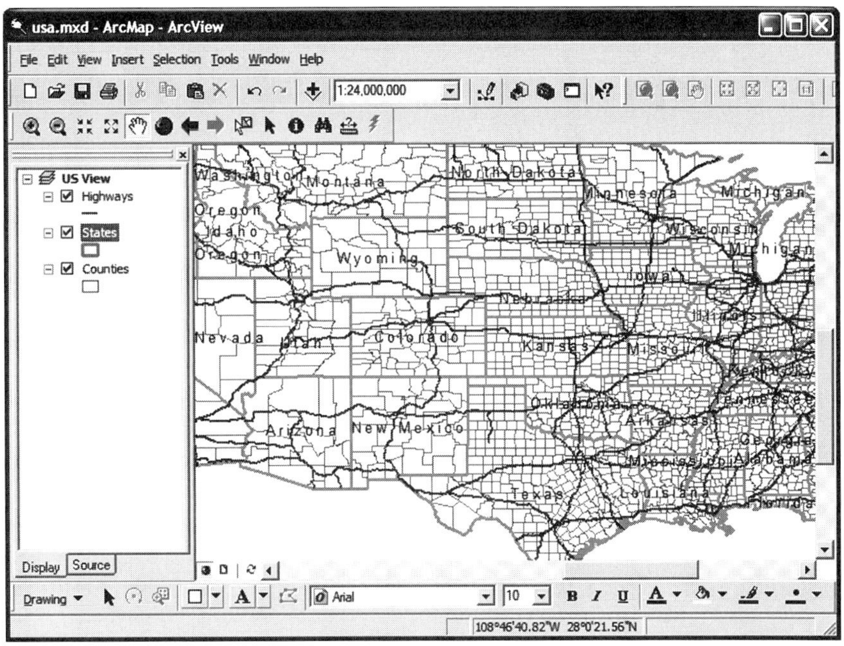

Figure 4–1. A data view of the US View *map document (*usa.mxd*).*

5 Click on the Layout View button to switch to layout view. Then add a legend, North arrow, scale bar, one line of text on top, and another line of text on the bottom (Figure 4–2).

6 Select File/Save As to save the project as *usa*.

7 Navigate to the folder *USA* of ESRI data such as *ESRI/USA* and select two shapefiles from that directory: *States* and *Counties*. Load the *Counties* layer first and then the *States* layer.

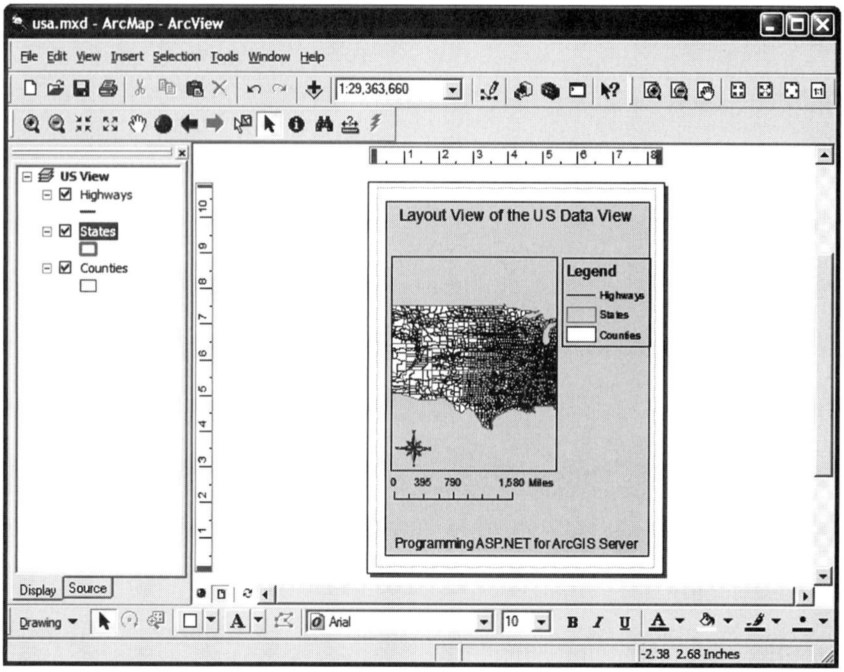

*Figure 4–2. A layout view of the US View map document (*usa.mxd*).*

8 Set the *Counties* layer using light yellow color, but do not fill the *States* polygons. Set the map frame name to *US Overview.*

9 Select File/Save As to save the project as *usa_locator.*

10 Navigate to the folder *USA* of ESRI data such as *ESRI/Virginia* and select *tgr51013rd* for the map document only. Set the map frame name to *Arlington, VA*.

11 Select File/Save As to save the project as *Arlington_rd*.

We have just created three map documents based on local layers. We will create another map server object for the map document using an ArcIMS image service from the Geography Network.

✔ **NOTE:** *The ESRI Geography Network is a free resource web site that offers ArcIMS image and feature services and Open GIS Consortium WMS image services. You can view available services using ArcGIS Desktop 9.x or ArcExplorer 9.x by entering* www.geographynetwork.com.

To create a map document based on the Geography Network in ArcMap:

1 Launch ArcMap.

2 Click on the Add Data button (shows a + sign) from the toolbar.

3 Navigate to the folder *GIS Servers* and double click on Add ArcIMS Server (Figure 4–3).

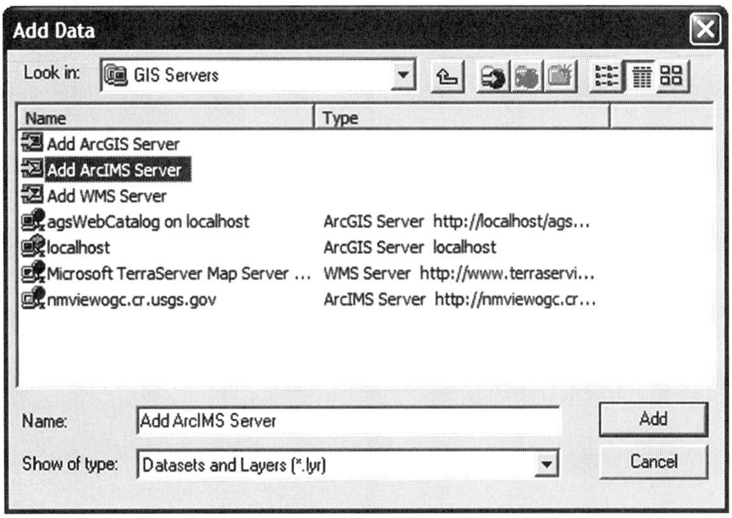

Figure 4–3. The Add Data dialog box with the Add ArcIMS Server option highlighted.

4 The Add ArcIMS Server dialog box appears (Figure 4–4). Enter *http://www.geographynetwork.com* for the URL of the server and then click on OK. The Geography Network will then be added as a new entry in the *GIS Servers* folder (Figure 4–5).

5 Highlight *www.geographynetwork.com* and then click on Add.

6 A list of available images and features from the Geography Network displays (Figure 4–6). Scroll down the list until you see the *CENSUS_TIGER2000* image service and select it. Click on Add. The *CENSUS_TIGER2000* image service was built based on the 2000 Census TIGER line features.

7 A view of the entire United States displays in the map window (Figure 4–7). Set the map frame name to *GNET* and save the map document as *gnet.mxd*.

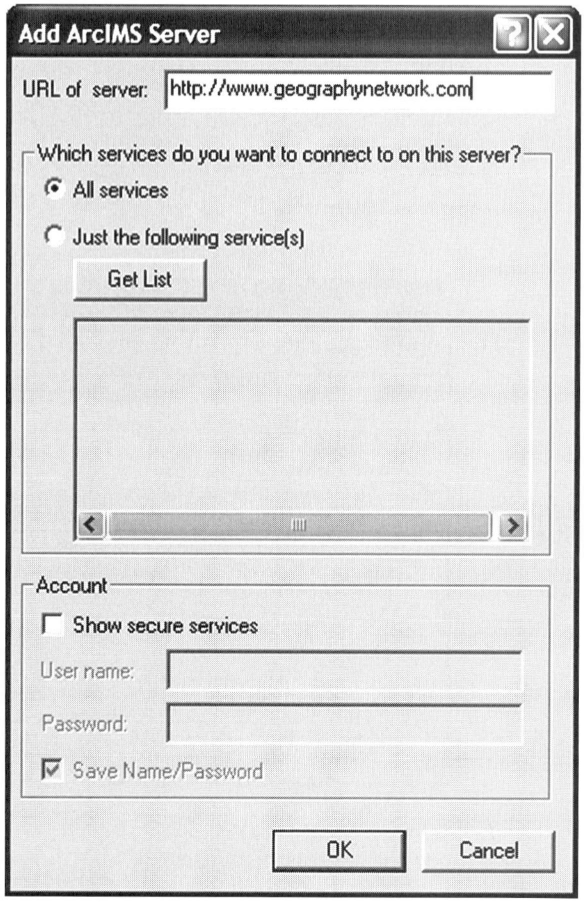

Figure 4–4. The Add ArcIMS Server dialog box.

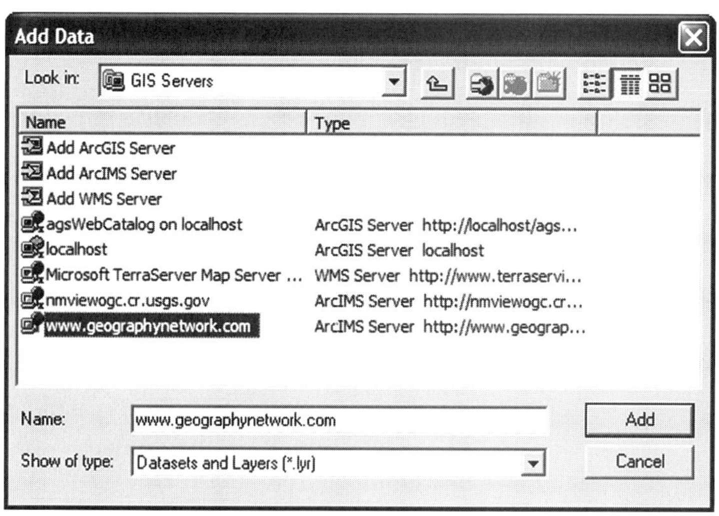

Figure 4–5. Newly added Geography Network entry is shown in the Add Data dialog box.

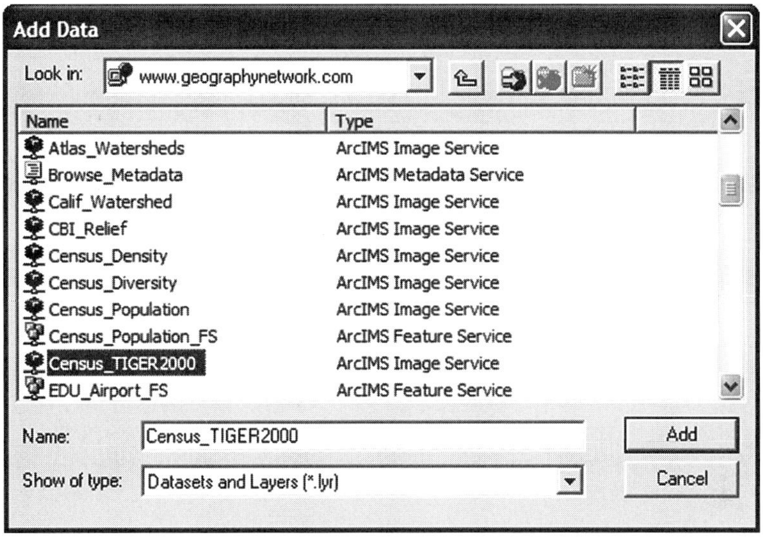

Figure 4–6. Available ArcIM image and feature services from the Geography Network web site.

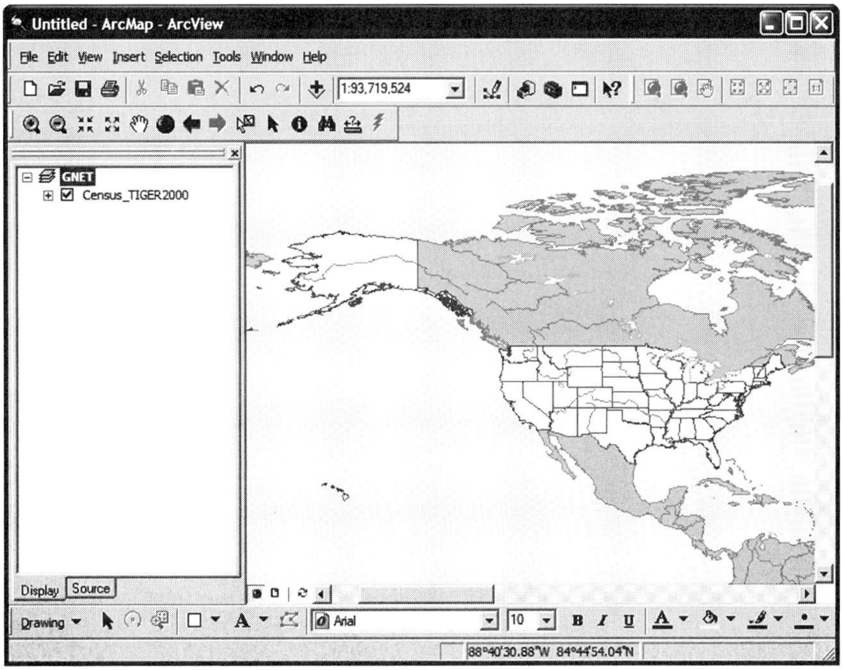

Figure 4–7. The GNET map document is shown in ArcMap.

We have just created four map documents based on local layers and remote ArcIMS image services, which will be used throughout the book. You can add more layers to each of the map documents created based on local layers and make the map display fancy, but make sure to keep those layers specified. Next, we will create an address locator, which is needed for creating a geocode server object.

Creating an Address Locator

Creating an address locator is relatively new. However, the address locator name itself has become standardized throughout the GIS and mapping communities. An address locator is essentially a geocoding engine. It has *.loc* as the file extension.

To create an address locator in ArcCatalog:

1 Launch ArcCatalog.

2 Click on the *Address Locators* folder in ArcCatalog.

3 Double click on Create New Address Locator. The Create New Address Locator dialog box appears (Figure 4–8) with a list of address locator styles displayed.

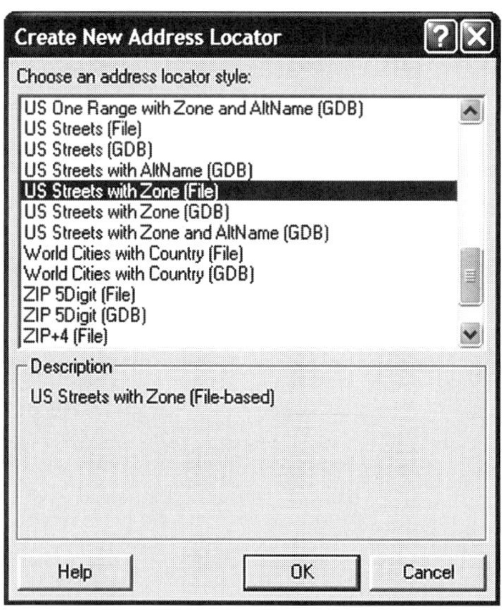

Figure 4–8. The Create New Address Locator dialog.

4 Scroll down the list and select the *US Streets (File)* style and click on OK. The road layer for Arlington, VA has the *US Streets* style. If your road or street layer has a different style, you would choose a style from the list that is appropriate to your layer.

5 The New US Streets (File) Address Locator dialog box displays (Figure 4–9). Type *ArlingtonLocator* in the Name text box.

6 Click on the Browse button on the Primary table tab. Navigate to the folder *USA* of ESRI data such as *ESRI/Virginia*, select the *tgr51013rd* shapefile for the *Arlington County Roads*, and click on Add. Leave everything else at the default settings. The required address elements have a bold field name.

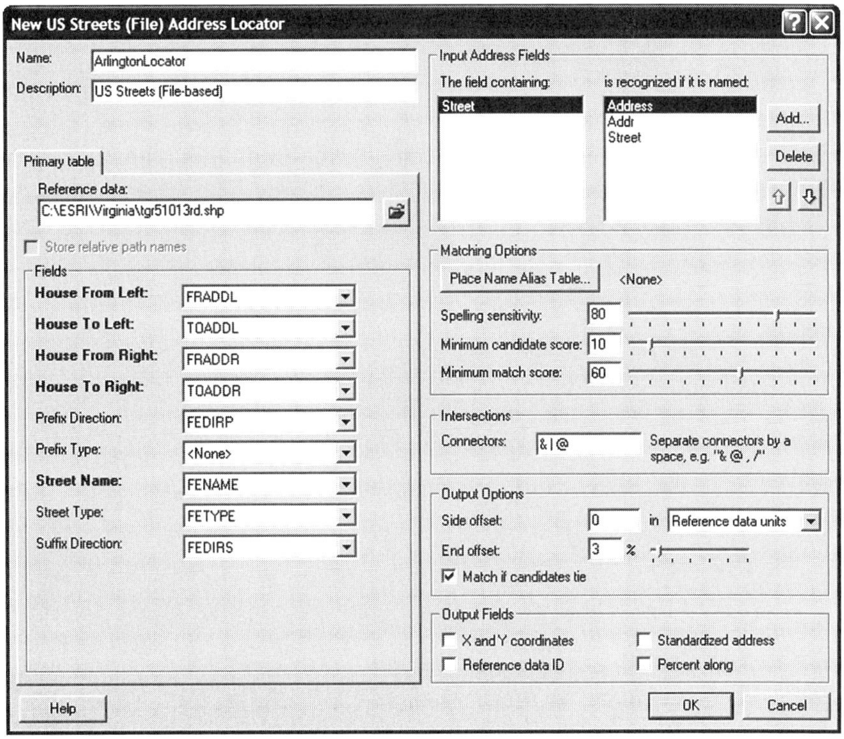

Figure 4–9. The New US Streets Address Locator dialog.

7 Click on OK to create the new address locator. It may take a minute to create the *ArlingtonLocator* address locator. Once it is created, *ArlingtonLocator* will be listed under the Address Locators in ArcCatalog (Figure 4–10).

8 Copy *ArlingtonLocator* from the *Address Locators* folder to the folder where all other map document files (*.mxd*) are. You need to make the address locator available from a publicly available directory. This is required for building a geocode server object. Otherwise, you would get a warning as shown in Figure 4–11.

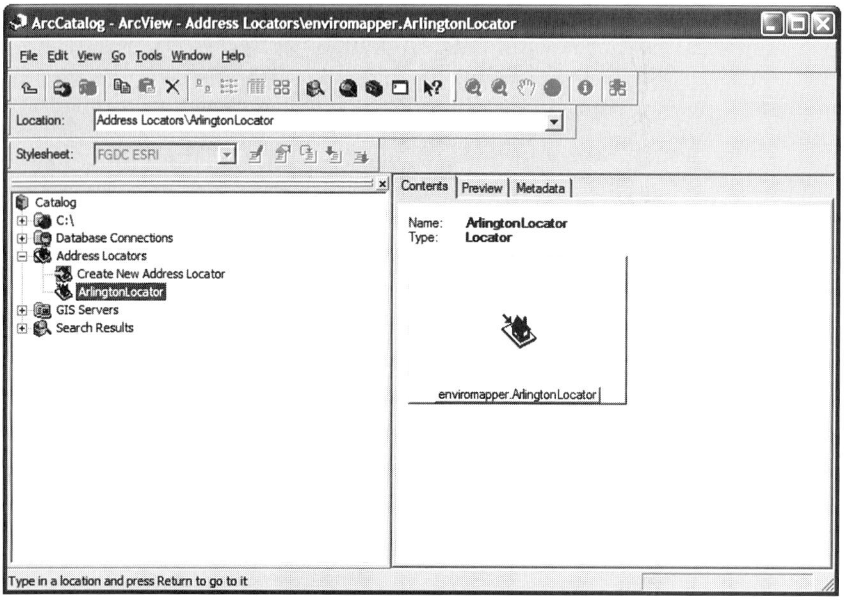

Figure 4–10. An address locator in ArcCatalog.

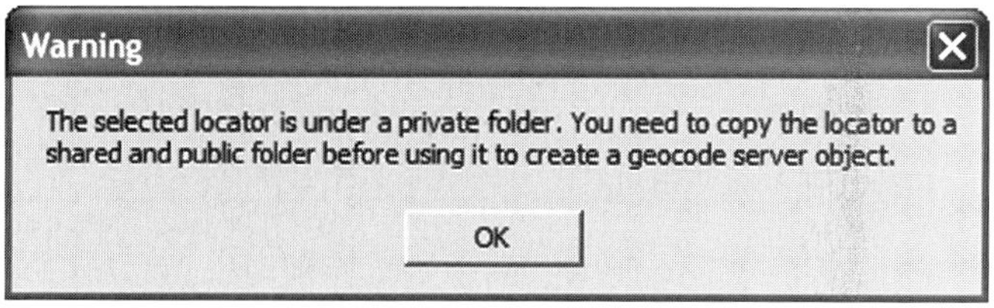

Figure 4–11. A warning message for the address locator creation when the locator is not located in a public directory.

Administrating Map and Geocode Server Objects

You need to set up ArcGIS Server objects before you can start building any ArcGIS Server application. As we discussed before, you can author two types of server objects: Map Server and Geocode Server. A map server object is a map document published through an ArcGIS server, whereas a geocoding server object is an address locator served through an ArcGIS server. If no ArcGIS server exists, you need to create it using ArcCatalog before you can start anything else. You also create both the map server and geocode server objects using ArcCatalog.

Creating Map Server Objects

A map server object can be pooled or non-pooled. When an application makes stateless use of the GIS server, a pooled map server object should be used. When an application makes state use of the GIS server, a non-pooled map server object should be used. The following shows you how to create pooled and non-pooled map server objects.

To create a map server object in ArcCatalog:

1. Launch ArcCatalog.
2. Expand the *GIS Servers* folder and double click on Add ArcGIS Server.
3. The Add ArcGIS Server dialog box appears. Click on the Local Server radio button, enter *localhost* for the Server text box and click on OK. *localhost* will be listed as an ArcGIS server under GIS Servers.
4. Expand the *localhost* folder and double click on Add Server Object. The Add Server Object Wizard dialog box appears (Figure 4–12). Enter *usa* for the Name of the server object, and select *MapServer* for the Type of the server object. Enter *USA state, county, and highways* for the Description and click on Next.
5. Click on the Browser button, navigate to the folder where those three map document projects are saved, and then select *usa* (Figure 4–13). Click on Next.
6. Click on Next two more times.
7. Click on Finish. We will start the map server object manually.

8 Highlight *usa* under the *localhost* folder, right-mouse click on it, and then select Start to start the map server object.

Repeat steps 1 through 7 to create two more map server objects. Name the first *us_locator* and the second *gnet*. We have created three pooled map server objects. We will create one non-pooled map server object so that we can use it for the thematic template (discussed in Chapter 6).

To create a non-pooled map server object in ArcCatalog:

1 Launch ArcCatalog.
2 Expand the *GIS Servers* folder and the *localhost* folder.
3 Double click on Add Server Object and the Add Server Object Wizard dialog box appears. Enter *usa_nonpool* for the Name of the server object, select *MapServer* for the Type of the server object, type *Non-pooled map server object for USA* for the Description, and then click on Next.
4 Click on the Browser button, navigate to the folder where those three map document projects are saved, and select *usa*. Click on Next.
5 Click on the Not Pooled radio button (Figure 4–14) and click on Next two times.
6 Click on Finish. We will start the map server object manually.
7 Highlight *usa_nonpool* under the *localhost* folder; right-mouse click on it, and select Start to start the non-pooled map server object.

Creating a Geocode Server Object

Earlier in this chapter, we created an address locator of *ArlingtonLocator* for Arlington County, VA, and the address locator was copied to the folder where all other *mxd* documents are located. Now we are going to build a geocode server object using *ArlingtonLocator.loc*.

To Create a geocode server object in ArcCatalog:

1 Launch ArcCatalog.
2 Expand the *GIS Servers* folder and the *localhost* folder.

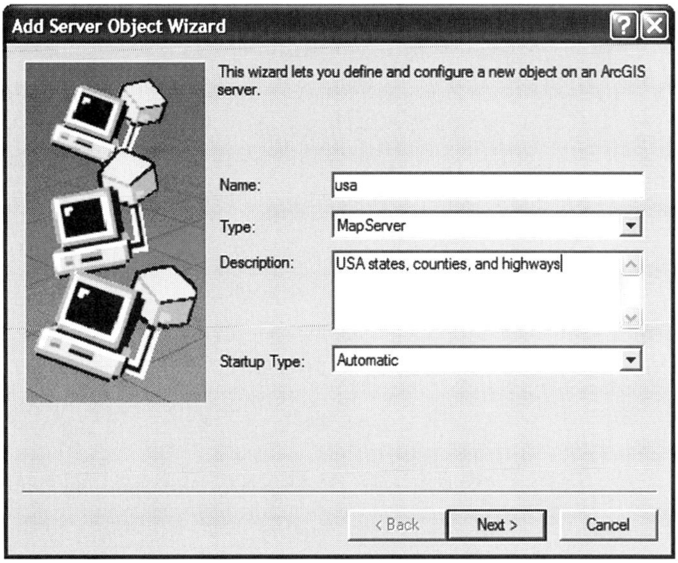

Figure 4–12. The Add Server Object Wizard dialog box.

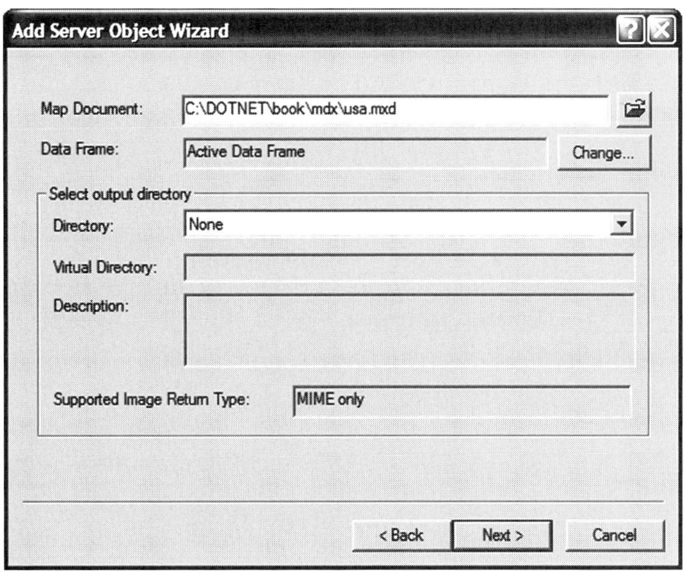

Figure 4–13. Specifying the location of the usa.mxd *map document on the Add Server Object wizard.*

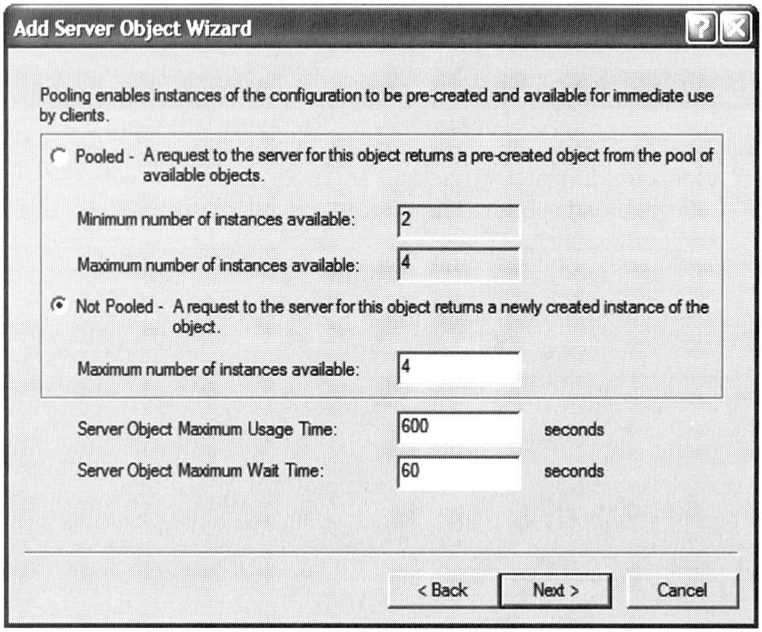

Figure 4–14. Specifying the non-pooled option for a map server object.

3 Double click on Add Server Object and the Add Server Object Wizard dialog box appears (Figure 4–15). Enter *ArlingtonAddress* for the Name of the server object, select *GeocodeServer* for the Type of the server object, enter *Geocoding Server for Arlington County, VA* for the Description, and click on Next.

4 Click on the Browser button, navigate to the public folder where *ArlingtonLocator* is located, and select *ArlingtonLocator*. Click on Next.

5 Click on Next two more times.

6 Click on Finish. We will start the geocode server object manually.

7 Highlight *ArlingtonAddress* under the *localhost* folder, right-mouse click on it, and select Start to start the geocode server object.

Now you have just created an ArcGIS Server geocode server object. It is ready for you to consume in an ASP.NET web application. Figure 4–16 shows the geocode server object in ArcCatalog.

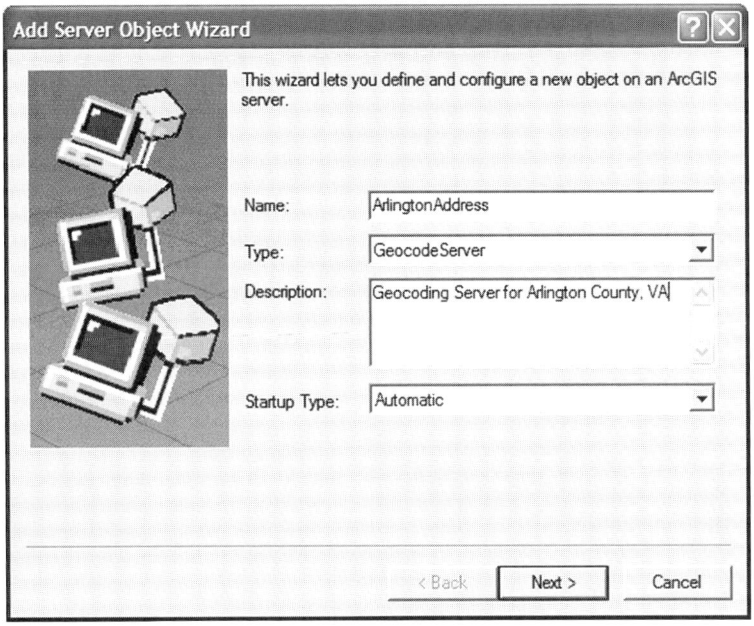

Figure 4–15. Adding a geocode server via the Add Server Object wizard.

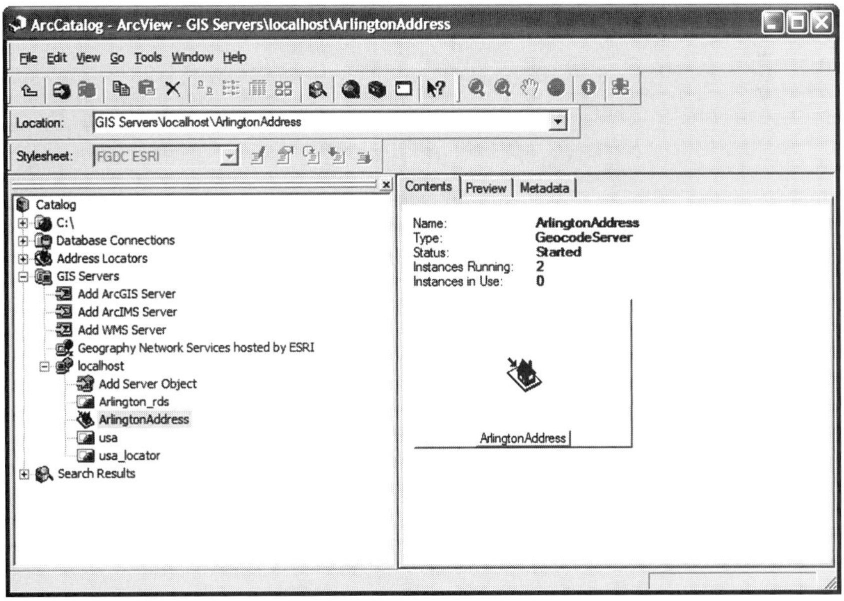

Figure 4–16. Geocode server object in ArcCatalog.

Building an ArcGIS Server Application Using Visual Studio .NET

Once you have created server objects, you are ready to start building web applications. The ArcGIS Server has two application programming interfaces for ASP.NET:

- The Server API
- The .NET web controls

We will focus on the server API in this chapter and discuss the ArcGIS Server web controls in Chapter 6. The server API is a collection of object libraries that contains the ArcObjects necessary to connect to the GIS server and access to server objects. The ArcGIS Server also includes a set of web controls and web templates as part of the application development framework (ADF) for .NET. We will discuss programming with the ADF and the web controls in Chapter 6.

Connecting to the GIS Server

Before you can gain access to either a map server object or a geocode server object on the GIS server, you need to connect to it first. Connections to the GIS server are made through the `ServerConnection` object, which is part of the *ESRI.ArcGIS. Server. WebControls* namespace. For example, to connect to the *localhost* GIS Server you write the Visual Basic .NET code as follows.

```
Dim svrConnection As ESRI.ArcGIS.Server.WebControls.ServerConnection
svrConnection = New ESRI.ArcGIS.Server.WebControls.ServerConnection
svrConnection = "localhost"
svrConnection.Connect()
```

If you build a web application using the ArcGIS Server web controls or templates, you do not have to write code to connect to a GIS server. You can connect to the server by setting the *Server* property on the web control.

Getting Server Objects from the GIS Server

Once you have a server connection object, you can easily retrieve the server objects from that server. You use the `GetServerObjectNames` method of the `ServerConnection` object to retrieve all types of servers or just obtain a certain type of server by specifying whether it is a Map Server or Geocode Server. For example, you can retrieve server objects and list them in a *ListBox* control (discussed in Chapter 5) as follows.

```
'retrieve all server objects
lstServerObjs.DataSource = GetServerObjectNames()
'retrieve map server objects
lstServerObjs.DataSource = GetServerObjectNames("MapServer")
'retrieve geocode server objects
lstServerObjs.DataSource = GetServerObjectNames("GeocodeServer")
```

Getting Data Frames of a Map Server Object

A map document can have multiple data frames. To retrieve all data frames out of a map document, you need to use the `GetData Frames-FromMapServerObject` method of the `Server Connection` object for a given map server object. For example, you can retrieve the data frame of *US View* for the *usa* map server object created previously and list it in a *ListBox* control using the following code segment.

```
'retrieve all data frames for a map server object
lstDataFrames.DataSource = _
      srvConnection.GetDataFramesFromMapServerObject("usa")
```

You can go back and add one more data frame to it. After that, you refresh the map server and run the same code again. You should see both the *US View* and the newly added data frames in the list.

Getting Layers and Columns

There is no method in the `ServerConnection` object to list all layers for a given data frame. You need to use the `WebMap` object. The *WebMap* class exposes the `MapServer` method, which returns a map server object. You can then use the `GetServerInfo` method of this object, and eventually use it to access the *IMapServerInfo* interface. The *IMapServer-Info* interface has a `MapLayerInfos` property that returns a list of layers for a given data frame. You can select a layer from the list and get all fields for that layer using the `Fields` property.

```
Dim webmap As WebMap = New WebMap(svrContext, host, dataFrame)
Dim msi As IMapServerInfo =
   webmap.MapServer.GetServerInfo(webmap.DataFrame)
Dim layers As IMapLayerInfos = msi.MapLayerInfos
Dim fields As IFields = layers.Element(1).Fields
```

Creating an Application

Now we will create a web application that accesses a GIS server via the server API. Once the application connects to the GIS server, it will retrieve all map server objects, data frames for a selected map server object, layers for a chosen data frame, and fields for a selected layer. The web form contains four *ListBox* controls, four *Label* controls, and an *Impersonation* control. The first *ListBox* control is used to display all map server object on a GIS server. The second *ListBox* control is used to display all data frames for a selected map server object. The third *ListBox* control is used to display all layers in the selected data frame. The last *ListBox* control is used to display the attribute fields of the selected layer. Four *Label* controls are used to label each list box. We will set the `AutoPostBack` property to *True* for the map server, data frame, and layer *ListBox* controls so that when a selection is made in one list the other list boxes will be updated accordingly. The first item in each list box will be set to *Selected* by default. For example, selecting a map server object in the map-server list box will trigger an update of the data-frame list box.

To create a simple application using the server API:

1. In Visual Studio .NET, create a new Visual Basic .NET project using the ASP.NET Web Application template and give a name of *agsServerAPI* to the application.
2. Drag and drop an *Impersonation* control from the ESRI tab of the toolbox onto the web form.
3. From the Properties window, click on the *Identity* property and an Ellipsis button displays. Click on the button and the Impersonation1 Identity dialog box appears. Enter a valid user name and password, and then click on OK. You can leave the Domain or Machine Name box blank. You will notice right away that the identity information you just entered is encrypted for security reasons.
4. Drag a *Label* control from the Web Forms tab of the toolbox onto the web form and set the *Text* property to Map Server Objects.
5. Drag a *ListBox* control from the Web Forms tab of the toolbox onto the web form and place it under the label. Set the *ID* property to *lstServerObjs*, the *AutoPostBack* property to *True*, the *Height* property to *80px*, and the *Width* property to *200px*.

6 Drag another *Label* control from the Web Forms tab of the toolbox onto the web form and place it under the previous *ListBox* control. Set the *Text* property to *Data Frames*.

7 Drag another *ListBox* control from the Web Forms tab of the toolbox onto the web form and place it under the data frame label. Set the *ID* property to *lstDataFrames*, the *AutoPostBack* property to *True*, the *Height* property to *80px*, and the *Width* property to *200px*.

8 Drag another *Label* control from the Web Forms tab of the toolbox onto the web form and place it under the previous *ListBox* control. Set the *Text* property to *Layers*.

9 Drag another *ListBox* control from the Web Forms tab of the toolbox onto the web form and place it under the data frame label. Set the *ID* property to *lstLayers*, the *AutoPostBack* property to *True*, the *Height* property to *80px*, and the *Width* property to *200px*.

10 Drag another *Label* control from the Web Forms tab of the toolbox onto the web form and place it under the previous *ListBox* control. Set the *Text* property to *Fields*.

11 Drag another *ListBox* control from the Web Forms tab of the toolbox onto the web form and place it under the data frame label. Set the *ID* property to *lstFields*, the *AutoPostBack* property to *True*, the *Height* property to *80px*, and the *Width* property to *200px*.

12 Click on the View Code button, and the code-behind appears. Add the following namespaces before the class.

```
Imports ESRI.ArcGIS.Server
Imports ESRI.ArcGIS.Server.WebControls
Imports ESRI.ArcGIS.Carto
Imports ESRI.ArcGIS.Geodatabase
```

13 Add the following code to the `Page_Load` event.

```
If Not Page.IsPostBack Then
  If Session.IsNewSession Then
   'Set host machine and save in session state.
   Dim host As String = "localhost"
   Session.Add("host", host)
   'Create a server connection.
   Dim svrConnection As ServerConnection = New ServerConnection(host,True)
   'Populate server objects and data frames list boxes.
   Get_ServerObjects(svrConnection)
   Get_DataFrames(svrConnection)
```

```
    Dim mapServerObj As String = lstServerObjs.SelectedValue
    Dim dataFrame As String = lstDataFrames.SelectedValue
    'Create WebMap and populate layers and fields list boxes.
    Dim webmap As WebMap =
       Get_WebMap(svrConnection, host, mapServerObj, dataFrame)
    Get_Layers(webmap)
    Get_Fields(webmap)
    'Keep server context for non-pooled objects.
    Keep_ServerContext(webmap)
  End If
End If
```

14 Add the `Get_ServerObjects` as a sub after the `Page_Load` event.

```
Private Sub Get_ServerObjects(ByVal svrConnection As ServerConnection)
  'Populate list of map server objects on server.
  lstServerObjs.DataSource =
   svrConnection.GetServerObjectNames("MapServer")
  lstServerObjs.DataBind()
  'set the first item to Selected
  lstServerObjs.SelectedIndex = 0
 End Sub
```

The sub uses the `GetServerObjectNames` method of the `ServerConnection` object to retrieve all map server objects on the *localhost* GIS server.

15 Add the `Get_DataFrames` as a sub after `Get_ServerObjects`.

```
Private Sub Get_DataFrames(ByVal svrConnection As ServerConnection)
  'Populate list of data frames of selected map server object
  Dim mapServerObj As String = lstServerObjs.SelectedValue
  lstDataFrames.DataSource = _
   svrConnection.GetDataFramesFromMapServerObject(mapServerObj)
  lstDataFrames.DataBind()
  'set the first item to Selected
  lstDataFrames.SelectedIndex = 0
End Sub
```

The sub uses the `GetDataFramesFromMapServerObject` method of the `ServerConnection` object to retrieve all data frames for a select map server object, which is specified by the current selected item in the *lstServerObjs* list box via `lstServerObjs.SelectedValue`.

16 Add the `Get_Layers` as a sub after `Get_DataFrames`.

```
Private Sub Get_Layers(ByVal webmap As WebMap)
  'Populate list of layers in selected data frame
```

```
  Dim arrLayers As ArrayList = New ArrayList
  Dim map As IMapServer = webmap.MapServer
  Dim msi As IMapServerInfo = map.GetServerInfo(webmap.DataFrame)
  Dim layers As IMapLayerInfos = msi.MapLayerInfos
  If layers.Count <> 0 Then
    Dim i As Integer
    Dim oneLayer As IMapLayerInfo
    For i = 0 To layers.Count - 1
       oneLayer = layers.Element(i)
       arrLayers.Add(oneLayer.Name)
    Next
  Else
    arrLayers.Add("No layers in data frame.")
  End If
  lstLayers.DataSource = arrLayers
  lstLayers.DataBind()
  'set the first item to Selected
  lstLayers.SelectedIndex = 0
End Sub
```

17 Add the `Get_Fields` as a sub after `Get_Layers`.

```
Private Sub Get_Fields(ByVal webmap As WebMap)
  'populate list of fields
  Dim arrFields As ArrayList = New ArrayList
  Dim layerIndex As Integer = lstLayers.SelectedIndex
  Dim map As IMapServer = webmap.MapServer
  Dim msi As IMapServerInfo = map.GetServerInfo(webmap.DataFrame)
  Dim layers As IMapLayerInfos = msi.MapLayerInfos
  If layers.Count <> 0 Then
    Dim oneLayer As IMapLayerInfo = layers.Element(layerIndex)
    If oneLayer.IsFeatureLayer Then
     Dim fields As IFields = oneLayer.Fields
     Dim i As Integer
      Dim oneField As IField
      For i = 0 To fields.FieldCount - 1
        oneField = fields.Field(i)
        arrFields.Add(oneField.Name)
      Next
    Else
      arrFields.Add("No fields for this layer.")
    End If
  Else
    arrFields.Add("No fields available.")
  End If
  lstFields.DataSource = arrFields
  lstFields.DataBind()
  'set the first item to Selected
```

```
  lstFields.SelectedIndex = 0
  End Sub
```

18 Add the Get_WebMap as a sub after Get_Fields.

```
Private Function Get_WebMap(ByVal svrConnection As ServerConnection, _
          ByVal host As String, ByVal mapServerObj As String,_
          ByVal dataFrame As String) As WebMap
  Dim Context As IServerContext = Session("serverContext")
  If IsDBNull(Context) Then 'for nonpooled objects
   Dim svrContext As IServerContext = Session("serverContext")
   Dim webmap As WebMap = New WebMap(svrContext, host, dataFrame)
     Return webmap
  Else ' for pooled objects
     Return New WebMap(svrConnection, mapServerObj, dataFrame)
  End If
End Function
```

This is a function that creates the *WebMap* object of webmap. For pooled objects, the Get_WebMap function returns webmap based on a server connection. The GIS server generates a new server context from the pool and returns it. For non-pooled objects, the function uses an existing server context extracted from session state to create webmap. In this case, the server returns the same server context the application used previously.

19 Add the Keep_ServerContext as a sub after Get_WebMap.

```
Private Sub Keep_ServerContext(ByVal webmap As WebMap)
  'Save ServerContext object in session state for non-pooled object.
  If Not webmap.IsPooled Then
   Session.Add("serverContext", webmap.ServerContext)
  End If
  webmap.Dispose()
End Sub
```

This sub saves the server context in session state for a non-pooled map server object through Session.Add. It then disposes the web map object for both pooled and non-pooled map server objects through webmap.Dispose().

```
If Not webmap.IsPooled Then
 Session.Add("serverContext", webmap.ServerContext)
End If
webmap.Dispose()
```

20 Double click on the *lstServerObjs ListBox* control and add the following code to the event.

```
Private Sub lstServerObjs_SelectedIndexChanged(ByVal sender As _
      System.Object, ByVal e As System.EventArgs)_
      Handles lstServerObjs.SelectedIndexChanged
  'New Map Server Object selected. Update list boxes.
  Dim host As String = Session("host")
  Dim svrConnection As ServerConnection = New ServerConnection(host, True)
  Get_DataFrames(svrConnection)
  Dim mapServerObj As String = lstServerObjs.SelectedValue
  Dim dataFrame As String = lstDataFrames.SelectedValue
  'Remove any ServerContext in session state and release server context.
  Dim svrContext As IServerContext = Session("serverContext")
  If IsDBNull(svrContext) Then
    svrContext.ReleaseContext()
    Session.Remove("serverContext")
  End If
  Dim webmap As WebMap = Get_WebMap(svrConnection, host, mapServerObj,_
    dataFrame)
  Get_Layers(webmap)
  Get_Fields(webmap)
  Keep_ServerContext(webmap)
End Sub
```

When a new map server is selected, for which this event triggered, this sub makes a new connection to the GIS server and populates the data frame list by calling `Get_DataFrames`. This sub continues to release the server context if it exists and creates a new web map object. Finally, this sub calls `Get_Layers` and `Get_Fields` to populate the layer and field lists, respectively.

21 Double click on the *lstDataFrames ListBox* control and add the following code to the event.

```
Private Sub lstDataFrames_SelectedIndexChanged(ByVal sender As _
      System.Object, ByVal e As System.EventArgs)_
      Handles lstDataFrames.SelectedIndexChanged
  'New data frame selected. Update list of layers and fields.
  Dim host As String = Session("host")
  Dim mapServerObj As String = lstServerObjs.SelectedValue
  Dim dataFrame As String = lstDataFrames.SelectedValue
  Dim svrConnection As ServerConnection =
     New ServerConnection(host, True)
  Dim webMap As WebMap = Get_WebMap(svrConnection, host, mapServerObj,_
    dataFrame)
```

```
 Get_Layers(webMap)
 Get_Fields(webMap)
 Keep_ServerContext(webMap)
End Sub
```

When a new map server is selected, for which this event triggered, this sub makes a new connection to the GIS server and creates a new `webmap` object. This sub then calls `Get_Layers` and `Get_Fields` to populate the layer and field lists, respectively.

22 Double click on the *lstLayers ListBox* control and add the following code to the event.

```
Private Sub lstLayers_SelectedIndexChanged(ByVal sender As _
      System.Object, ByVal e As System.EventArgs) _
      Handles lstLayers.SelectedIndexChanged
 'New layer selected. Update list of fields.
 Dim host As String = Session("host")
 Dim mapServerObj As String = lstServerObjs.SelectedValue
 Dim dataFrame As String = lstDataFrames.SelectedValue
 Dim svrConnection As ServerConnection = New ServerConnection(host, True)
 Dim webMap As WebMap = Get_WebMap(svrConnection, host, mapServerObj,_
   dataFrame)
 Get_Fields(webMap)
 Keep_ServerContext(webMap)
End Sub
```

When a new map server is selected, for which this event is triggered, this sub makes a new connection to the GIS server and creates a new `webmap` object. This sub then calls `Get_Fields` to populate the field list.

23 Click on the Start button or press F5 to run the application.

When the web page displays, select *usa* for the map server object. Because the *AutoPostBack* property on the map-server list box was set to *True*, the data frame list box will be then updated automatically to show the *US View* data frame. Both the layer and field list boxes will also be updated automatically. Select the *States* layer from the layer list box. The field list box will be updated because the *AutoPostBack* property on the layer list box was also set to *True*. The field list box will then show all field names for the *States* layers, shown in Figure 4–17.

Figure 4–17. A simple application was created to show how to access the server API.

Summary

In this chapter, we discussed how to create a map document and address locator. A map document is a *.mxd* file created in ArcMap, whereas an address locator is a *.loc* file created in ArcCatalog. A map document can be published through a GIS server, whereupon it becomes an ArcGIS Server map object. An address locator can be served through a GIS server, whereupon it turns into an ArcGIS Server geocode server. A server object can be pooled or non-pooled. When an application makes stateless use of the GIS server, a pooled map server object should be used. When an application makes state use of the GIS server, a non-pooled map server object should be used.

The ArcGIS Server has two application programming interfaces for ASP.NET: the Server API and the .NET web controls. The server API is a collection of object libraries that contains the ArcObjects necessary to connect to the GIS server and access to server objects. The very first thing you need to do in ArcGIS Server programming is to gain access to either a map server object or a geocode server object on a GIS server. You need to connect to a GIS server first. Connections to the GIS server are made through the `ServerConnection` object, which is part of the *ESRI.ArcGIS.Server.WebControls* namespace. For pooled objects, the GIS server generates a new server context from the pool. For non-pooled objects, the GIS server extracts an existing server context from the session state. You can also obtain the data frames by using the `GetDataFrames FromMapServerObject` method of the `ServerConnection` object for a given map server object. However, there is no method in the `ServerConnection` object to list all layers for a given data frame. You need to use the `WebMap` object, which is also from the *ESRI.ArcGIS. Server.WebControls* namespace.

Several web controls were used to build that simple application based on the server API. If you somehow felt puzzled by those web controls and their associated events, you do not need to worry about it. We did not go into the details about those web controls because we will discuss most commonly used web controls in detail in Chapter 5. Tips such as when to set the *AutoPostBack* property to *True* or *False* will also be addressed. We will also discuss the ArcGIS Server ADF for .NET in Chapter 6.

Chapter 5

Web Forms, Web Controls, and Web Templates

ASP.NET WEB PAGES CONSIST OF WEB FORMS, web controls, code, markup tags, and text. Based on the request, the server controls generate a proper markup language. The ASP.NET runtime then combines the HTML content of all web controls and sends them to the client as a page to display in a web browser.

One of the major achievements of ASP.NET was to bring to the Web a rich application development environment similar to that used in traditional desktop development. It also tries to bring as many features as possible to ASP.NET web pages. ASP.NET achieves its goal by providing a flexible development methodology that allows the use of multiple .NET-compatible languages for the same ASP.NET application development. Furthermore, you can have access to the entire .NET framework via any .NET-compatible language.

ASP.NET web applications begin with web forms. In this chapter we will start with web forms, and then go on to discuss in detail web controls, validation controls, and template controls.

Window Forms, HTML Forms, and Web Forms

Visual Basic became one of the primary programming languages primarily due to its famous Windows forms. A Visual Basic application is a form-based application. That allows you to place buttons, labels, or text boxes on top of forms. Windows forms are used for Windows-based applications. When Internet use became popular, HTML forms were created. An HTML form is a section of a document containing normal content, markup, special elements called controls (e.g., check boxes, radio buttons, and inputs), and labels on those controls (W3C Consortium). A form is an HTML area that can contain form elements, as indicated by the following.

```
<FORM name="myForm" action="myFormAction.asp" method="get">
<LABEL for="username">User name: </LABEL>
<INPUT type="text" name="user">
<INPUT type="submit" value="Submit"> <INPUT type="reset">
</FORM>
```

HTML forms made publishing dynamic HTML possible. They are supported by all web-based development tools. However, HTML forms have a limitation: they can only go back to the server as a form action but not with each individual HTML control such as a button or an input box.

With the birth of the .NET framework, Microsoft was able to create an ASP.NET version of Windows forms called web forms. With web forms, you can design a web application in the same way you do with Windows applications. An ASP.NET application can contain one or many web pages. Each web page is a web form. You can place buttons, labels, check boxes, and other types of controls on the form. From a web user perspective, every ASP.NET page is basically an HTML form. Everything is rendered in standard HTML, and it works with either Internet Explorer or Netscape. As covered in Chapter 2, Visual Studio .NET provides an IDE to build ASP.NET applications. An ASP.NET application is essentially a Visual Studio .NET project. It can contain one or many web forms. You can drag and drop elements such as buttons and labels, which are called web controls, onto a web form. In the following sections, we will discuss web forms, web controls, template web controls, and web templates.

Web Forms

An ASP.NET page is essentially a web form whose operation is twofold: it runs on both the server side and client side. A web form is a web-based implementation of the Visual Basic Windows form. Most of the elements placed on the form can interact with the server independently. Another dimension is its client-side operation. A web form is rendered as an HTML form inside a web browser, with all elements on the form rendered as HTML form elements. However, each of the elements can communicate with the server individually. Note that every web control tag contains a special attribute of `runat="server"`. So does the form tag. If you leave it out, a web control becomes a regular HTML form element.

```
<form id="Form1" method="post" runat="server">
 <asp:Label id="Label1" runat="server">User name:</asp:Label>
 <asp:TextBox id="TextBox1" runat="server" MaxLength="12"
     Columns="12"></asp:TextBox>
 <asp:Label id="Label2" runat="server">Password:</asp:Label>
 <asp:TextBox id="TextBox2" runat="server" MaxLength="12"
     Columns="12" TextMode="Password"></asp:TextBox>
 <asp:Button id="Button1" runat="server" Text="Login"></asp:Button>
</form>
```

You can drag and drop a web control onto a web form in the same way as you do with Visual Basic form, and you can design the presentation of web forms for a browser in a similar way to that you do with HTML. Web forms take advantage of a robust set of web controls exposed by the .NET framework namespaces, including HTML controls, web controls, validation controls, and mobile controls. Each web control on a web form has a set of properties, methods, and events that can be manipulated on the server side. Web forms also support custom controls, which are controls you create by inheriting features from other controls just mentioned (see also Chapter 7). Web forms produce browser-independent code that can be run inside Internet Explorer or Netscape.

Web Controls

Web controls are those elements that can be placed on a web form. They are specially designed ASP.NET tags that can only be interpreted by the server. The currently available web controls were listed in Table 2-4.

Layout and Formatting Controls: Label, Panel, and Table

There are three categories of web controls that can be used for page layout and formatting:

- *Label* control
- *Panel* control
- *Table* control, *TableRow* control, and *TableCell* control

The *Label* control is used to display any HTML text with style and within a predefined area. So it does more than just HTML text. You can set font, color, location, and size for it. Once a *Label* control is placed on a web form, it has all of the previously listed already. All of those properties can be set at design time with Visual Studio .NET. Labels are static text, meaning that users cannot edit the text once the HTML page is rendered. There is no handler event associated with the *Label* control. After you have placed the *Label* control on a web form, you can resize it to suit your needs.

To add a Label *control to an ASP.NET web form:*

1 Drag a *Label* control from the Web Forms tab of the toolbox onto the web form.

2 Set the *Text* property to *"Hello ArcGIS!"*.

3 Set Font/Bold to *True*, Font/Name to *Comic Sans MS*, Font/Size to *X-Large*, *Font*/Overline to *True*, and Font/Underline to *True*.

4 Set FontColor to Red.

5 Set BackColor to Yellow.

6 Click on the Start button or press F5 to run the application. A web page with the label displays (Figure 5–1).

Figure 5–1. Example of a Label *control.*

Listing 5–1 is the *.aspx* file for the web form with a *Label* control.

Listing 5–1. *Label* control *.aspx* file.

```
<form id="Form1" method="post" runat="server">
 <asp:Label id="Label1" style="Z-INDEX: 101; LEFT: 48px;
    POSITION: absolute; TOP: 32px" runat="server"
    Font-Bold="True"
    Font-Names="Comic Sans MS"
    Font-Size="X-Large"
    Font-Overline="True"
    Font-Underline="True"
```

```
        ForeColor="Red"
    BackColor="Yellow">Hello ArcGIS!
 </asp:Label>
</form>
```

The *Panel* control is used as a container for other controls. It is rendered as an HTML `<DIV>` element. It can be used to group a set of controls. It is especially useful for displaying and hiding that group of controls. By setting the *BackImageUrl* property, you can display an image in the background of the *Panel* control in the same way as you do for a `<DIV>` tag. You can specify the horizontal alignment of the items contained in the *Panel* control by using the *HorizontalAlignment* property.

To add a Panel *control to an ASP.NET web form:*

1 Drag a *Panel* control from the Web Forms tab of the toolbox onto the web form.

2 Set the *Height* property to 200px, the *Width* property to 150px, the *BorderStyle* property to *inset*, and the *BorderWidth* to 5px.

3 Drag another *Panel* control from the Web Forms tab of the toolbox onto the web form.

4 Set the *Height* property to 200px, the *Width* property to 150px, the *BorderStyle* property to *outset*, and the *BorderWidth* to 5px.

5 Click on the Start button or press F5 to run the application. A web page with two panels displays (Figure 5–2).

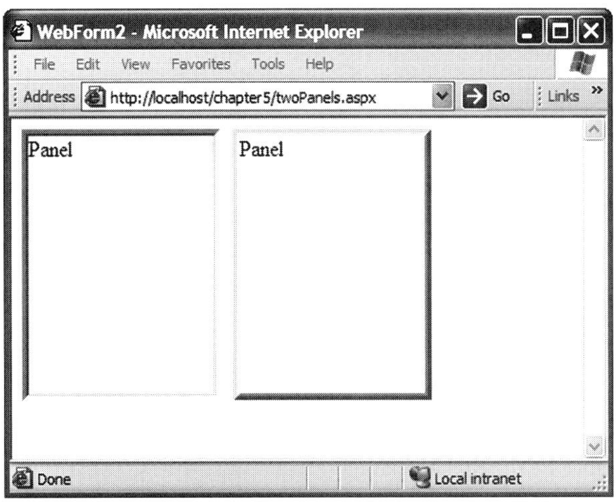

Figure 5–2. Example of two Panel *controls.*

Listing 5–2 is the *.aspx* file for the web form with two *Panel* controls.

Listing 5–2. *.aspx* file for web form with two *Panel* controls.

```
<form id="Form1" method="post" runat="server">
 <asp:Panel id="Panel1" style="Z-INDEX: 101; LEFT: 8px;
    POSITION: absolute; TOP: 8px" runat="server"
    Width="150px" Height="200px"
    BorderStyle="Inset" BorderWidth="5px">Panel</asp:Panel>
```

```
<asp:Panel id="Panel2" style="Z-INDEX: 102; LEFT: 168px;
   POSITION: absolute; TOP: 8px" runat="server"
      Width="150px" Height="200px"
   BorderStyle="Outset" BorderWidth="5px">Panel</asp:Panel>
</form>
```

Tables are widely used in web applications. A table is commonly used to format tabular data or a set of HTML form controls because it is supported by all HTML browsers. There are three controls used to create a table: *Table* control, *TableRow* control, and *TableCell* control. In the following example, we will build a table for the federal web site dedicated to providing geospatial data. The table has two columns: Website and URL.

To add a table to an ASP.NET web form:

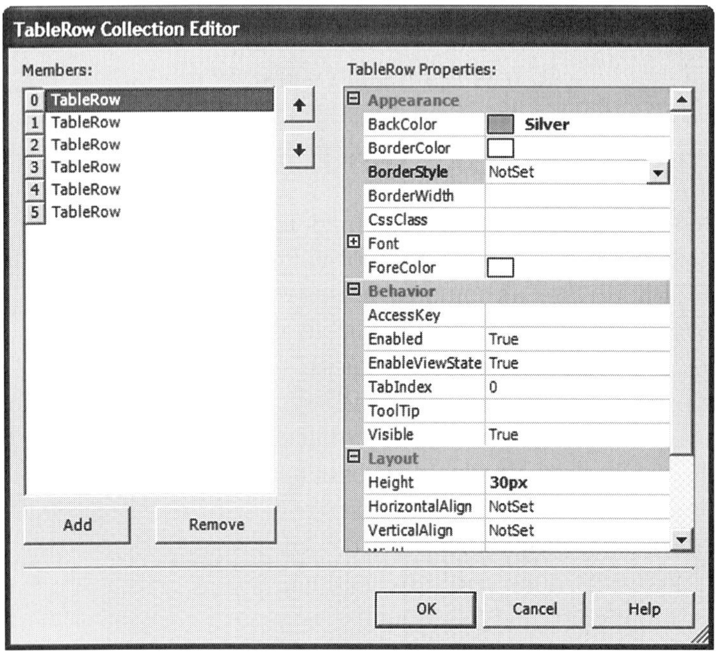

Figure 5–3. TableRow Collection Editor dialog.

1 Drag a *Table* control from the Web Forms tab of the toolbox onto the web form.

2 Set the *BorderWidth* property to 5px, the *BorderStyle* property to Double, and the *BorderColor* property to Silver.

3 From the Properties window, click on the Ellipsis button of the *Rows* property, and then the TableRow Collection Editor dialog box appears (Figure 5–3). You use the TableRow Collection Editor to manually define the rows for the table.

4 Click on Add to add a *TableRow* control. The TableRow Properties pane appears. Set the *BackColor* property to Silver and the *Height* property to 30px. We will use this row for the table caption.

5 From the TableRow Collection Editor, click on the Ellipsis button of the *Cells* property and the TableCell Collection Editor dialog box appears (Figure 5–4).

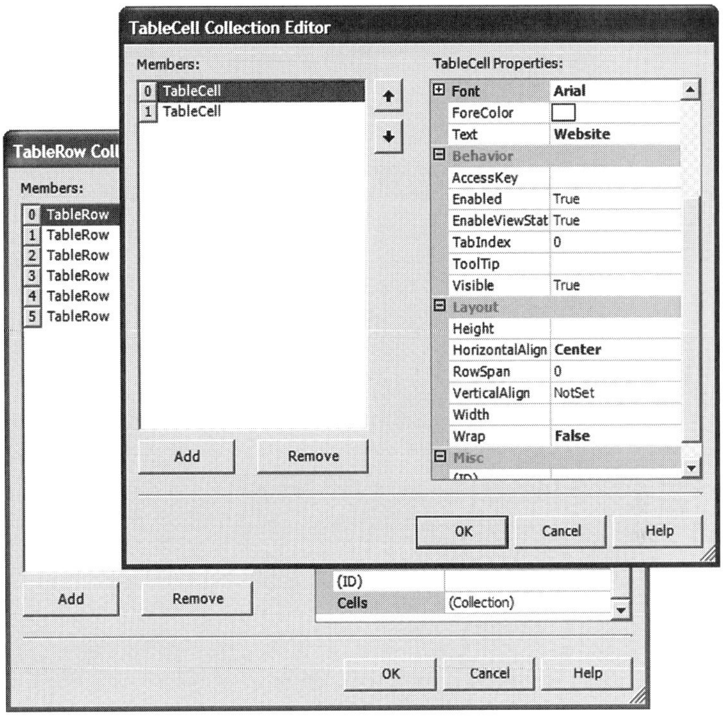

Figure 5–4. TableCell Collection Editor dialog on top of the TableRow Collection Editor dialog.

6 Click on Add to add a *TableCell* control. The TableCell Properties pane appears. Set the *Text* property to Website, the *Font/Name* property to Arial, the *Font/Bold* property to *True*, the *HorizontalAlign* property to Center, and the *Wrap* property to *False*. This sets up the Website column. Repeat the same step to add another *TableCell* control for the URL column. Now we have defined the table header row.

7 Repeat Steps 4 through 6 to add five *TableRow* controls for five federal geospatial data web sites named Geospatial One-Stop, GeoCommunicator, EnviroMapper, National Atlas, and National Map. Leave all properties at the default settings but set the *BackColor* property to #E0E0E0 for every other row. Add two *TableCell* controls for each row and leave all properties at the default settings. Figure 5–5 shows the completed table at design time.

8 Click on the Start button or press F5 to run the application. A web page with the table displays (Figure 5–6).

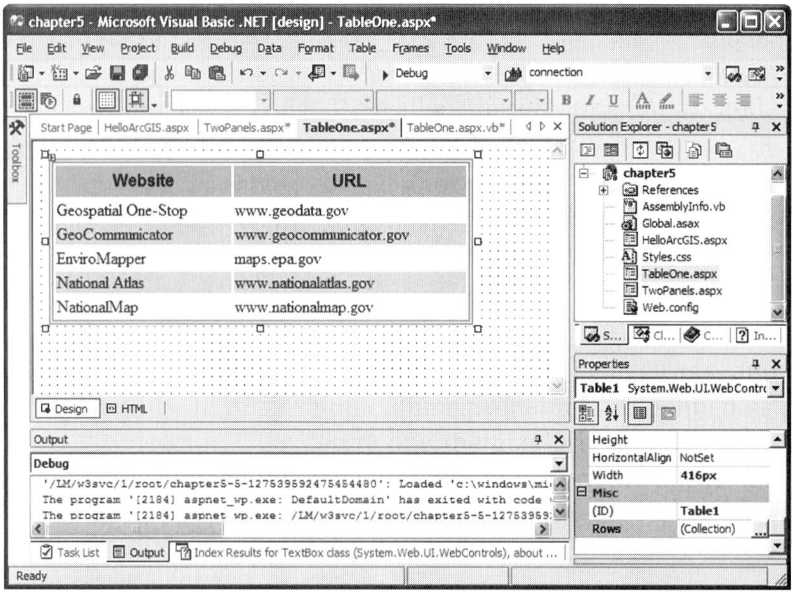

Figure 5–5. Example of the Table control at design time.

Figure 5–6. Example of the Table control at runtime.

Listing 5–3 is the *.aspx* file for the web form with a *Table* control, six *TableRow* controls, and twelve *TableCell* controls.

Listing 5–3. *.aspx* file for the web form with a *Table* control, six *TableRow* controls, and twelve *TableCell* controls.

```
<form id="Form1" method="post" runat="server">
<asp:Table id="Table1" style="Z-INDEX: 101; LEFT: 16px;
   POSITION: absolute; TOP: 16px" runat="server"
   Width="416px" BorderColor="Silver"
```

```
   BorderStyle="Double" BorderWidth="5px">
<asp:TableRow BackColor="Silver" Height="30px">
   <asp:TableCell Font-Names="Arial" Font-Bold="True"
      HorizontalAlign="Center" Wrap="False"
      Text="Website"></asp:TableCell>
   <asp:TableCell Font-Names="Arial" Font-Bold="True"
      HorizontalAlign="Center" Wrap="False"
      Text="URL"></asp:TableCell>
</asp:TableRow>
<asp:TableRow>
   <asp:TableCell Text="Geospatial One-Stop"></asp:TableCell>
   <asp:TableCell Text="www.geodata.gov"></asp:TableCell>
</asp:TableRow>
<asp:TableRow BackColor="#E0E0E0">
   <asp:TableCell Text="GeoCommunicator"></asp:TableCell>
   <asp:TableCell Text="www.geocommunicator.gov"></asp:TableCell>
</asp:TableRow>
<asp:TableRow>
   <asp:TableCell Text="EnviroMapper"></asp:TableCell>
   <asp:TableCell Text="maps.epa.gov"></asp:TableCell>
</asp:TableRow>
<asp:TableRow BackColor="#E0E0E0">
   <asp:TableCell Text="National Atlas"></asp:TableCell>
   <asp:TableCell Text="www.nationalatlas.gov"></asp:TableCell>
</asp:TableRow>
<asp:TableRow>
   <asp:TableCell Text="NationalMap"></asp:TableCell>
   <asp:TableCell Text="www.nationalmap.gov"></asp:TableCell>
</asp:TableRow>
</asp:Table>
</form>
```

You can manually type in the table if you are familiar the *Table* control, *TableRow* control, and *TableCell* control tags in the same way you write an HTML table. However, we strongly suggest that you drag and drop those controls if you are not fluent with the syntax of those control tags.

TextBox Control

The *TextBox* control is an input control that lets the user enter text. By default, the *TextMode* property is set to *SingleLine*, which creates a text box with only one line for things such as user name and password. You can also set the property to MultiLine or Password. The MultiLine option lets you create a text box with multiple lines of text. The Password option lets you create a single-line text box by hiding the value entered by the user (only asterisks showing inside the Password text box). The size of the text box is determined by its *Columns* property. If the text box is a MultiLine *TextBox* control, the number of lines for the text is determined by the *Rows* property. A SingleLine *TextBox* control is the counterpart of the HTML Text field for the `<input>` tag. A MultiLine *TextBox* control is that of the `<textarea>` tag.

The content of the text is entered via the *Text* property of the *TextBox* control. You can limit the number of characters that one can enter by setting the *MaxLength* property. Set the *Wrap* property to *true* to specify that the content within the *TextBox* control should automatically continue on the next line when the end of the text box is reached. In the following example, we will build a web form with a single-line *TextBox* control for the use name, a password *TextBox* control for the password, and a multi-line *TextBox* control for the feedback textbox.

To add a TextBox *control to an ASP.NET web form:*

1. Drag a *Label* control from the Web Forms tab of the toolbox onto the web form and set the *Text* property to "User Name".
2. Drag a *TextBox* control from the Web Forms tab of the toolbox onto the web form and set the *TextMode* property to SingleLine and the *Columns* property to 30.
3. Drag another *Label* control from the Web Forms tab of the toolbox onto the web form and set the *Text* property to "Password".
4. Drag another *TextBox* control from the Web Forms tab of the toolbox onto the web form and set the *TextMode* property to Password and the *Columns* property to 12.
5. Drag the third *Label* control from the Web Forms tab of the toolbox onto the web form and set the *Text* property to "Feedback".
6. Drag the third *TextBox* control from the Web Forms tab of the toolbox onto the web form and set the *TextMode* property to MultiLine, the *Rows* property to 10, and the *Columns* property to 50.
7. Click on the Start button or press F5 to run the application. A web page with the table displays (Figure 5–7).

Listing 5–4 is the *.aspx* file for the web form with three *Label* controls and three *TextBox* controls.

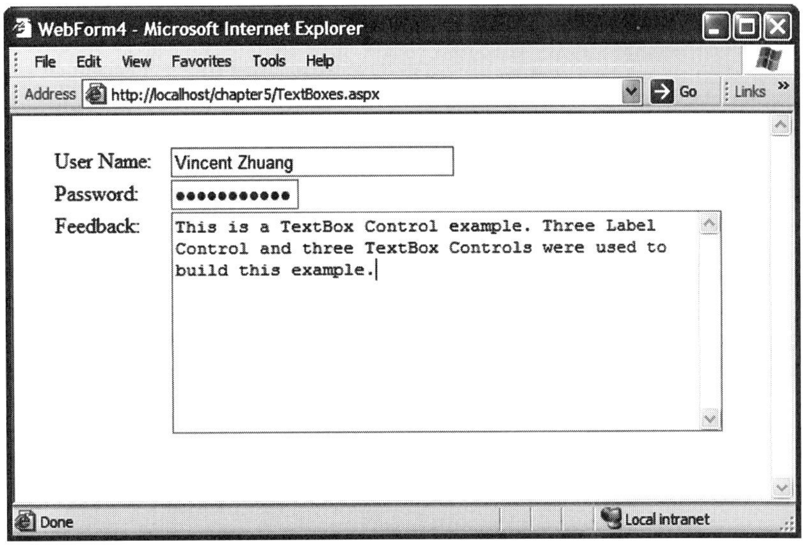

Figure 5–7. Example with TextBox controls in SingleLine, Password, and MultiLine, modes, respectively.

Listing 5–4. *aspx* file for the web form with three *Label* controls and three *TextBox* controls.

```
<form id="Form1" method="post" runat="server">
<asp:Label id="Label1" style="Z-INDEX: 102; LEFT: 32px;
   POSITION: absolute; TOP: 24px"
   runat="server">User Name:</asp:Label>
<asp:TextBox id="TextBox1" style="Z-INDEX: 101; LEFT: 120px;
   POSITION: absolute; TOP: 24px" runat="server"
   Columns="30"></asp:TextBox>
<asp:Label id="Label2" style="Z-INDEX: 103; LEFT: 32px;
   POSITION: absolute; TOP: 48px"
   runat="server">Password:</asp:Label>
<asp:TextBox id="TextBox2" style="Z-INDEX: 105; LEFT; 120px;
   POSITION: absolute; TOP: 48px" runat="server"
   TextMode="Password" Columns="12"></asp:TextBox>
<asp:Label id="Label3" style="Z-INDEX: 104; LEFT: 32px;
   POSITION: absolute; TOP: 72px"
   runat="server">Feedback:</asp:Label>
<asp:TextBox id="TextBox3" style="Z-INDEX: 106; LEFT: 120px;
   POSITION: absolute; TOP: 72px"
   runat="server" TextMode="MultiLine"
   Rows="10" Columns="50"></asp:TextBox>
</form>
```

If you do not want a text box to be modified, you can set the *ReadOnly* property to *True*. You can even make it invisible by setting the *Visible* property to *False*. Another very useful property is *ToolTip*. Once it is set, the tip message displays when you move the mouse on top of the *TextBox* control. These properties could be very useful if you really want to create some dynamics for the *TextBox* control.

Three Types of Button Controls: Button, ImageButton, and LinkButton

There are three types of web controls that can be used to generate a post-back action to the server:

- Button control
- ImageButton
- LinkButton

The *Button* control renders a regular push button. The push button can be a *submit* button or a *command* button. By default, the *Button* control is a submit button. A submit button does not have a command name and posts the web page back to the server when it is clicked on, similar to the HTML form element `<input type=submit>`. The submit button can have an event handler to control the actions performed when it is clicked. A command button has a command name. You can have multiple command buttons on a web form, similar to the HTML form element `<input type=button>`. The command button can also have an event handler to control the actions performed when it is clicked.

The *ImageButton* control renders like a push button but with a graphic image for the button instead of text similar to the HTML form element `<input type=image>`. The graphic image can be set via the *ImageUrl* property. The *ImageButton* can have an even handler in the same way as the *Button* control. The *LinkButton* control renders a hyperlink in the same way as an HTML hyperlink, but unlike HTML it can post the web page back to the server instead of simply navigating to another URL. In the following example, we will create a web form with a summit button, a command button, an image button, and a hyperlink button.

To add a Button *control, an* ImageButton *control, and a* LinkButton *control to an ASP.NET web form:*

1 Drag a *Button* control from the Web Forms tab of the toolbox onto the web form and set the *Text* property to "Submit" and the *ToolTip* property to "This is a submit button".

2 Drag a *Label* control from the Web Forms tab of the toolbox onto the web form and place it next to the "Submit" button. Clear the default value for the *Text* property.

3 Drag another *Button* control from the Web Forms tab of the toolbox onto the web form and set the *Text* property to "Go", the *CommandName* property to "Go", the *CommandArgument* property to "Next", and the *ToolTip* property to "This is a command button".

4 Drag the second *Label* control from the Web Forms tab of the toolbox onto the web form and place it next to the "Go" button. Clear the default value for the *Text* property.

5 Drag an *ImageButton* control from the Web Forms tab of the toolbox onto the web form and set the *ImageUrl* property to "images/color/happy.bmp" and the *ToolTip* property to "This is an image button".

6 Drag the third *Label* control from the Web Forms tab of the toolbox onto the web form and place it next to the image button. Clear the default value for the *Text* property.

7 Drag a *LinkButton* control from the Web Forms tab of the toolbox onto the web form and set the *Text* property to "Next > " and the *ToolTip* property to "This is a hyperlink button".

8 Drag the fourth *Label* control from the Web Forms tab of the toolbox onto the web form and place it next to the hyperlink button. Clear the default value for the *Text* property.

9 Add the events specified in Table 5–1 to each of button clicks.

Table 5–1: Visual Basic .NET Codes for Button-click Events

Button Click	VB.NET Code
Button1_Click	Label1.Text = Button1.ToolTip
Button2_Click	Label2.Text = Button2.ToolTip
ImageButton1_Click	Label3.Text = ImageButton1.ToolTip
LinkButton1_Click	Label4.Text = LinkButton1.ToolTip

10 Click on the Start button or press F5 to run the application. A web page with the table displays (Figure 5–8).

Click on all four buttons on the web page. The tooltip message for each button appears next to the button where we placed the label for each button. Figure 5–8 also shows the tooltip when the mouse moves over the *LinkButton* control. Listing 5–5 is the *.aspx* file for the web form with four *Label* controls, two *Button* controls, an *ImageButton* control, and a *LinkButton* control.

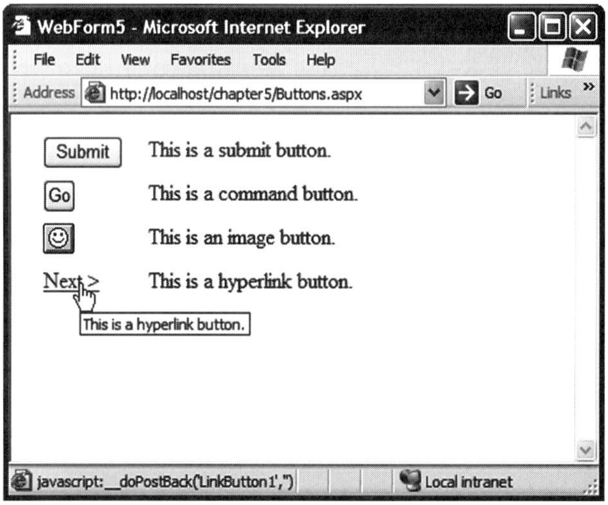

Figure 5–8. Four types of button controls.

Listing 5–5. *.aspx* file for the web form with four *Label* controls, two *Button* controls, an *ImageButton* control, and a *LinkButton* control.

```
<form id="Form1" method="post" runat="server">
<asp:Button id="Button1" style="Z-INDEX: 101; LEFT: 24px;
   POSITION: absolute; TOP: 16px" runat="server"
   Text="Submit" ToolTip="This is a submit button."></asp:Button>
<asp:Label id="Label1" style="Z-INDEX: 102; LEFT: 104px;
   POSITION: absolute; TOP: 16px"
   runat="server"></asp:Label>
<asp:Button id="Button2" style="Z-INDEX: 103; LEFT: 24px;
   POSITION: absolute; TOP: 48px" runat="server"
   Text="Go" CommandName="Go" CommandArgument="Next"
   ToolTip="This is a command button."></asp:Button>
<asp:Label id="Label2" style="Z-INDEX: 104; LEFT: 104px;
   POSITION: absolute; TOP: 48px" runat="server"></asp:Label>
<asp:ImageButton id="ImageButton1" style="Z-INDEX: 105; LEFT: 24px;
   POSITION: absolute; TOP: 80px"
   runat="server" ImageUrl="images/color/happy.bmp"
   ToolTip="This is an image button.">
</asp:ImageButton>
<asp:Label id="Label3" style="Z-INDEX: 107; LEFT: 104px;
   POSITION: absolute; TOP: 80px" runat="server"></asp:Label>
<asp:LinkButton id="LinkButton1" style="Z-INDEX: 106; LEFT: 24px;
```

```
   POSITION: absolute; TOP: 112px"
   runat="server" ToolTip="This is a hyperlink button.">
   Next &gt;</asp:LinkButton>
<asp:Label id="Label4" style="Z-INDEX: 108; LEFT: 104px;
   POSITION: absolute; TOP: 112px"
   runat="server"></asp:Label>
</form>
```

As you can see, both the submit button and the command are the ASP.NET *Button* control. The only difference is that the *CommandName* and *CommandArgument* properties for the command button have been set.

The *ImageButton* control is used most in web-based GIS applications. Let's expand the previous example to illustrate how to use an *ImageButton*. We will add another *ImageButton* control and *Label* control to the previous web form.

To add an ImageButton *control and a* Label *control to an ASP.NET web form:*

11 Drag an *ImageButton* control from the Web Forms tab of the toolbox onto the web form and set the *ImageUrl* property to "images/color/dc.png" and the *ToolTip* property to "Click map to show the coordinates". The image file is a map of Washington, D.C.

12 Drag a *Label* control from the Web Forms tab of the toolbox onto the web form and place it next to the image button. Clear the default value for the *Text* property.

13 Add the following code to the *ImageButton2_Click* event.

```
Private Sub ImageButton2_Click(ByVal sender As System.Object, _
 ByVal e As System.Web.UI.ImageClickEventArgs) _
 Handles ImageButton2.Click
    Label5.Text = "Coordinates: " & e.X & "," & e.Y
End Sub
```

14 Click on the Start button or press F5 to run the application.

A web page similar to that shown in Figure 5-8 displays. This time, a map of Washington, D.C., was added as an *ImageButton* control. Click anywhere on the map and you will see a pair of coordinates displayed next to the map (Figure 5-9). They are the screen's X and Y coordinates. In web-based GIS, we often capture a pair of screen coordinates and then convert them into a pair of real-world coordinates in which the underlying GIS data are stored. After the conversion, we can do feature identification on points of the map. Also, if we convert two diagonal pairs of screen coordinates of the map we would get the geographic window the current map is in.

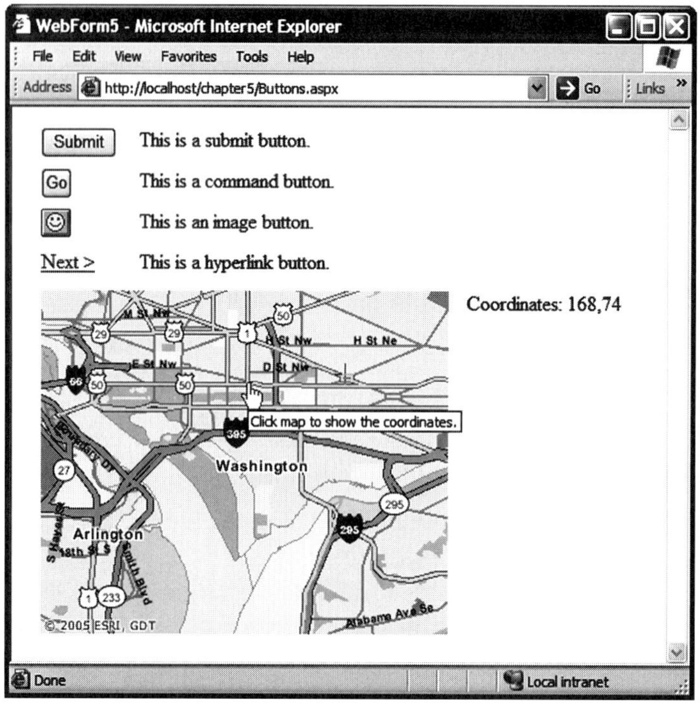

Figure 5–9. Example of the ImageButton *control showing a pair of coordinates.*

Single-choice Controls: CheckBox and RadioButton

There are two types of single-choice controls.

- *CheckBox* control
- *RadioButton* control

The *CheckBox* control renders a check box on the web form that allows the user to switch between an *On* and *Off* or a *True* and *False* state. It is similar to the HTML form `element <input type=checkbox>`. You can specify the caption to display in the control by setting the *Text* property. The caption can appear either on the right or left of the check box. Set the *TextAlign* property to specify which side you would like the caption to appear on. The *CheckBox* control also has the *ToolTip* property.

If you want the *CheckBox* control to be checked, set the *Checked* property to *True*. When a *CheckBox* control is switched from checked to unchecked, or vise versa, the *CheckedChanged* event can be raised to send the web page back to the server. However, the *CheckBox* control does not automatically post the web form to the server when it is clicked. To enable automatic posting, set the *AutoPostBack* property to *True*.

The *RadioButton* control renders a radio button on the web form and works the same way as the *CheckBox* control. You can specify the caption to display in the control by setting the *Text* property. The caption can appear either on the right or left of the check box. Set the *TextAlign* property to specify which side you would like the caption to appear on. The *RadioButton* control also has the *ToolTip* property.

It is common to use a group of radio buttons by setting the *GroupName* for each *RadioButton* control. When working in group, the *RadioButton* control allows a mutually exclusive selection from the group. In other words, only one member of the group can be selected at a time. *CheckBox* controls are also often used by group. Instead of setting each individual *CheckBox* control or *RadioButton* control, you can use a *CheckBoxList* control and a *RadioButtonList* control, discussed in the next section.

Multiple Choice Controls: CheckBoxList and RadioButtonList

There are two types of multiple choice controls:

- *CheckBoxList* control
- *RadioButtonList* control

The *CheckBoxList* control renders a group of check boxes that can be dynamically generated using data binding. You can use multiple *CheckBox* controls to achieve a similar result. But the *CheckBoxList* control is easier for creating a set of check boxes, especially when they need to use data binding, whereas the individual *CheckBox* control gives you greater control over layout such as color and font type. In the following, we will create a web form with a *CheckBoxList* control of five items and a *Label* control.

To add a CheckBoxList *control to an ASP.NET web form:*

1 Drag a *CheckBoxList* control from the Web Forms tab of the toolbox onto the web form and set the *ID* property to "chkLst1" and the *AutoPostBack* property to *True*.

2 From the Properties window, click on the *Items* property and an Ellipsis button displays. Click on the button and the ListItem Collection Editor dialog box appears (Figure 5–10). You use the ListItem Collection Editor to manually define each item for the *CheckBoxList* control.

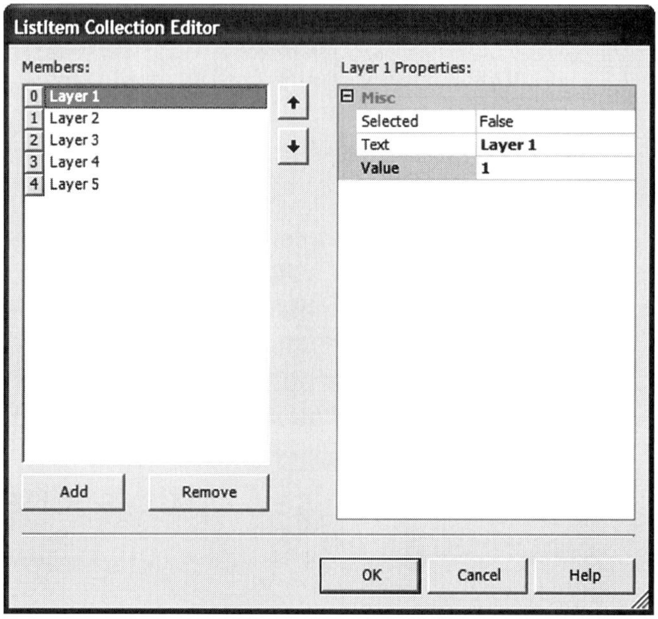

Figure 5–10. ListItem Collection Editor dialog.

3 Click on Add to add a new item to the control. The ListItem Properties window appears. Set the *Text* property to "Layer 1" and the *Value* property to "1". Leave the *Selected* property to its default value of *False*. After you specify the *Text* property, the window title switches to Layer 1 Properties.

4 Repeat step 3 to add layers 2 through 5.

5 Click on OK to close out the ListItem Collection Editor dialog.

6 Drag a *Label* control from the Web Forms tab of the toolbox onto the web form, place it beneath the *CheckBoxList* control, and clear the default value for the *Text* property.

7 Add the following code to the *chkLst1_Selected IndexChanged* event to show which layer is selected.

```
Dim i
Label1.Text = "Selected layer(s):<BR>" & vbCrLf
For i = 0 To chkLst1.Items.Count - 1
```

```
  If chkLst1.Items(i).Selected Then
   Label1.Text += chkLst1.Items(i).Text + "<BR/>" & vbCrLf
  End If
Next
```

8 Click on the Start button or press F5 to run the application. When a web page displays, click on those five check boxes one by one. Every time you click a check box, the web page is refreshed to show which layer is turned on via the *SelectedIndexChanged* event of the *CheckBoxList* control (Figure 5–11).

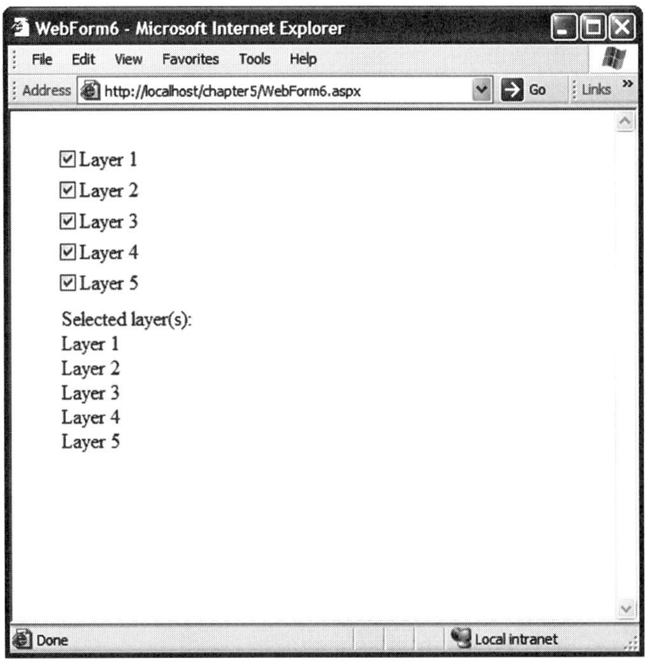

Figure 5–11. Example of the CheckBoxList *control.*

If we actually used this *CheckBoxList* Control for a real application, it would create traffic between the client and the server every time the user clicks on a check box. Also, if a layer took time to draw, the application would be very slow. To mitigate problems associated with increased traffic, set the *AutoPostBack* property of the *CheckBoxList* control to its default value of *False*.

Now we need another control to centralize the click actions on the individual check boxes. Let's add a *Button* control to the web form to control traffic sent back to the server. The web page will not be redrawn until this button is clicked. This means that the user decides when the post-back to the server is executed, regardless of how many check boxes are selected.

9 Set the *AutoPostBack* property to *False* for the *CheckBoxList* control.

10 Drag a *Button* control from the Web Forms tab of the toolbox onto the web form and place it between the *CheckBoxList* control and the *Label* control. Set the *Text* property to "Submit" and the *ToolTip* property to "Show selected layers".

11 Cut the code for *chkLst1_SelectedIndexChanged* and paste it under the *Button1_Click* event.

12 Click on the Start button or press F5 to run the application. When the web page appears, click on a couple of check boxes. You will notice that nothing happens to the page because there is no post to the server. Click on the submit button. The web page will be refreshed with the selected layers listed under the button (Figure 5–12), which also displays the tooltip information.

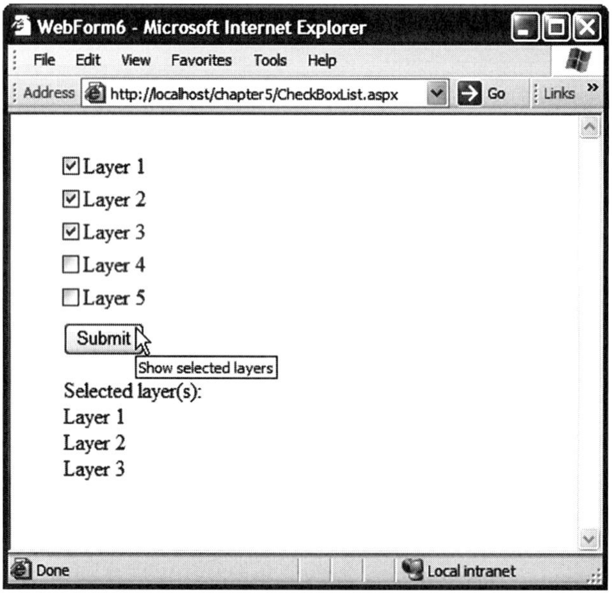

Figure 5–12. Example of the CheckBoxList *control without post-back.*

Listing 5–6 is the *.aspx* file for the web form with a *CheckBoxList* control, *Label* control, and *Button* control.

Listing 5–6. *.aspx* file for the web form with a *CheckBoxList* control, *Label* control, and *Button* control.

```
<form id="Form1" method="post" runat="server">
<asp:CheckBoxList id="chkLst1" style="Z-INDEX: 101; LEFT: 32px;
   POSITION: absolute; TOP: 24px" runat="server"
   Width="88px" AutoPostBack="False">
   <asp:ListItem Value="1">Layer 1</asp:ListItem>
   <asp:ListItem Value="2">Layer 2</asp:ListItem>
   <asp:ListItem Value="3">Layer 3</asp:ListItem>
   <asp:ListItem Value="4">Layer 4</asp:ListItem>
   <asp:ListItem Value="5">Layer 5</asp:ListItem>
</asp:CheckBoxList>
<asp:Label id="Label1" style="Z-INDEX: 102; LEFT: 40px;
   POSITION: absolute; TOP: 192px" runat="server"></asp:Label>
```

```
<asp:Button id="Button1" style="Z-INDEX: 103; LEFT: 40px;
   POSITION: absolute; TOP: 152px" runat="server"
   Text="Submit" ToolTip="Show selected layers">
</asp:Button>
</form>
```

The *RadioButtonList* control renders a group of radio buttons that can be dynamically generated using data binding as if you had several HTML radio buttons with the same name. You can also use multiple *RadioButton* controls to achieve a similar result. You should use the *RadioButtonList* control, however, when you need to create a set of radio buttons, especially when they use data binding. On the other hand, an individual *RadioButton* control gives you greater control over layout such as color and font type. In the following example, we will create a web form with a *RadioButtonList* control of six items and a *Label* control.

To add a RadioButtonList *control to an ASP.NET web form:*

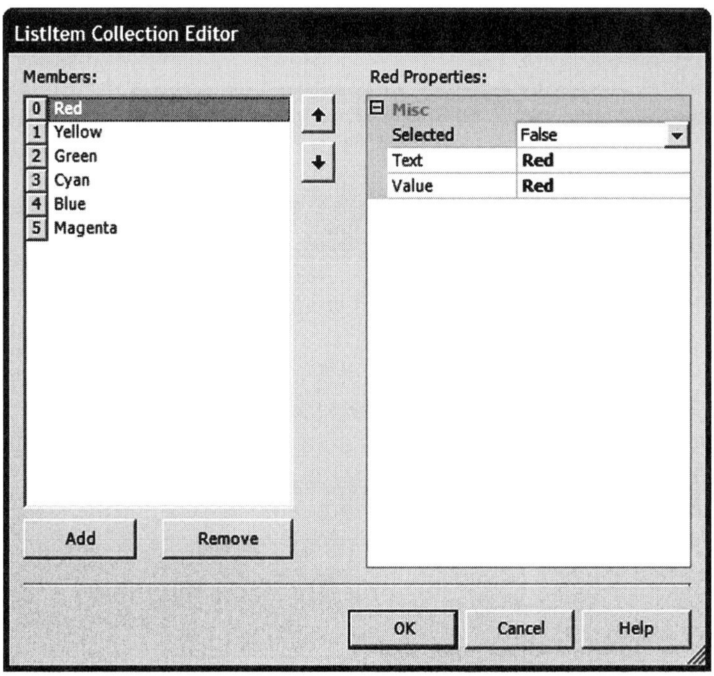

Figure 5–13. ListItem Collection Editor dialog for the RadioButtonList *control.*

1. Drag a *RadioButtonList* control from the Web Forms tab of the toolbox onto the web form and set the *ID* property to "rdLst1" and the *AutoPostBack* property to *True.*
2. From the Properties window, click on the *Items* property and an Ellipsis button displays. Click on the button, and then the ListItem Collection Editor dialog box appears (Figure 5–13). You use the ListItem Collection Editor dialog to manually define each item for the *RadioButtonList* control.

3 Click on Add to add a new item to the control. The ListItem Properties dialog box appears. Set the *Text* property to "Red" and the *Value* property to "Red". Leave the *Selected* property to its default value of *False*. After you specify the *Text* property, the dialog title switches to Red Properties.

4 Repeat step 3 to add five more colors: Yellow, Green, Cyan, Blue, and Magenta.

5 Click on OK to close out the ListItem Collection Editor dialog.

6 Drag a *Label* control from the Web Forms tab of the toolbox onto the web form, place it beneath the *RadioButtonList* control, and clear the default value for the *Text* property.

7 Add the following code to the *rdLst1_SelectedIndexChanged* event to show which color is selected.

```
Dim myColor As System.Drawing.Color
If rdLst1.SelectedIndex > -1 Then
 Label1.Text = "Selected color: " & rdLst1.SelectedItem.Text
 Label1.ForeColor = myColor.FromName(rdLst1.SelectedValue)
End If
```

Figure 5–14. Example of the RadioButtonList *control.*

8 Click on the Start button or press F5 to run the application.

When a web page displays, click on each of the six radio buttons one by one. Every time you click on a radio button, the web page is refreshed to show which color is selected via the *SelectedIndexChanged* event of the *RadioButtonList* control (Figure 5–14). A more advanced enhancement was made to the *Label* control. We created a color object called `myColor` and used the `FromName` method to generate a color. When a radio button was selected, we fed its value, which is a color name in this case, to the `FromName` method of the color object. Then the foreground color of the *Label* control was set by the color object, which is the color specified by the checked radio button.

Listing 5–7 is the *.aspx* file for the web form with a *RadioButtonList* control and a *Label* control.

Listing 5–7. *.aspx* file for the web form with a *RadioButtonList* control and a *Label* control.

```
<form id="Form1" method="post" runat="server">
<asp:RadioButtonList id="rdLst1" style="Z-INDEX: 101; LEFT: 32px;
   POSITION: absolute; TOP: 24px" runat="server"
   AutoPostBack="True">
   <asp:ListItem Value="Red">Red</asp:ListItem>
   <asp:ListItem Value="Yellow">Yellow</asp:ListItem>
   <asp:ListItem Value="Green">Green</asp:ListItem>
   <asp:ListItem Value="Cyan">Cyan</asp:ListItem>
   <asp:ListItem Value="Blue">Blue</asp:ListItem>
   <asp:ListItem Value="Magenta">Magenta</asp:ListItem>
</asp:RadioButtonList>
<asp:Label id="Label1" style="Z-INDEX: 102; LEFT: 40px;
   POSITION: absolute; TOP: 184px" runat="server"
   Font-Bold="True"></asp:Label>
</form>
```

List Controls: ListBox and DropDownList

There are two types of list controls:

- *ListBox* control
- *DropDownList* control

The *ListBox* control is used to create a list control that allows single or multiple item selection. It renders the HTML form selection box that contains a list of text items. You can specify the height of the control by setting the *Rows* property. By default, the *SelectionMode* property is set to *Single*. To enable multiple item selection, set the *SelectionMode* property to *Multiple*, which is equivalent to the keyword *MULTIPLE* for the HTML form `<select>` tag. In the following example, we will create a web form with a *ListBox* control of five items, a *Button* control, and a *Label* control.

To add a ListBox *control to an ASP.NET web form:*

1 Drag a *ListBox* control from the Web Forms tab of the toolbox onto the web form and set the *ID* property to "lst1", the *AutoPostBack* property to *False*, the *SelectionMode* property to *Multiple*, and the *Rows* property to 4. We do not want to send the page back to the server every time we click on a list item.

2 From the Properties window, click on the *Items* property and an Ellipsis button displays. Click on the button and the ListItem Collection Editor dialog box appears, which is the same as that shown in Figure 5-10. You use the ListItem Collection Editor dialog to manually define each item for the *ListBox* control.

3 Click on Add to add a new item to the control. The ListItem Properties pane appears. Set the *Text* property to "Layer 1" and the *Value* property to "1". Leave the *Selected* property to its default value *False*. After you specify the *Text* property, the pane title switches to Layer 1 Properties.

4 Repeat step 3 to add layers 2 through 5.

5 Click on OK to close out the ListItem Collection Editor dialog.

6 Drag a *Button* control from the Web Forms tab of the toolbox onto the web form and place it beneath the *ListBox* control. Set the *Text* property to "Submit" and the *ToolTip* property to "Show selected layers".

7 Add the following code to the *Button1_Click* event to show selected layers.

```
Dim i As Integer
Dim myLyrs As String
myLyrs = "Selected layer(s):" & "<BR>" & vbCrLf
For i = 0 To lst1.Items.Count - 1
 If lst1.Items(i).Selected = True Then
  myLyrs += "Layer " & lst1.Items(i).Value & "<BR>" & vbCrLf
 End If
Next
Label1.Text = myLyrs
```

8 Drag a *Label* control from the Web Forms tab of the toolbox onto the web form and place it beneath the *Button* control, and clear the default value for the *Text* property.

9 Click on the Start button or press F5 to run the application.

When a web page appears, you will notice that four of five layers are visible in the box because the height of the control was set to 4 via the *Rows* property.

Because we set the *SelectionMode* property to *Multiple*, we can select more than one item. Hold down the Ctrl key and click layers 1, 2 and 4. Click on the submit button. The *Label* control will display those selected layers, which are layers 1, 2, and 4 in this case (Figure 5-15).

Because the *SelectionMode* property was set to *Multiple,* we were able to iterate the items in the *ListBox* control and determine which item was selected. We detected the selected items in the control by testing the *Selected* property of each item in the list, as shown in the Visual Basic .Net code. If the *SelectionMode* property is set to *Single,* we can use the *SelectedIndex* property directly to determine the selected item.

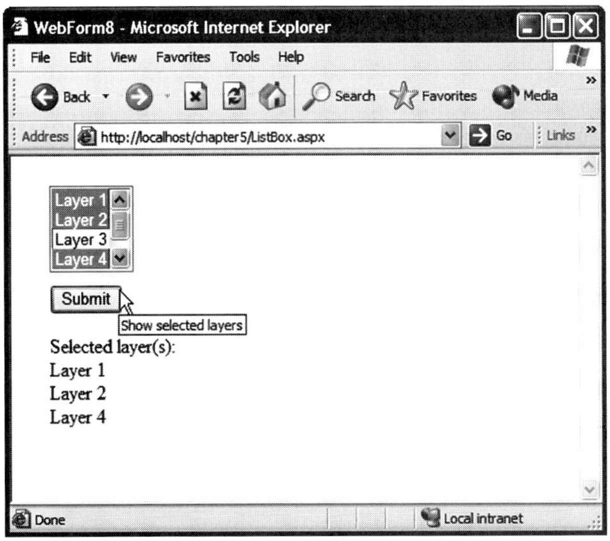

Figure 5–15. Example of the ListBox *control.*

The steps to manually define a *ListBox* control are almost identical to those to specify a *CheckBoxList* control. They are all *ListItem* collections. That is the reason they both use the ListItem Collection Editor dialog. So does the *RadioButtonList* control.

The *ListBox* control also supports data binding. To bind the control to a data source, you first create a data source (such as an ArrayList or ADO.NET DataTable) that contains the items to be displayed in the control. Next, you use the *DataBind* method to bind the data source to the *ListBox* control. You use the *DataTextField* and *DataValueField* properties to specify which field in the data source to bind to the *Text* and *Value* properties, respectively, of each list item in the control. The *ListBox* control will then display the information from the data source.

Let's repeat the previous example. But this time we will use dynamic data binding. We create an Access table called *layers* that contains two columns: `layer_name` and `layer_ID`, where column `layer_name` stores "Layer1", "Layer2", "Layer 3", Layer 4", and "Layer 5" and column `layer_ID` stores those values for each layer: 1, 2, 3, 4, and 5. We design exactly the same web form as before: a *ListBox* control, a *Button* control, and a *Label* control. Instead of manually defining the items for the list, we bind the Access table to the list at runtime.

To dynamically bind a ListBox *control with a database table:*

1 Drag a *ListBox* control from the Web Forms tab of the toolbox onto the web form and set the *ID* property to "lstBnd1", the *AutoPostBack* property to *False*, the *SelectionMode* property to *Multiple*, and the

Rows property to 4. We do not want to send the page back to the server every time we click on a list item.

2 Add the following code to the *Page_Load*.

```
If Not IsPostBack Then
 Dim dbConn As OleDbConnection = New OleDbConnection( _
   "Provider=Microsoft.Jet.OLEDB.4.0; data source=" & _
   Server.MapPath("data/chapter5.mdb"))
 Dim sql As String
 Dim dbComm As OleDbCommand = dbConn.CreateCommand()
 dbComm.CommandText = "SELECT * FROM layers"
 dbConn.Open()
 Dim dbRead As OleDbDataReader = dbComm.ExecuteReader()
 lstBnd1.DataSource = dbRead
 lstBnd1.DataBind()
 dbRead.Close()
 dbConn.Close()
End If
```

The code basically defines an OLE DB connection to the Access database, retrieves all layer records from the database table, and reads them into an OLE DB reader. It then assigns the reader to the *DataSource* of the *ListBox* control and follows by a data binding between the control and the database using the *DataBind* method. But it only executes the code the first time the page is loaded by testing whether `IsPostBack` is *not* true.

✔ **NOTE:** *OLE DB was often used in classic ASP applications to connect any database whether it is Access, SQL Server, or Oracle. The advantage of OLE DB is that it makes a connection on the fly. After being connected, SQL queries can be issued through the* OleDbCommand. *The* OleDbDataReader *can then be used to read a stream of records from a data source. Once the data is read into a* DataReader, *they can be bound to a qualified web control such as a* ListBox *control. Make sure you always have a correct OLE DB provider for your data source. Check with the data source web site for the latest provider. The following are the providers for Access, SQL Server, and Oracle.*

- Access: *Microsoft.Jet.OLEDB.4.0 (Microsoft OLE DB Provider for Microsoft Jet)*
- SQL Server: *SQLOLEDB (Microsoft OLE DB Provider for SQL Server)*
- Oracle: *OraOLEDB.Oracle (Oracle OLE DB Provider for Oracle 9.x)*

3 Add the *Imports* statement for the OLEDB namespace.

```
Imports System.Data.OleDb
```

4 Set the *DataTextField* property to `layer_name` and the *DataValue-Field* property to `layer_ID`. The layer name will be used for item text displayed in the list, whereas the layer ID will be used as the item value of each list item.

5 Drag a *Button* control from the Web Forms tab of the toolbox onto the web form and place it beneath the *ListBox* control. Set the *Text* property to "Submit" and the *ToolTip* property to "Show selected layers".

6 Add the following code to the *Button1_Click* event to show selected layers.

```
Dim i As Integer
Dim myLyrs As String
myLyrs = "Selected layer(s):" & "<BR>" & vbCrLf
For i = 0 To lstBnd1.Items.Count - 1
 If lstBnd1.Items(i).Selected = True Then
  myLyrs += "Layer " & lstBnd1.Items(i).Value & "<BR>" & vbCrLf
 End If
Next
Label1.Text = myLyrs
```

7 Drag a *Label* control from the Web Forms tab of the toolbox onto the web form and place it beneath the *Button* control, and clear the default value for the *Text* property.

8 Click on the Start button or press F5 to run the application.

When a web page appears, it will appear the same as the previous example. There is no visible difference in terms of the web page. The only difference was in the design. The list items were dynamically generated at runtime through data binding. Actually, dynamic data binding is often used in ASP.NET applications, especially when the data are stored in the database.

The *DropDownList* control is used to create a list control that only allows single item selection. It renders the HTML form selection box that contains a list of text items but with one item displayed at a time. You can use this control for state and county selections. You first create a *DropDownList* control to hold all 50 states and then define another *DropDownList* control to dynamically hold all counties for a selected state. Every time a new state is chosen, the county list gets refreshed automatically with all counties for that state. The rule of thumb is that you use the *DropDownList* control if you will need to select one from several choices and only the selected one is visible in the list.

In the following example, we will create an application that portrays population by region, as shown in Table 5–2. We will create a web form with a *DropDownList* control of four items and a *Label* control.

Table 5–2: Population by Geographical Region

ID	Region	Population	Male	Female
1	Northeast	53594378	25897327	27697051
2	Midwest	64392776	31555438	32837338
2	South	100236820	49057320	51179500
3	West	63197932	31543478	31654454

To create a DropDownList *control with a dynamic binding:*

1 Drag a *DropDownList* control from the Web Forms tab of the Toolbox onto the Web Form. Set the ID property to "ddLst1", the AutoPostBack property to *True*, the DataTextField property to "Region", and the DataValueField property to "Population".

We want to send the page back to the server every time the *DropDownList* item is selected. Also we use the Region column for the text in the list box and the Population column for the value of each item in the list.

2 Add the following code to the *Page_Load*.

```
If Not IsPostBack Then
 Dim dbConn As OleDbConnection = New OleDbConnection( _
   "Provider=Microsoft.Jet.OLEDB.4.0; data source=" & _
   Server.MapPath("data/chapter5.mdb"))
 Dim sql As String
 Dim dbComm As OleDbCommand = dbConn.CreateCommand()
 dbComm.CommandText = "SELECT * FROM Region"
 dbConn.Open()
 Dim dbRead As OleDbDataReader = dbComm.ExecuteReader()
 ddLst1.DataSource = dbRead
 ddLst1.DataBind()
 ddLst1.Items.Insert(0, "Select a Region")
 dbRead.Close()
 dbConn.Close()
End If
```

This code is almost the same as the one used for the *ListBox* control. The difference is that this one reads from the Region table. After the list control is bound with the data, a caption line of "Select a Region" was inserted on top of the list to the *DropDownList* control using the `insert` method from the *DropDownList* control. So, the *DropDownList* control has five list items, including the four regions shown in Table 5–2 and the caption.

3 Add the *Imports* statement for the OLEDB namespace.

```
Imports System.Data.OleDb
```

4 Add the following code to *ddLst1_SelectedIndexChanged*.

```
Label1.Text = "Population: " & ddLst1.SelectedItem.Value
```

5 Drag a *Label* control from the Web Forms tab of the toolbox onto the web form and place it beneath the *DropDownList* control, and then clear the default value for the *Text* property. This *Label* control is used to display selected *DropDownList* items.

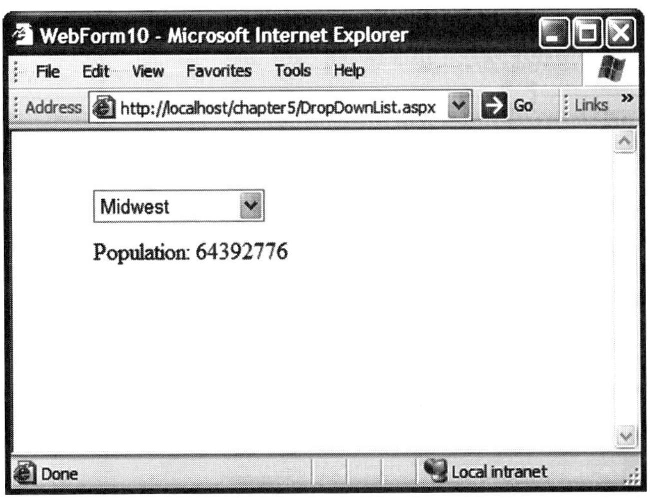

Figure 5–16. Example of the DropDownList *control.*

6 Click on the Start button or press F5 to run the application.

When a web page appears, click on a region. The regional population number will appear underneath the *DropDownList* control where the *Label* control is (Figure 5–16). The population number for a selected region is dynamically assigned to the *Text* property of the *Label* control.

Listing 5–8 is the *.aspx* file for the web form with a *DropDownList* control and a *Label* control.

Listing 5–8. *.aspx* file for the web form with a *DropDownList* control and a *Label* control.

```
<form id="Form1" method="post" runat="server">
  <asp:DropDownList id="ddLst1" style="Z-INDEX: 101; LEFT: 56px;
```

```
        POSITION: absolute; TOP: 40px" runat="server"
        DataTextField="Region" DataValueField="Population"
        AutoPostBack="True"></asp:DropDownList>
    <asp:Label id="Label1" style="Z-INDEX: 102; LEFT: 56px;
        POSITION: absolute; TOP: 72px" runat="server"></asp:Label>
</form>
```

Validation Controls

Validation controls are a collection of controls for validating an input type of server control, such as a *TextBox*, and displaying a custom message when validation fails. A validation control tests the user's inputs and if the data fails any of the validations the server sends the page back to the client with a predefined error message. Each validation control performs a specific type of validation. For example, you can validate against a specific value or a range of values by using the *CompareValidator* and *RangeValidator* controls, respectively. You can even define your own validation criteria by using the *CustomValidator* control. All validation controls are listed in Table 2–5. Because the error message is displayed in the validation control, you can place a validation control next to a *Web* control. You can also display a summary of the results from all validation controls on the page by using the *ValidationSummary* control.

RequiredFieldValidator Control

The *RequiredFieldValidator* control is used to make an input control, such as a *TextBox* control, a mandatory field. The *TextBox* control fails validation if the value it contains does not change from its initial value when validation is performed. The initial value can be an empty string ("") or instructions to the user. For example, you can use a *RequiredFieldValidator* control for a ZIP code entry. Instead of placing a label of "ZIP Code" in front of the *TextBox* control, you can simply set the *Text* property of the *TextBox* control to "ZIP Code". The *RequiredFieldValidator* is very useful for validating passwords. A user must enter a valid password before continuing.

In the following example, we will create a web form that contains a ZIP code field and a password field. We will put the instructions inside the ZIP code entry without using any label control, but we must use a *Label* control for the caption of the password field.

To add a RequiredFieldValidator *control to an ASP.NET web form:*

1 Drag a *TextBox* control from the Web Forms tab of the toolbox onto the web form and set the *ID* property to "txtZip" and the *Text* property to "ZIP Code".

2 Drag a *RequiredFieldValidator* control from the Web Forms tab of the toolbox onto the web form, and set the *ID* property to "vldtrTxtZip", the *InitialValue* property to "ZIP Code", and the *ErrorMessage* property to "Please enter a ZIP code".

✔ **NOTE:** *If you do specify an initial value for the Text property of the* TextBox *control as the instruction, you need to set the* InitialValue *property of the* RequiredFieldValidator *to the same initial value.*

3 From the Properties window, click on the *ControlToValidate* property and a drop-down list appears. Select the ZIP code text box "txtZip" from the list.

4 Drag a *Label* control from the Web Forms tab of the toolbox onto the web form and set the *Text* property to "Password".

5 Drag another *TextBox* control from the Web Forms tab of the toolbox onto the web form and set the *ID* property to "txtPasswd" and the *TextMode* property to *Password*.

6 Drag another *RequireFieldValidator* control from the Web Forms tab of the toolbox onto the web form and set the *ID* property to "vldtrTxtPasswd" and the *ErrorMessage* property to "Password is required". Leave the *InitialValue* property blank.

7 Drag a *Button* control from the Web Forms tab of the toolbox onto the web form and set the *Text* property to "Validate".

8 Click on the Start button or press F5 to run the application.

When a web page displays, click on the Validate button without changing the initial values for two text boxes on the web page. Two error messages are displayed in red: one for the ZIP code entry and the other for the password field appear on the web page (Figure 5–17). As you can see, some well-thought-out engineering was put into the validation controls. The error message has been set to red by default. Now you enter a ZIP code in the text box and a password in the field and then click on the Validate button. The error messages disappear.

Figure 5–17. Example of the RequiredFieldValidator *control.*

Listing 5–9 is the *.aspx* file for the web form with the *RequiredFieldValidator* control.

Listing 5–9. *.aspx* file for the web form with the *RequiredFieldValidator* control.

```
<form id="Form1" method="post" runat="server">
<asp:TextBox id="txtPasswd" style="Z-INDEX: 101; LEFT: 120px;
   POSITION: absolute; TOP: 96px"
   runat="server" TextMode="Password"></asp:TextBox>
<asp:RequiredFieldValidator id="vldtrTxtPasswd"
   style="Z-INDEX: 102; LEFT: 120px;
   POSITION: absolute; TOP: 128px"
   runat="server"
   ErrorMessage="Password is required!"
   ControlToValidate="txtPasswd">
</asp:RequiredFieldValidator>
<asp:Label id="Label1" style="Z-INDEX: 103; LEFT: 40px;
   POSITION: absolute; TOP: 96px"
   runat="server">Password:</asp:Label>
<asp:TextBox id="txtZip" style="Z-INDEX: 104; LEFT: 40px;
   POSITION: absolute; TOP: 16px"
   runat="server">ZIP Code</asp:TextBox>
<asp:RequiredFieldValidator id="vldtrTxtZip"
   style="Z-INDEX: 105; LEFT: 40px;
   POSITION: absolute; TOP: 40px"
```

```
   runat="server"
   ErrorMessage="Please enter a ZIP code!"
   ControlToValidate="txtZip"
   InitialValue="ZIP Code">
</asp:RequiredFieldValidator>
<asp:Button id="Button1" style="Z-INDEX: 106; LEFT: 48px;
   POSITION: absolute; TOP: 152px" runat="server"
   Text="Validate"></asp:Button>
</form>
```

CompareValidator Control

The *CompareValidator* control is used to compare the value entered by the user into an input control, such as a *TextBox* control, with the value entered into another input control, or with a predefined value. For example, a registration web site that requires retyping the e-mail address and user-selected password could use the *CompredValidate* to verify that the information is correct. The relationship of the *TextBox* controls to compare can be specified by a comparison operator of *Equal NotEqual*, *GreaterThan*, *GreaterThanEqual*, *LessThan*, *LessThanEqual*, and *DataTypeCheck*. By setting the *Operator* property to *DataTypeCheck*, you can also use the *CompareValidator* control to determine whether the value entered into an input control can be converted to the data type specified by the *Type* property. Listing 5–10 is the *.aspx* file for the web form with two *CompareValidator* controls to verify a password input.

Listing 5–10. *.aspx* file for the web form with two *CompareValidator* controls to verify a password input.

```
<form id="Form1" method="post" runat="server">
<asp:Label id="Label1" style="Z-INDEX: 101; LEFT: 104px;
   POSITION: absolute; TOP: 32px"
   runat="server">Password:</asp:Label>
<asp:TextBox id="txtPasswd1" style="Z-INDEX: 102; LEFT: 184px;
   POSITION: absolute; TOP: 32px"
   runat="server" TextMode="Password"></asp:TextBox>
<asp:Label id="Label2" style="Z-INDEX: 103; LEFT: 56px;
   POSITION: absolute; TOP: 80px"
   runat="server">Re-type password:</asp:Label>
<asp:TextBox id="txtPasswd2" style="Z-INDEX: 104; LEFT: 184px;
   POSITION: absolute; TOP: 80px"
   runat="server" TextMode="Password"></asp:TextBox>
<asp:CompareValidator id="vldtrCmp" style="Z-INDEX: 105; LEFT: 184px;
   POSITION: absolute; TOP: 104px"
   runat="server"
```

```
   ErrorMessage="Inputs are not identical!"
   ControlToValidate="txtPasswd2"
   ControlToCompare="txtPasswd1"
   Operator="Equal">
</asp:CompareValidator>
<asp:Button id="Button1" style="Z-INDEX: 106; LEFT: 184px;
   POSITION: absolute; TOP: 136px" runat="server"
   Text="Validate"></asp:Button>
</form>
```

The comparison uses two *textbox* controls. The first is for the initial entry of the password, whereas the second is for the reentry. The *CompareValidator* control is applied to the second *TextBox* control by setting its *ControlToValidate* property to "txtPasswd2". The comparison of the reentry of the password to the initial password is defined via the *ControlToCompare* property and the *Operator* property. You may not see the *Operator* attribute as shown in this code if you choose the *Equal* operator, because it is the default value.

RangeValidator Control

The *RangeValidator* control is used to check whether the value entered by the user is within a range specified by a minimum and a maximum value. The range value can be a pair of numbers, alphabetic characters, and dates. Both the minimum and maximum values are expressed as constants. Listing 5–11 is the *.aspx* file for the web form with a *RangeValidator* control and a *CompareValidator* control to verify a numeric text input.

Listing 5–11. *.aspx* file for the web form with a *RangeValidator* control and a *CompareValidator* control to verify a numeric text input.

```
<form id="Form1" method="post" runat="server">
<asp:TextBox id=txtRange style="Z-INDEX: 101; LEFT: 72px;
   POSITION: absolute; TOP: 32px" runat="server"
   AutoPostBack="True"></asp:TextBox>
<asp:RangeValidator id=RangeValidator1
   style="Z-INDEX: 102; LEFT: 72px;
   POSITION: absolute; TOP: 80px" runat="server"
   ErrorMessage="Value must between 1 - 10"
   ControlToValidate="txtRange"
   MaximumValue="10" MinimumValue="1"
   Type="Integer"></asp:RangeValidator>
<asp:CompareValidator id=CompareValidator1
   style="Z-INDEX: 103; LEFT: 72px;
   POSITION: absolute; TOP: 56px" runat="server"
   ErrorMessage="Numbers only!"
```

```
   ControlToValidate="txtRange"
   Type="Integer" Operator="DataTypeCheck">
</asp:CompareValidator>
</form>
```

The reason to use a *CompareValidator* control is to make the text entry numeric only. Thus, the *Type* property and the *Operator* property are set to *Integer* and *DataTypeCheck*, respectively, for the *CompareValidator* control. The *RangeValidator* is used to check whether the numeric text entry is between 1 and 10. The *AutoPostBack* property of the *TextBox* control is set to *True* because otherwise a *Button* control is needed.

RegularExpressionValidator Control

The *RegularExpressionValidator* control is used to determine whether the value of an input control matches a pattern defined by a regular expression. As a matter of fact, each of other types of validation controls can be achieved by a *RegularExpressionvalidator* control via a pattern that matches that specific validation. The *RegularExpressionValidator* control is effective for checking predictable sequences of characters, such as those in Social Security numbers, e-mail addresses, ZIP codes, and latitude and longitude coordinates. Basically, if you can come up a pattern expression you can use the *RegulareExpressionValidator* control. Some commonly used regular expressions are listed in Table 3–10.

In the following example, we will create a web form that contains a ZIP code field and an e-mail field. Instead of using the *Label* control for denoting each of the text boxes we will set the initial value of the *Text* property for the *TextBox* control to the instruction of itself. We will set the *Text* property to "ZIP Code" for the ZIP code entry and the *Text* property to "e-mail address" for the e-mail entry. Then we will add a *RegularExpressionValidator* control to each *TextBox* control.

To add a RegularExpressionValidator *control to an ASP.NET web form:*

1. Drag a *TextBox* control from the Web Forms tab of the toolbox onto the web form and set the *ID* property to "txtZip" and the *Text* property to "ZIP Code".
2. Drag a *RegularExpressionValidator* control from the Web Forms tab of the toolbox onto the web form and place it next to the *TextBox* control. Set the *ID* property to "vldtrRegexZip", the *ErrorMessage* property to "Invalid ZIP Code", and the *ControlToValidate* property to "txtZip".

3 From the Properties window, click on the *ValidationExpression* property and an Ellipsis button displays. Click on the button, and then the Regular Expression Editor dialog box appears (Figure 5–18) with a list of predefined regular expressions. Select "U.S. ZIP code" from the list and then click on OK.

4 Drag another *TextBox* control from the Web Forms tab of the toolbox onto the web form and set the *ID* property to "txtEmail" and the *Text* property to "e-mail address".

5 Drag another *RegularExpressionValidator* control from the Web Forms tab of the toolbox onto the web form and place it next to the *TextBox* control. Set the *ID* property to "vldtrEmail", the *ErrorMessage* property to "Invalid e-mail address", and the *ControlToValidate* property to "txtEmail".

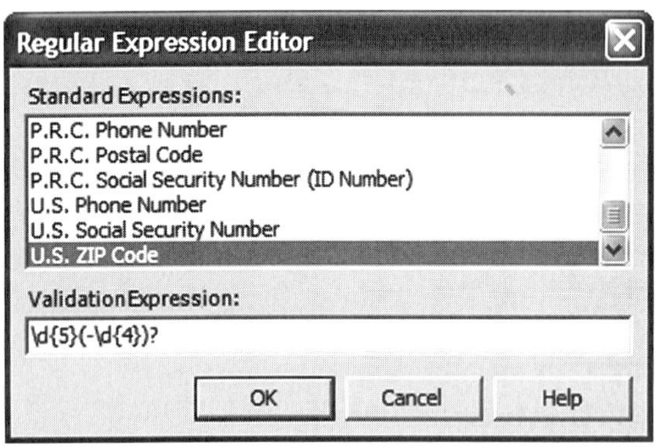

Figure 5–18. Regular Expression Editor dialog box.

6 From the Properties window, click on the *ValidationExpression* property and an Ellipsis button displays. Click on the button, and then the Regular Expression Editor dialog box appears with a list of predefined regular expressions, as shown in Figure 5–18. Select "Internet E-mail Address" from the list and then click on OK.

7 Drag a *Button* control from the Web Forms tab of the toolbox onto the web form and set the *Text* property to "Validate".

8 Click on the Start button or press F5 to run the application.

When a web page appears, click on the Validate button and then you should see both "Invalid ZIP code" and "Invalid e-mail address" displaying on the web page (Figure 5–19). If you enter *"22042"* and *"myself@aol.com"* for the ZIP code entry and e-mail entry, respectively, both error messages disappear (Figure 5–20). You can also use the regular expressions listed in Table 3–10 for the ZIP code and e-mail validation. As you can see, the key to successful use of the *RegularExpressionValidator* control is the pattern expression. Once you have a valid regular expression, you have a valid validation control.

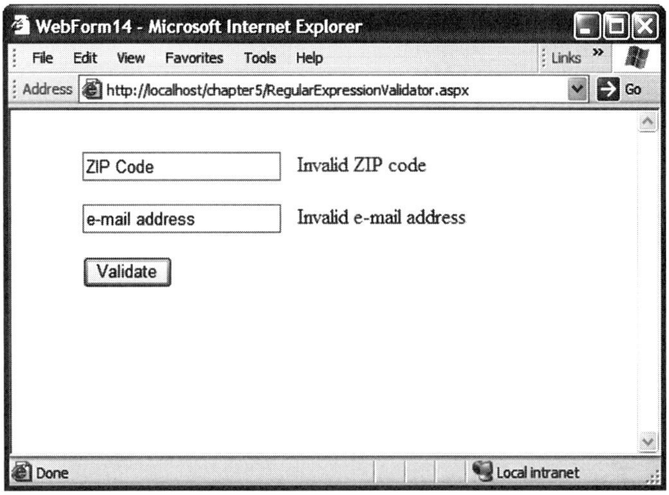

Figure 5–19. Example of the RegularExpressionValidator *control. Two* RegularExpressionValidator *controls were used to catch the input errors.*

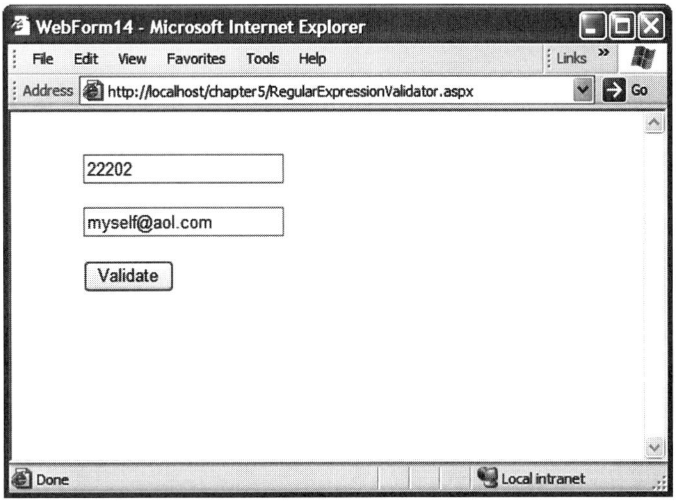

Figure 5–20. No error found when the inputs were entered correctly.

Listing 5–12 is the *.aspx* file for the web form with two *RegularExpressionValidator* controls.

Listing 5–12. *.aspx* file for the web form with two *RegularExpression Validator* controls.

```
<form id="Form1" method="post" runat="server">
<asp:TextBox id="txtZip" style="Z-INDEX: 101; LEFT: 56px;
   POSITION: absolute; TOP: 32px"
   runat="server">ZIP Code</asp:TextBox>
<asp:TextBox id="txtEmail" style="Z-INDEX: 102; LEFT: 56px;
   POSITION: absolute; TOP: 72px"
   runat="server"
   Columns="20">e-mail address</asp:TextBox>
<asp:RegularExpressionValidator id="vldtrRegexZip"
   style="Z-INDEX: 104; LEFT: 224px;
   POSITION: absolute; TOP: 32px" runat="server"
   ControlToValidate="txtZip" ErrorMessage="Invalid ZIP code"
   ValidationExpression="\d{5}(-\d{4})?">
</asp:RegularExpressionValidator>
<asp:RegularExpressionValidator id="vldtrRegexEmail"
   style="Z-INDEX: 105; LEFT: 224px; POSITION: absolute; TOP: 72px"
   runat="server" ControlToValidate="txtEmail"
   ErrorMessage="Invalid e-mail address"
   ValidationExpression="\w+([-+.]\w+)*@\w+([-.]\w+)*\.\w+([-.]\w+)*">
</asp:RegularExpressionValidator>
<asp:Button id="Button1" style="Z-INDEX: 103; LEFT: 56px;
   POSITION: absolute; TOP: 112px" runat="server"
   Text="Validate"></asp:Button>
</form>
```

ValidationSummary Control

The *ValidationSummary* control is used to summarize the error messages from all validation controls on a web page in a single location. It is a web control for validation summary. Depending on the value of the *DisplayMode* property, the validation summary can be displayed as a list, a bulleted list, or single paragraph. The error message displayed in the *ValidationSummary* control for each validation control on the same page is specified by the *ErrorMessage* property of each individual validation control. If the *ErrorMessage* property of the validation control is not set, no error message is displayed in the *ValidationSummary* control for that validation control. You can also specify a custom heading for the validation summary in the *ValidationSummary* control by setting its *HeaderText* property.

You can control whether to show or hide the validation summary by setting the *ShowSummary* property of the *ValidationSummary* control. That summary can also be displayed in a pop-up dialog box by setting the *ShowMessageBox* property to *True*.

In the following example, we will copy the previous web form and then add a *ValidationSummary* control to the web form to show the validation summary. Because each individual *RegularExpressionValidator* control has already been set to show the validation, we need to turn it off to avoid duplicating the error message. It can be easily done. We only need to set the *Text* property to an asterisk (*) for each *RegularExpressionValidator* control.

To add a ValidationSummary *control to an ASP.NET web form:*

1 Repeat steps 1 through 7 of the example for the *RegularExpressionValidator* control.

2 Set the *Text* property to an asterisk (*) for both *RegularExpressionValidator* controls.

3 Drag a *ValidationSummary* control from the Web Forms tab of the toolbox onto the web form and set the *HeaderText* property to "Error(s)", the *BorderColor* property to Red, the *BorderStyle* property to solid, and the *BorderWidth* property to "1px".

4 Click on the Start button or press F5 to run the application. When the web page appears, click on the Validate button. You should see the same error messages as shown in Figure 5–19, but the messages are listed in bullet style and shown in a red box (Figure 5–21).

5 Stop the application.

6 Set the *ShowSummary* property to *False* and the *ShowMessageBox* property to *True* for the *ValidationSummary* control.

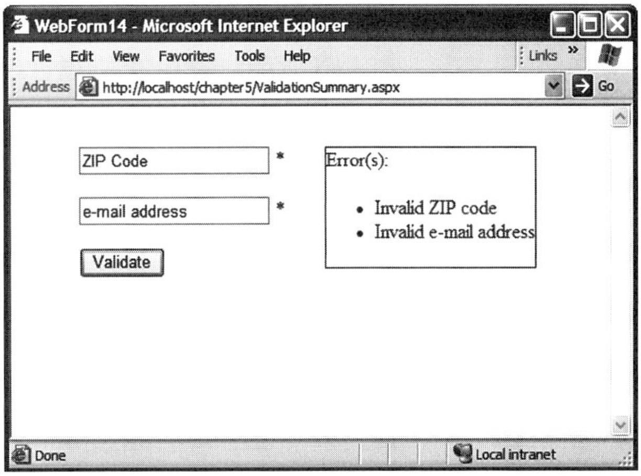

Figure 5–21. Example of the ValidationSummary *control.*

7 Click on the Start button or press F5 to run the application again.

When the web page appears, click on the Validate button. You should see the error messages shown in a pop-up message box (Figure 5–22). Basically, ASP.NET automatically generates a JavaScript alert to show the error messages.

Figure 5–22. Validation summary message showing in a pop-up message box.

Listing 5–13 is the *.aspx* file for the web form with the *ValidationSummary* control.

Listing 5–13. *.aspx* file for the web form with the *ValidationSummary* control.

```
<HTML>
<HEAD>
<title>WebForm14</title>
   <meta name="GENERATOR" content="Microsoft Visual Studio .NET 7.1">
   <meta name="CODE_LANGUAGE" content="Visual Basic .NET 7.1">
   <meta name="vs_defaultClientScript" content="JavaScript">
   <meta name="vs_targetSchema"
         content="http://schemas.microsoft.com/intellisense/ie5">
</HEAD>
<body MS_POSITIONING="GridLayout">
<form id="Form1" method="post" runat="server">
<asp:TextBox id="txtZip" style="Z-INDEX: 101; LEFT: 56px;
   POSITION: absolute; TOP: 32px"
   runat="server">ZIP Code</asp:TextBox>
<asp:TextBox id="txtEmail" style="Z-INDEX: 102; LEFT: 56px;
   POSITION: absolute; TOP: 72px" runat="server"
   Columns="20">e-mail address</asp:TextBox>
<asp:Button id="Button1" style="Z-INDEX: 103; LEFT: 56px;
   POSITION: absolute; TOP: 112px" runat="server"
   Text="Validate"></asp:Button>
<asp:RegularExpressionValidator id="vldtrRegexZip"
   style="Z-INDEX: 104; LEFT: 216px; POSITION: absolute; TOP: 32px"
   runat="server" ControlToValidate="txtZip"
   ErrorMessage="Invalid ZIP code"
   ValidationExpression="\d{5}(-\d{4})?">*
</asp:RegularExpressionValidator>
<asp:RegularExpressionValidator id="vldtrRegexEmail"
   style="Z-INDEX: 105; LEFT: 216px; POSITION: absolute; TOP: 72px"
   runat="server" ControlToValidate="txtEmail"
   ErrorMessage="Invalid e-mail address"
```

```
   ValidationExpression="\w+([-+.]\w+)*@\w+([-.]\w+)*\.\w+([-.]\w+)*">
*</asp:RegularExpressionValidator>
<asp:ValidationSummary id="ValidationSummary1"
   style="Z-INDEX: 106; LEFT: 256px; POSITION: absolute; TOP: 32px"
   runat="server" HeaderText="Error(s):" BorderColor="Red"
   BorderStyle="Solid" BorderWidth="1px" ShowSummary="False"
   ShowMessageBox="True"></asp:ValidationSummary>
</form>
</body>
</HTML>
```

Template Web Controls and Web Templates

ASP.NET provides a versatile feature called templates that allow the separation of control data from its presentation. A template control does not itself provide a user interface (UI). The UI for the control is supplied by a page developer through inline templates that allow a page developer to customize the UI for the control. For example, the *Repeater* control, *DataGrid* control, and *DataList* control are three template web controls.

Repeater Control

The *Repeater* control is used to create a simple list but with a user-specified template for each row. The list can be a data-bound list. The *Repeater* control has no built-in layout or styles, but it comes with five templates (Table 5–3) for its layout. You must explicitly declare all HTML layout, formatting, and style tags for each template you are using.

Table 5–3: Templates Available for the *Repeater* Control

Template	Description
HeaderTemplate*	HTML to render once before any data-bound rows in the list. A typical use is to begin a container element, such as <table>.
ItemTemplate	HTML to render each data-bound row in the list.
AlternatingItem Template	HTML to render for every other row in the list.
SeparatorTemplate*	HTML to render between rows in the list.
FooterTemplate*	HTML to render once after all data-bound rows in the list. A typical use is to close the container opened in the HeaderTemplate item, such as </table>.

* HeaderTemplate, SeparatorTemplate, and FooterTemplate cannot be data bound.

The *Repeater* control is different from other data list controls in that you need to explicitly specify HTML fragments in its templates. For example, you can use a HTML table to display a list for a *Repeater* control. You start with the table by placing the *<table>* tag in the *HeaderTemplate*. Next, you create the rows and columns of the table by placing *<tr>* tags, *<td>* tags, and data-bound items in the *ItemTemplate*. If you want a different appearance for alternating items in the table, you create an *AlternatingItemTemplate* with the same content as the *ItemTemplate*, except with a different style specified. Finally, you complete the table by placing the *</table>* tag in the *FooterTemplate*. In the following example, we will create a web form that contains a *Repeater* control to show the regional population listed Table 5–2.

To add a Repeater *control to an ASP.NET web form:*

1 Drag a *Repeater* control from the Web Forms tab of the toolbox onto the web form.

2 Click on the HTML tab and add the following code for *<asp:Repeater>*.

```
<HeaderTemplate>
 <table width="100%" cellspacing=1 cellpadding=3 style="font:8pt verdana">
  <tr style="background-color:#ddc">
    <th>ID</th>
    <th>Region</th>
    <th>Population</th>
    <th>Male</th>
    <th>Female</th></tr>
    </HeaderTemplate>
<ItemTemplate>
  <tr>
    <td><%#Container.DataItem("ID") %></td>
    <td><%#Container.DataItem("Region") %></td>
    <td><%#Container.DataItem("Population") %></td>
    <td><%#Container.DataItem("Male") %></td>
    <td><%#Container.DataItem("Female") %></td>
  </tr>
</ItemTemplate>
<FooterTemplate>
 </table>
</FooterTemplate>
```

A table is used inside the *Repeater* control to hold the regional population information. First, the *HeaderTemplate* is used for the table header, then the *ItemTemplate* is used to show all records, and finally the *FooterTemplate* is used to close out the HTML table tag. Inline scripting is used to populate each table cell using *Container.DataItem*. The *ItemTemplate* must be defined in order to show the repeated list.

3 Add the reference to the OLE DB namespace before the page class.

```
Imports System.Data.OleDb
```

4 Add the following code to the *Page_Load*.

```
If Not IsPostBack Then
 Dim dbConn As OleDbConnection = New OleDbConnection( _
   "Provider=Microsoft.Jet.OLEDB.4.0; data source=" & _
   Server.MapPath("data/chapter5.mdb"))
 Dim sql As String
 Dim dbComm As OleDbCommand = dbConn.CreateCommand()
 dbComm.CommandText = "SELECT * FROM Region"
 dbConn.Open()
 Dim dbRead As OleDbDataReader = dbComm.ExecuteReader()
 Repeater1.DataSource = dbRead
 Repeater1.DataBind()
 dbRead.Close()
 dbConn.Close()
End If
```

5 Click on the Start button or press F5 to run the application. A web page displays with regional population information (Figure 5–23).

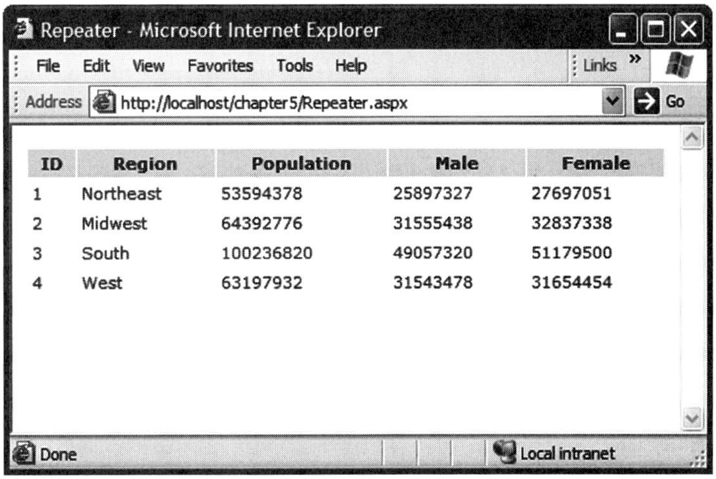

Figure 5–23. Example of the Repeater *control.*

Listing 5–14 is the *.aspx* file for the web form with a *Repeater* control.

Listing 5–14. *.aspx* file for the web form with a *Repeater* control.

```
<%@ Page Language="vb" AutoEventWireup="false"
Codebehind="Repeater.aspx.vb" Inherits="chapter5.Repeater"%>
<HTML>
<HEAD>
   <title>Repeater</title>
</HEAD>
<body MS_POSITIONING="GridLayout">
<form id="Form1" method="post" runat="server">
<asp:Repeater id="Repeater1" runat="server">
  <HeaderTemplate>
  <table width="100%" cellspacing="1" cellpadding="3"
    style="font:8pt verdana">
    <tr style="background-color:#ddc">
      <th>ID</th>
      <th>Region</th>
      <th>Population</th>
      <th>Male</th>
      <th>Female</th></tr>
   </HeaderTemplate>
   <ItemTemplate>
     <tr>
       <td><%#Container.DataItem("ID") %></td>
       <td><%#Container.DataItem("Region") %></td>
       <td><%#Container.DataItem("Population") %></td>
       <td><%#Container.DataItem("Male") %></td>
       <td><%#Container.DataItem("Female") %></td>
     </tr>
   </ItemTemplate>
   <FooterTemplate>
   </table>
   </FooterTemplate>
</asp:Repeater>
</form>
</body>
</HTML>
```

In this example, an HTML table was used. The *HeaderTemplate* was used to render the table header. Because that template cannot be data bound, the table header had to be manually entered. The *ItemTemplate* was used to render each row of the table. Although you did not see any looping, as you might have seen with classic ASP, it was all handled by the *DataSource* in acknowledge of ASP.NET. The *footerTemplate* was only used to close out the *<table>* tag. As shown in this example, you can display the row-based data in tabular form easily with the *Repeater* control.

DataGrid Control

The *DataGrid* control is used to display structured data in table format. It displays a database table where each column of the table holds a field and each row of the table represents a record. People also call this type of display a grid format. That is the reason this control is called *DataGrid*. The *DataGrid* control is ideal for displaying query results from a geospatial layer. It can also allow the user to edit the data in the grid. In the following example, we will create a web form that contains a *DataGrid* control to show the regional population listed in Table 5–2.

To add a DataGrid *control to an ASP.NET web form:*

1 Drag a *DataGrid* control from the Web Forms tab of the toolbox onto the web form, and then enter the following property settings.

- Click on the *AlternatingItemStyle* property, set the *BackColor* property to "#C0C0FF", and then click on the *Font property* folder and set the *Name* property to *Arial*.
- Click on the *ItemStyle* property, set the *BackColor* property to "#8080FF", and then click on the *Font property* folder, set the *Name* property to *Arial*, and set the *Size* property to 10pt.
- Click on the *HeaderStyle* property, and then click on and set the *BackColor* to "#8080FF".

2 Add the reference to the OLE DB namespace before the page class.

```
Imports System.Data.OleDb
```

3 Add the following code to the *Page_Load*.

```
If Not IsPostBack Then
 Dim dbConn As OleDbConnection = New OleDbConnection( _
   "Provider=Microsoft.Jet.OLEDB.4.0; data source=" & _
   Server.MapPath("data/chapter5.mdb"))
 Dim sql As String
 Dim dbComm As OleDbCommand = dbConn.CreateCommand()
 dbComm.CommandText = "SELECT * FROM Region"
 dbConn.Open()
 Dim dbRead As OleDbDataReader = dbComm.ExecuteReader()
 DataGrid1.DataSource = dbRead
 DataGrid1.DataBind()
```

```
    dbRead.Close()
    dbConn.Close()
End If
```

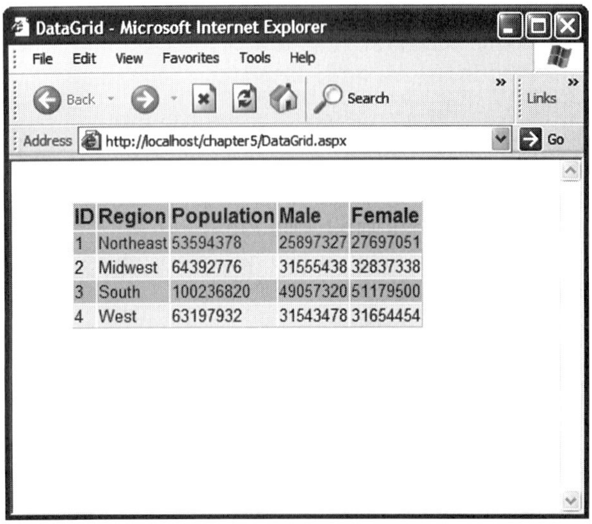

Figure 5–24. Example of the DataGrid *control.*

4 Click on the Start button or press F5 to run the application. A web page displays with regional population information (Figure 5–24).

In this example, three templates (*AlternatingItemStyle*, *ItemStyle*, and *HeaderStyle*) were used. Instead of manually entering HTML tags for the templates, the styles were set through the properties, which are much easier and flexible compared to the *Repeator* control. Unlike the *Repeator* control, the *DataGrid* control is more like a table control. An HTML table is used to render the result of the *DataGrid* control. The style of the header row of the table was specified through the *HeaderStyle* property. The data row was specified via the *ItemStyle* and *AlternatingItemStyle* properties. The *AlternatingItemStyle* was used primarily to alter the color for every other row so that the table is more readable, especially when a table has more than five rows. The *DataGrid* is designed smartly so that the *AlternatingItemStyle* is independent from the data content itself, as you can see in Listing 5–15.

Listing 5–15. *.aspx* file for the Web Form with a *DataGrid* control.

```
<%@ Page Language="vb" AutoEventWireup="false"
Codebehind="DataGrid.aspx.vb" Inherits="chapter5.DataGrid"%>
<HTML>
<HEAD>
   <title>DataGrid</title>
</HEAD>
<body MS_POSITIONING="GridLayout">
<form id="Form1" method="post" runat="server">
<asp:DataGrid id="DataGrid1" style="Z-INDEX: 101; LEFT: 48px;
   POSITION: absolute; TOP: 32px" runat="server">
   <AlternatingItemStyle BackColor="#C0C0FF">
   </AlternatingItemStyle>
   <ItemStyle Font-Size="10pt" Font-Names="Arial"
      BackColor="#8080FF">
```

```
    </ItemStyle>
    <HeaderStyle Font-Names="Arial" Font-Bold="True"
      BackColor="#8080FF"></HeaderStyle>
</asp:DataGrid>
</form>
</body>
</HTML>
```

If you want to display a database table or a query result, the *DataGrid* control formats the content quickly. With a few clicks, the resultant table can be made much more readable than its classical ASP counterpart. The key part to use the *DataGrid* control is to bind a *DataSource*. The Visual Basic .NET code for the `Page_Load` can be used with any database table once you use a proper OLE DB provider. Listing 5-15 is the *.aspx* file for the Web Form with a *DataGrid* control.

DataList Control

The *DataList* control is, like the *Repeater* control, used to display a repeated list of items that are bound to the control. However, the *DataList* control adds a table around the data items by default. Instead of typing HTML within each set of tags, you can just simply customize each template for the *DataList* control through its properties in a similar fashion to other web controls. The *DataList* control comes with seven templates (Table 5-4) and seven styles (Table 5-5) for its layout. You can customize those templates and the corresponding styles through the property setting approach. The *DataList* control may be bound to a database table, an XML file, or another list of items.

In the following example, we will create a web form that contains a *DataList* control to show the regional population listed Table 5-2. Instead of typing HTML tags for each tag, we will define the styles through the Properties window.

To add a DataList *control to an ASP.NET web form:*

1. Drag a *DataList* control from the Web Forms tab of the toolbox onto the web form, and then establish the following property settings
    - Set the *BackColor* property to *Navy*, the *BorderStyle* property to Solid, and the *BorderWidth* property to 5px.
    - Set the *GridLines* property to *Horizontal*, the *CellPadding* property to 2, and the *CellSpacing* property to 2.

- Click on the *AlternatingItemStyle* property, set the *BackColor* property to "#C0C0FF", and then click on the *Font property* folder and set the *Name* property to *Arial Narrow.*
- Click on the *ItemStyle* property, set the *BackColor* property to "#8080FF", and then click on the *Font property* folder and set the *Name* property to *Arial Narrow.*

2 Add the reference to the OLE DB namespace before the page class.

```
Imports System.Data.OleDb
```

3 Add the following code to the *Page_Load*.

```
If Not IsPostBack Then
 Dim dbConn As OleDbConnection = New OleDbConnection( _
   "Provider=Microsoft.Jet.OLEDB.4.0; data source=" & _
   Server.MapPath("data/chapter5.mdb"))
 Dim sql As String
 Dim dbComm As OleDbCommand = dbConn.CreateCommand()
 dbComm.CommandText = "SELECT * FROM Region"
 dbConn.Open()
 Dim dbRead As OleDbDataReader = dbComm.ExecuteReader()
 DataList1.DataSource = dbRead
 DataList1.DataBind()
 dbRead.Close()
 dbConn.Close()
End If
```

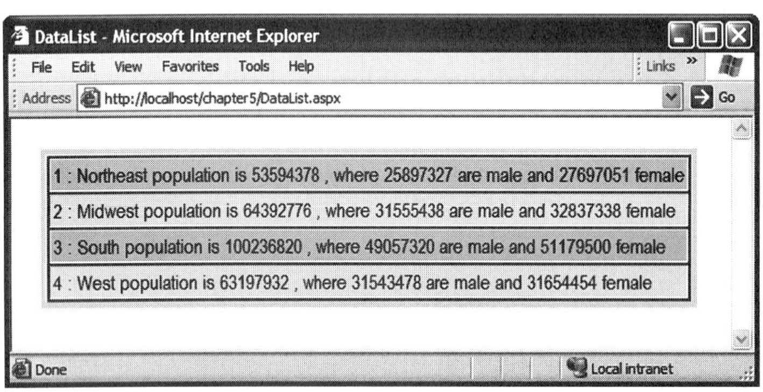

4 Click on the Start button or press F5 to run the application. A web page displays with regional population information (Figure 5–25).

Figure 5–25. Example of the DataList *control.*

Listing 5–16 is the *.aspx* file for the web form with a *DataList* control.

Listing 5–16. *.aspx* file for the web form with a *DataList* control.

```
<%@ Page Language="vb" AutoEventWireup="false"
Codebehind="DataList.aspx.vb"
Inherits="chapter5.DataList"%>
```

```
<HTML>
<HEAD>
   <title>DataList</title>
</HEAD>
<body MS_POSITIONING="GridLayout">
<form id="Form1" method="post" runat="server">
<asp:datalist id="DataList1" style="Z-INDEX: 101; LEFT: 24px;
   POSITION: absolute; TOP: 24px" runat="server"
   GridLines="Horizontal" BorderStyle="Solid"
   BorderWidth="5px" CellPadding="2" CellSpacing="2"
   BackColor="Navy">
   <AlternatingItemStyle Font-Names="Arial Narrow"
      BackColor="#C0C0FF">
   </AlternatingItemStyle>
   <ItemStyle Font-Names="Arial Narrow" BackColor="#8080FF">
   </ItemStyle>
   <ItemTemplate>
   <%#Container.DataItem("ID")%>
   :
   <%#Container.DataItem("Region")%>
   population is
   <%#Container.DataItem("Population")%>
   , where
   <%#Container.DataItem("Male") %>
   are male and
   <%#Container.DataItem("Female") %>
   female.
   </ItemTemplate>
   <AlternatingItemTemplate>
   <%#Container.DataItem("ID")%>
   :
   <%#Container.DataItem("Region")%>
   population is
   <%#Container.DataItem("Population")%>
   , where
   <%#Container.DataItem("Male") %>
   are male and
   <%#Container.DataItem("Female") %>
   female.
   </AlternatingItemTemplate>
</asp:datalist></form>
</body>
</HTML>
```

In this example, only *AlternatingItemTemplate* and *ItemTemplate* were used. Instead of manually entering HTML tags for the templates, the styles were set through the properties. Unlike the *Repeater* control, the

DataList control is more like other list controls discussed in this chapter. You drag and drop a *DataList* control to a web form, customize it through its properties, and then set the data bound for it. The *AlternatingItemTemplate* was used to change the color of every other row of the table. It was defined in the same way as *ItemTemplate* because we wanted the same layout for the alternating rows. The styles for the *AlternatinItemTemplate* and *ItemTemplate* were defined in the *AlternatingItemStyle* and *ItemStyle*, respectively. Recall how much effort it took to achieve the same alternating row effect in classical ASP. ASP.NET has made alternating table rows trivial.

We all know that when a table has more than five rows it would very difficult for our eyes to capture all rows. Alternating rows makes rows more readable. In addition, each row was made as a sentence instead of cells to show the flexibility with the *DataList* control. For simplicity, the *HeadTemplate* and *FooterTemplate* were not used in this example, but you can add them to the example and make a nice-looking header and footer. You can do a lot more than what you can imagine with the *HeaderStyle* and *FooterStyle*. Remember that you do not need to manually enter any HTML tag—everything is done through IDE.

Web Templates

After we develop a few web applications, we often develop a preference for one or more of our web page designs and would like to make a page template. For example, almost every web-based GIS application has a geography entry page that allows users to enter a ZIP, city, county, state, region, watershed, and a pair of latitude and longitude coordinates. It is natural that we want to encapsulate those common geography entry components into a page template, which a new web page can consume or on which a new web page can be possibly customized. However, there is no easy way because ASP.NET 1.1 does not provide an intuitive way of structuring an application based on a page template. ASP.NET offers the *ITemplate* interface, which allows you to build a custom control with inline templates. You can then populate those templates for the custom control at design time. You have to be very ASP.NET savvy to program page templating with *ITemplate*. You need to know how to design a user control before you can develop a page template. We cover creation of user controls in Chapter 7. Basically, a user control is a control converted from an ASP.NET web page. Any ASP.NET web page can be converted into a user control. Thus, we discuss web templates in Chapter 7 as well.

At the time of writing, ASP.NET 2.0 is still in beta. One key feature is master pages, which provide a template for the look and implementation of a page. A master page is an ASP.NET page that provides a template for other pages, providing shared page-level layout and functionality. The master page defines placeholders for the content, which can be overridden by child pages. The resultant page is a combination of the master page and the child page. A master page is a new type of control, a sibling of the current *Page* and *UserControl* classes. As you can see, the templating solution from Microsoft is right around the corner. We suggest that you not spend too much time developing your own page templates. ASP.NET 2.0 is scheduled to release in the fall of 2005.

Summary

In this chapter, we have introduced web forms, the foundation of ASP.NET web pages. We compared them with the Windows form and HTML forms. There are differences and similarities.

We covered in great detail *Web* controls, *Validation* controls, and *Template* controls. Building ASP.NET applications with *Web* controls is just like building blocks with LEGOs. You can use your imagination; your options are limitless. *Web* controls are like Windows form elements. You can drag and drop them onto a web form using Visual Studio .NET and then customize them at design time. The *ImageButton* control is the one we use most in web-based GIS application because it first holds an image and provides a pair of coordinates of a clickable point. Using list controls can be difficult at first, but once you use them a couple of times you will find them very flexible. If a data binding is required, you just make sure to define a *DataSource* object.

There is no specific section in this chapter on ADO.NET. However, we have used OLE DB a few times throughout this chapter. OLE DB is the most powerful data access component Microsoft has ever produced. It is easy, powerful, and painless. However, you still need to hunt down the right version of the *Provider* string. We also provided the OLE DB provider strings for Microsoft Access, SQL Server, and Oracle—the three most commonly seen databases. You need to use *OleDbCommand* and

OleDbDataReader to define an SQL *SELECT* statement and load in the "recordset", respectively. Finally, do not forget to bind the data to the list control.

Validation controls are gracious gifts from Microsoft. They relieve us of writing lines and lines of client-side JavaScript code for validation as we did with classic ASP applications. *RegularExpressionValidator* works like a cordless drill/screwdriver combo kit. It operates in many ways and can be used to do many things as long as you find a proper drill bit or driver bit. In addition, Visual Studio .NET comes with a handy set of predefined patterns. This book provides a list of commonly used patterns in Table 3–10.

Be patient for the release of ASP.NET 2.0. You will be rewarded with the advantageous web templating mechanism in ASP.NET 2.0. In addition, Visual Studio .NET 2005 will be released as well so that you can author templates with the new version of the IDE.

Now you have learned most of the ASP.NET *Web* controls but have not seen any ArcGIS *Server* control yet. In Chapter 6 we explore in detail ArcGIS Server *Web* controls and web templates used to build web applications or web services.

CHAPTER 6

ArcGIS Server Web Templates and Web Controls

ARCGIS SERVER CONSISTS OF TWO PRIMARY COMPONENTS: a GIS server and an Application Development Framework (ADF) for .NET and Java. The GIS server hosts ArcObjects that can be used for web, enterprise, and desktop applications. It includes the core ArcObjects libraries and provides a scalable environment for running ArcObjects on a server. The ArcGIS ADF allows you to build and deploy desktop applications, web applications, and web services that use ArcObjects but run on top of the GIS server. You manage the ArcGIS Server objects using ArcCatalog and author the ArcGIS Server objects (including mapping and geocoding) with ArcMap. ArcGIS Server provides templates, web controls, and convenience classes for you to build web applications.

ArcGIS Server Application Developer Framework

ArcGIS Server includes an ADF for .NET that provides GIS functionality to build ASP.NET web applications. The ArcGIS Server ADF for .NET is built on top of Microsoft's .NET Framework and includes a set of ESRI Web controls and templates you can use to build GIS web applications. You can start building your web application with one of several predefined templates, including the Map Viewer, Search, Page Layout, Geocoding, Thematic, Buffer Selection, and Web Service Catalog templates. You can also start developing your web application with a combination of ESRI *Web* controls and regular web controls.

The ArcGIS Server ADF for .NET basically provides you a channel to communicate with ArcObjects. The .NET ADF built on top of Microsoft's .NET Framework extends the .NET Framework class library with new classes that support a set of custom web controls and provide remote access to ArcGIS Server and thus ArcObjects (see also the *ArcGIS Server Administrator and Developer Guide*). According to ESRI, the ArcGIS Server ADF for .NET consists of:

- Visual Studio .NET templates
- Custom ASP.NET *Web* controls
- Convenience classes for building client applications that access a GIS server

The ADF provides a set of objects that can access a GIS server as ESRI *Web* controls and convenience classes. With the ADF, you can directly access ArcGIS Server objects and ArcObjects through ASP.NET. The ADF encapsulates the details of programming directly with ArcGIS Server objects and exposes a rich set of mapping-centric web controls you can place directly on your web forms.

The Visual Studio .NET templates help you start building your web applications. Each ArcGIS Server template was designed to accomplish a common GIS task such as map viewing and feature searching. If you need to develop a rapid prototype, the templates out of the box are your best option. All code for the templates is provided, and you can customize a template as you desire. You can also cut and paste code fragments from those templates into your own applications. The templates serve as a great starting point for developing your own web applications.

ArcGIS Server Web Templates

As we have seen many times so far, every time you create a new Visual Studio .NET project you are presented with a set of templates you can use as a starting point to create an application. We often use ASP.NET Web Application, ASP.NET Web Service, the Web Control Library, and so on. In the New Project dialog box, there is a folder that contains ArcGIS Server Projects as well. There are two subfolders of the set of templates distributed with the ArcGIS Server ADF for .NET: one for Visual Basic .NET and the other for C# (Figure 6-1). We will focus on those templates for Visual Basic .NET.

✔ **NOTE:** *If you do not see the ArcGIS Server templates in the New Project dialog box, they probably are not installed on your system.*

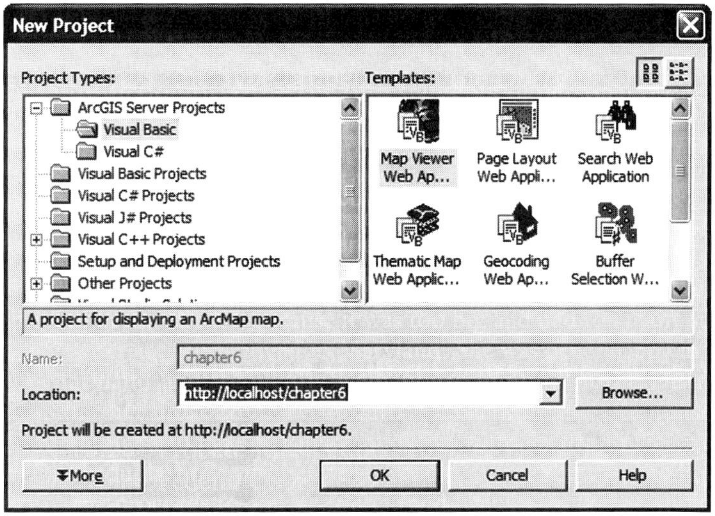

Figure 6–1. ArcGIS Server templates for Visual Studio .NET.

There are seven templates that are shipped with ArcGIS Server:

- Map Viewer
- Search
- Page Layout
- Geocoding
- Thematic
- Buffer Selection
- Web Service Catalog

Each of these server templates uses the set of *Web* controls that are part of ArcGIS Server ADF for .NET. For example, if you want to build an ASP.NET web application to show the census data, you may want to consider choosing the thematic template. It comes with *Web* controls that allow you to do dynamic classification of demographics. The primary advantage of building your web application with a predefined template is that much of the commonly used functionality has already been built into it, so you do not have to program it yourself. That is extremely helpful if you want to develop a prototype to show some new concept.

Predefined templates provide an easy starting point for your web application. You can then customize the layout of web controls on the template (such as the banner, fonts, and colors) to suit your needs. You can drop some unused controls from the template or add more controls to it. Each of the templates is discussed in the following sections.

Map Viewer Template

The Map Viewer template, which is similar to any other map viewer in MapObjects or ArcObjects, provides a basic map viewer to display and query spatial layers. This map viewer, however, runs within a web browser. It consists of a main map, an overview map (also called locator map), a table of contents (dynamic legend for turning layers on/off), a toolbar, a North arrow, and a scale bar (Figure 6–2). The toolbar offers tools and commands for zooming, panning, and querying.

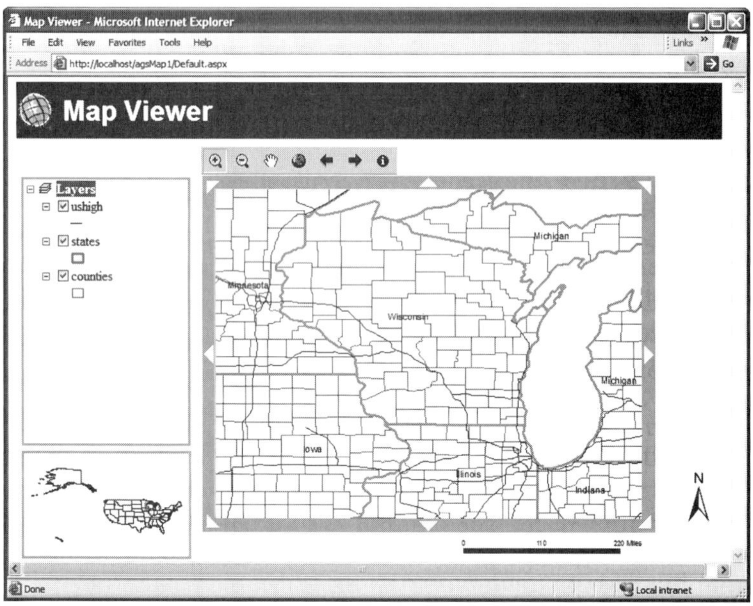

Figure 6–2. An example created based on the Map Viewer template.

The Map Viewer template is one of the best ways to build a rapid prototype. It is a good starting point. Once you get familiar with it, you can start to customize and modify it. You need to specify the *Host*, *ServerObject*, and *DataFrame* (if you want to use one data frame other than the default) for the main map and the overview map. You also need to set up the *BuddyControl* for the toolbar, the overview map, and the table of contents. A *BuddyControl* specifies what you want the current control to work with. Finally, you need to remember to set up the impersonation. You need to provide the user name and password of a user who is part of ArcGIS Server users group (*agsuser*).

Page Layout Template

The Page Layout template offers you a page layout that is the same as the layout view in ArcView or ArcMap. It shows all data frames on the map on the page layout as well as the map title, legend, North arrow, and scale bar. This template also comes with a toolbar, which allows you to pan and zoom each data frame on the map and pan and zoom around the page layout itself.

A web application built based on the Page Layout template is preferable for printing through the browsers. It provides a WYSIWYG look-and-feel interface. With the toolbar included in the template, you can easily adjust what you want to print. For example, you can zoom in precisely to an area to generate a map.

Search Template

You can build a geographic-centric search site similar to *maps.google.com*, using the Search template (Figure 6–3). The Search template provides a search interface for finding features and then tabulating and mapping the results. In addition to a traditional search textbox and a submit button, the Search template contains a *DataGrid* control for the tabulated results, a *Map* control for mapping the search results, and a toolbar. ESRI has enhanced the Search template with a live *Map* control so that you can directly zoom in and out and pan around to study the search results geographically.

A web application built based on the Search template searches for matching values in the attribute tables of all available layers from the map server object you specified. You can only search those features offered through the server object, and thus you need to make sure to publish the feature layer against which you want to conduct the search via the server object. Because a search is done on all layers on the map, you may want to refine your search criteria when building an application using this template.

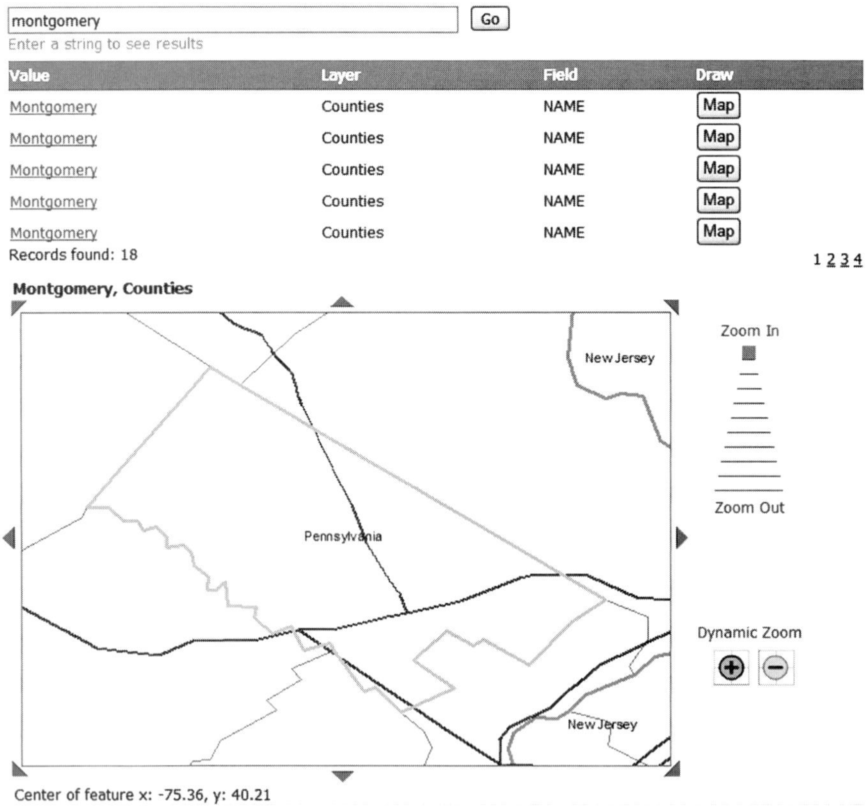

Figure 6–3. Example of using the Search template.

Geocoding Template

Often we are given an address and asked to put it on a map. What we need is address matching. The Geocoding template provides an interface for converting a valid urban address into a mappable point on the map. The Geocoding template contains a *DataGrid* control for the matched candidates, a *Map* control for mapping each individual candidate, and a toolbar. ESRI has enhanced the Geocoding template with a live *Map* control so that one can directly zoom in and out and pan around to study the individual geocoded addresses. By default, the Geocoding template assumes the address style to be US Streets or US Streets with Zone.

Thematic Template

If you ever need value rendering, the Thematic template is the best. The Thematic template provides thematic mapping symbolization. It looks like the Map View template but has an extra tool on the toolbar. When you click on it, it allows you to choose which layer to render and what field to render. The tool offers three algorithms to compute the class breaks: *Natural Breaks*, *Equal Interval*, and *Quantile*. You can define the number of classes for the thematic rendering. The tool also provides a color selection tool so that you can choose your own start and end colors for your value rendering. This template actually modifies the map server objects based on the user selection. Therefore, you need to use a non-pooled server object for the Thematic template. If you do not, you will get a warning message when you run the application built based on this template.

Buffer Selection Template

If you want to set up a new ATM machine near a certain intersection or conduct a corridor analysis for a freeway expansion, you need to conduct a buffer analysis. The buffer analysis allows you to find out, for example, how many residents are within 10 miles of that intersection and how many people live 5 miles along each side of the freeway. The Buffer Selection template allows you do that. It finds features in one layer based on the buffering distance relative to features in another layer. The Buffer Selection template offers options for finding features that completely fall within the buffer or intersect with the buffer.

Web Service Catalog Template

When you create your first ArcGIS Server object and are able to see it displayed in an application built based on the Map Viewer template, you probably say to yourself, "Gee, am I good or what?" After you create more than 10 server objects, you probably start scratching your head and wondering how you could manage them if the number of server objects continues to grow. The Web Service Catalog template can be used to address this concern. It provides a way of organizing related server objects into groups and makes them accessible to a certain group of people. The Web Service Catalog template makes server objects accessible as web services. As a result, any IT developer can build GIS web applications.

You create a project using the Web Service Catalog template first and then build a solution. You then publish a catalog service using ArcCatalog. The process is similar to publishing local server objects, but you specify the Internet server as the ArcGIS server you want to connect to and enter a URL address such as *http://localhost/agsWebService/default.aspx*. Upon creation, a web service catalog will become available in the *GIS Servers* folder in ArcCatalog (Figure 6–4). It is a true SOAP/WSDL web service. You can then write a program to consume it. ArcGIS desktop users can also connect to it and use it as a local data source. To reference the web service catalog, you use the following URL: *http://localhost/agsWebService/default.aspx?wsdl*.

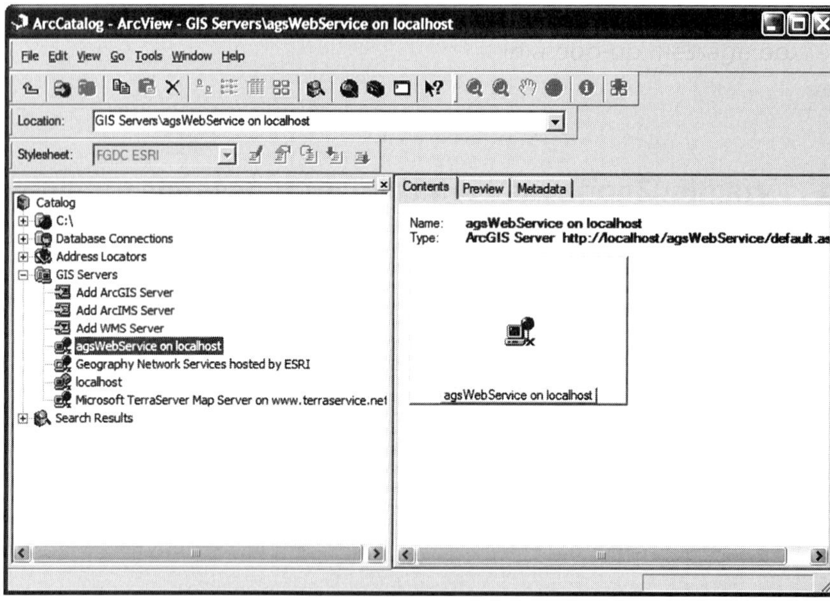

Figure 6–4. An ArcGIS Server web service shown in ArcCatalog.

Unlike other templates discussed previously, the Web Service Catalog template has no user interface for end users. But it does have an *Impersonation* control you need to set up before building any web service. When you create a web service using the template, you will be presented all server objects available on the server you specify. You can choose one, a few, or all server objects for your web service catalog. Once the catalog is created, it can be used by the ArcGIS desktop users or a program.

Creating Web Applications with ArcGIS Server Templates

The easiest way to create a rapid prototype is to use one of templates discussed previously. Once you see what it could look like, you can then customize it. In the following example, we are going to create a web application using the Thematic template.

1 Start Visual Studio .NET and click on the New Project button.

2 The New Project dialog box appears.

3 Double click on ArcGIS Server Projects and then click on Visual Basic.

4 From the Templates box, click on Thematic Map Web Application. Type in *http://localhost/agsThematic* for the Location text box and then click on OK.

As you can see, the web form contains a *Map* control, an *Overview Map* control, a *Table of Contents* control, a *Toolbar* control, a *Scale Bar* control, a *North Arrow* control, and an *Impersonation* control. There are also 10 hidden input variables.

5 Click on the *Impersonation* control, and from the Properties window click on the *Identity* property and an Ellipsis button displays. Click on the button, and then the Impersonation1 Identity dialog box appears. Enter a valid user name and password, and then click on OK. You can leave the "Domain or Machine Name" box blank.

You will notice right away that the identity information you just entered is encrypted for security reasons. Always remember to set up the *Impersonation* control before you run the application.

Figure 6–5. A runtime error for the Thematic template that does not use a non-pooled server object.

6 Click on the *Map* control, and then set the *Host* property to "localhost". Click on the *ServerObject* property. A drop-down list displays. Select the "usa_nonpool" server for the *ServerObject* property.

As discussed previously, you need to use a non-pooled server object. If you use the pooled server object of "usa", for example, you would get an error (Figure 6–5), to remind you to use a non-pooled server object.

7 Click on the *Overview Map* control, and then set the *BuddyControl* property to "Map1" and the *Host* property to "localhost". Click on the *ServerObject* property. A drop-down list displays. Select the "usa_locator" server for the *ServerObject* property.

8 Click on the *Table of Contents* control, and then set the *BuddyControl* property to "Map1".

9 Click on the *Toolbar* control, and then make sure the *BuddyControl* property is set to "Map1".

10 Click on the Start button or press F5 to run the application.

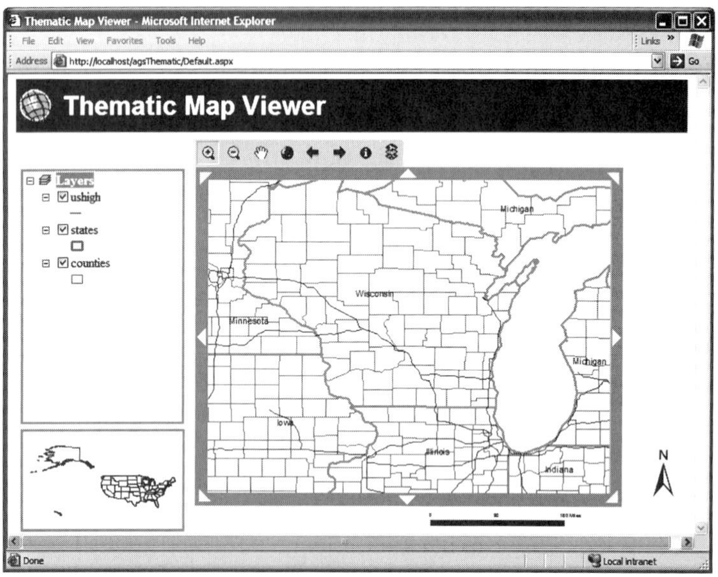

Figure 6–6. An example of using the Thematic template.

A web page displays (Figure 6–6). Click on the Classify Layer button (far right) and the Classify dialog box appears (Figure 6–7). Select "Counties" for the Layer, "POP1999" for the Value, Quantile for the Type, "10" for the Classes, dark green for the Start Color, and light green for the End Color. Click on the OK button. Click on a color box to access the Color Selector, and then select an appropriate color from the pop-up dialog box. The web page will be refreshed right away, with the population classification shown in the *Map* control and the list of 10 class breaks shown in the *Table of Contents* control (Figure 6–8).

As you can see, you can rapidly build a prototype using an ArcGIS Server template. You can try with other templates to see how they work and what functionality they provide. Each project built based on a template comes with a *readme* file that describes what you need to do to make an application run from the template. Read this file for more detailed information about the template you are using. As you can see, creating a web application based on a predefined ArcGIS template is easy and rapid. You will need to know more about ArcGIS Server *Web* controls before you can customize the application built based on a template. In the following section, we discuss ArcGIS Server *Web* controls.

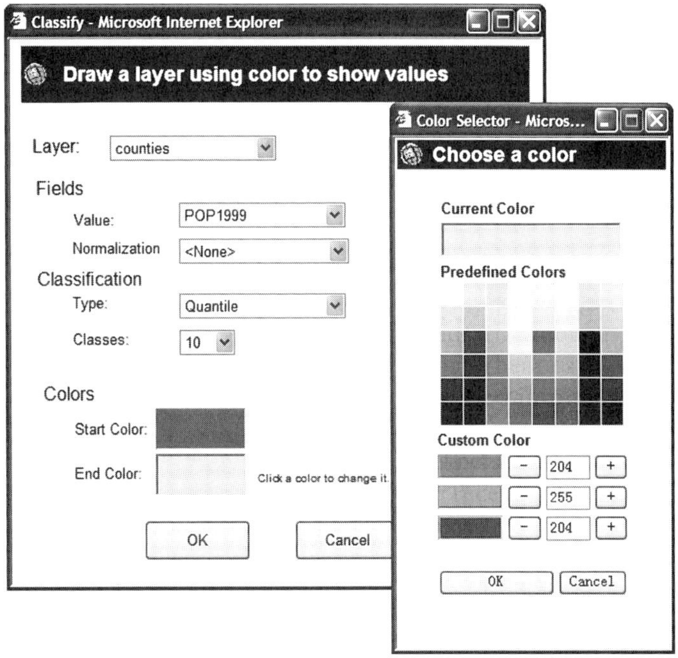

Figure 6–7. Classification dialog and color selection dialog.

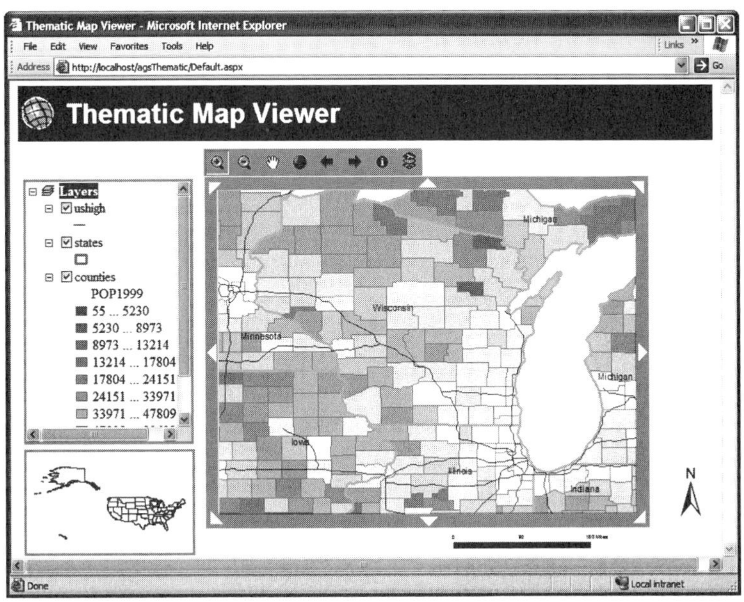

Figure 6–8. The Thematic Map Viewer shows the classified 1999 county population.

ArcGIS Server Web Controls

As mentioned earlier in this chapter, the ArcGIS Server ADF for .NET provides a set of *Web* controls and convenience classes you need for building web applications. Just like those *Web* controls discussed in Chapter 5, you can drag and drop the ArcGIS Server *Web* controls onto a web form using Visual Studio .NET. The ArcGIS Server offers eight *Web* controls, including *Impersonation, Map, Toolbar, Overview Map, North Arrow, Scale Bar, Table of Contents*, and *Page Layout* (Figure 6–9). It also provides a *GeocodeConnection* component for easing the complexity of geocoding (also called address matching).

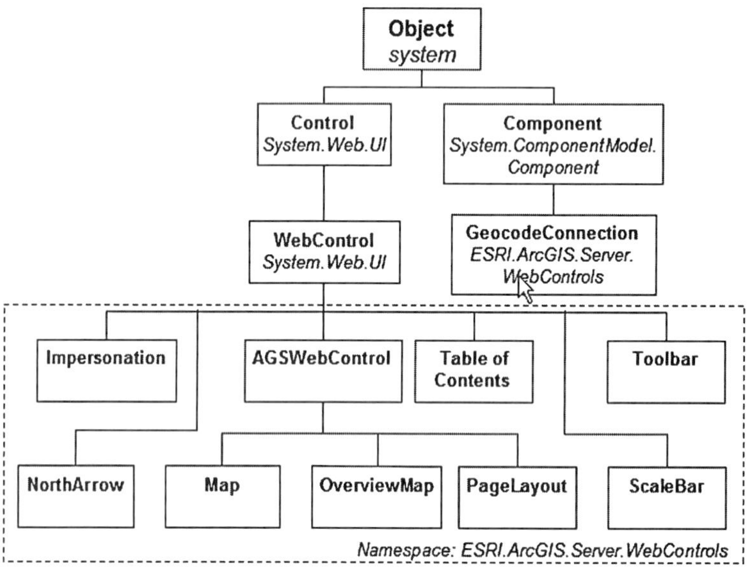

Figure 6–9. The conceptual object model diagram for the .NET ADF Web *controls.*

ArcGIS Server ADF for .NET provides three namespaces:

- *ESRI.ArcGIS.Server.WebControls*
- *ESRI.ArcGIS.Server.WebControls.Design*
- *ESRI.ArcGIS.Server.WebControls.Tools*

The *ESRI.ArcGIS.Server.WebControls* namespace contains the eight ArcGIS *Web* controls cited previously, as well as their associated convenience classes. You use those web controls and classes to build your ArcGIS Server web applications. The *ESRI.ArcGIS.Server.WebControls.Design* namespace consists of the classes that can be used to extend ArcGIS Server *Web* controls or to create new custom controls. The *ESRI.ArcGIS.Server.WebControls.Tools* namespace offers the classes that provide server-side actions for tools on a *Toolbar* control or for *TooItems* on a *Map* control or *Page Layout* control.

By combining the ArcGIS Server *Web* controls and those ASP.NET *Web* controls discussed in Chapter 5, you can easily build a GIS web application in any IDE. For example, you would use ArcGIS Server *Web* controls for map zooming and panning but use the *DataGrid* control for listing a geospatial query results. In the following, we will discuss all ArcGIS Server *Web* controls one by one.

Impersonation Control

The *Impersonation* control is used to set up the appropriate security credential for access to a GIS server on which your web application is built. This control allows the ArcGIS Server web application to impersonate the identity of a particular user and has all the privileges that identity is entitled to. The user to be impersonated needs to be part of the ArcGIS Server users group (*agsuser*). From now on, every single ArcGIS Server web application you are going to build needs an *Impersonation* control.

Map Control

The *Map* control is used to display a frame of the map. It provides the display functionality of the ArcMap data view for a map document published using ArcGIS Server. The *Map* control is the control that all web GIS applications must have. It accesses a single data frame at a time. You can set the *DataFrame* property to change the data frame for mapping. You can also program changes at runtime. Like other *Web* controls, the *Map* control has style and dimension. One of the most convenient properties is *CopyrightText*. You can enter your copyright text at design time and specify whether you want it to be left or right, top or bottom, or at one of the corners. You can even specify the direction the copyright is drawn, either horizontal or vertical. The convenience class for the *Map* control is *Web Map*. This class provides methods for panning and zooming the map, searching spatial features, and querying the attributes of a spatial feature.

✔ **NOTE:** *The vertical option for the* CopyrightText *does not work at the time of writing. A bug report on it has been filed according to ESRI.*

To add a Map control to an ASP.NET web form:

1 Drag a *Map* control from the ESRI tab of the toolbox onto the web form and set the *ID* property to "map1", the *Height* property to "300px", the *Width* property to "400px, and the *Host* property to "local host".

✔ **NOTE:** *You should notice that the height and width properties are arranged in reverse order. They should have the width first and height second. Instead,* Height *is before* Width.

2 Click on the *ServerObject* property. A drop-down list displays. Select the "usa" server for the *ServerObject* property.

If you run the application now, you will get an error message saying "Access is denied" because you have not set up an *Impersonation* control.

3 Expand the *CopyrightText* folder and set the *Text* property to "Programming ASP.NET for ArcGIS Server" and the *Halo* property to *True.*

4 Drag and drop an *Impersonation* control from the ESRI tab of the toolbox onto the web form.

5 From the Properties window, click on the *Identity* property and an Ellipsis button displays. Click on the button, and then the Impersonation1 Identity dialog box appears. Enter a valid user name and password and click on OK. You can leave the "Domain or Machine Name" box blank.

6 Click on the Start button or press F5 to run the application.

A web page displays with the map (Figure 6–10). Pay attention to the copyright text. It is so easy to add copyright information to the *Map* control. You can design your copyright text in many ways. You can even design a watermark for the maps you are publishing from your web site so that people cannot just simply use them without notifying you. Once they receive permission from you, you can then take the watermark off. Listing 6–1 is the *.aspx* file for the web form with the *Map* control.

Listing 6–1. *.aspx* file for the web form with the *Map* control.

```
<form id="Form1" method="post" runat="server">
<esri:Map id="Map1" style="Z-INDEX: 101; LEFT: 88px;
   POSITION: absolute; TOP: 24px" runat="server"
   Host="localhost" ServerObject="usa"
   Height="300px" Width="400px"
   CopyrightText-Font="Arial, 10pt"
```

```
   CopyrightText-Text="Programming ASP.NET for ArcGIS Server"
   CopyrightText-RightToLeft="False"
   CopyrightText-ForeColor="Black"
   CopyrightText-TextDirection="Horizontal"
   CopyrightText-TextAnchorPosition="BottomLeft"
   CopyrightText-TextAlignment="Left"
   CopyrightText-FillColor="Transparent"
   CopyrightText-Shadow="False"
   CopyrightText-Halo="True">
</esri:Map>
<esri:Impersonation id="Impersonation1"
   style="Z-INDEX: 102; LEFT: 80px;
   POSITION: absolute; TOP: 336px"
   runat="server"
   Identity="ENGMEMHMUMPMFMUNOMOMENUMPZGCWIYNIIKPKWUQWVECPZ">
</esri:Impersonation>
</form>
```

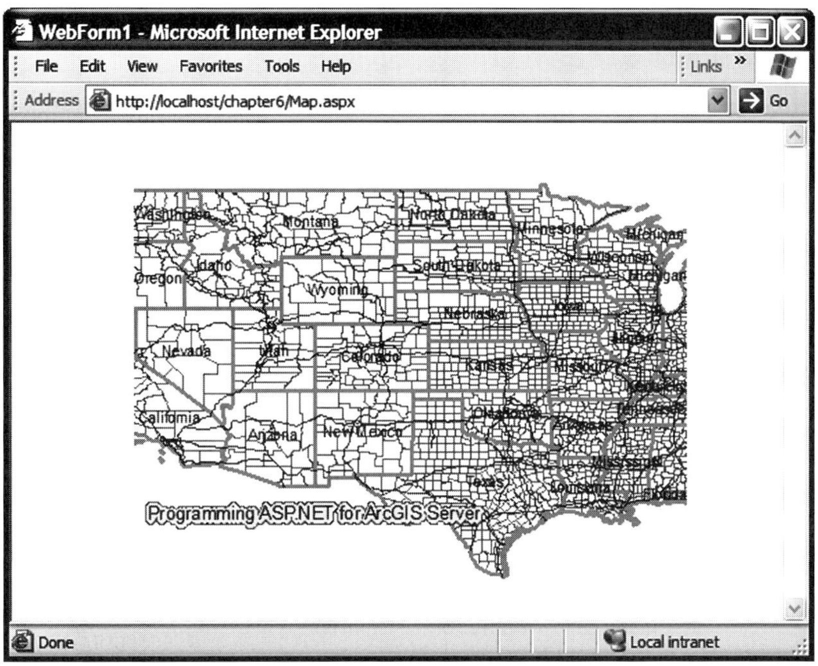

Figure 6–10. An example of using the Map *control.*

Toolbar Control

Once you have seen the map display, you may want to have some controls such as zoom or pan. You need to add a *Toolbar* control. This control is a collection of tools, commands, separators, and spaces. A tool interacts with the *Map* control and has a client-side action or a server-side action or both. For example, Zoom In, Zoom Out, and Identify are tools. A command does not interact with the *Map* control. Instead, it simply executes the code behind such as zooming to the full extent or previous extent. The *Toolbar* class implements all methods it needs, so there is no associated convenience class.

In the following example, we will add a *Toolbar* control based on the example for the *Map* control. We will add two tools (Zoom In and Zoom Out) and one command of full extent to the web form.

To add a Map control to an ASP.NET web form:

1 Drag a *Map* control from the ESRI tab of the toolbox onto the web form, and set the *ID* property to "map1", the *Height* property to "300px", the *Width* property to "400px, and the *Host* property to "localhost".

2 Click on the *ServerObject* property. A drop-down list displays. Select the "usa" server, which we just set up earlier for the *ServerObject* property.

If you run the application now, you will get an error message saying "Access is denied" because you have not set up an *Impersonation* control.

3 Expand the *CopyrightText* folder, and set the *Text* property to "Programming ASP.NET for ArcGIS Server" and the *Halo* property to *True*.

4 Drag and drop an *Impersonation* control from the ESRI tab of the toolbox onto the web form.

5 From the Properties window, click on the *Identity* property and an Ellipsis button displays. Click on the button, and then the Impersonation1 Identity dialog box appears. Enter a valid user name and password and click on OK. You can leave the "Domain or Machine Name" box blank.

So far, that is what you did with the example of the *Map* control. Now we will add a *Toolbar* control to it.

6 Drag a *Toolbar* control from the ESRI tab of the toolbox onto the web form and place it on top of the *Map* control. Set the *ID* property to "Toolbar1" and the *BuddyControlType* property to *Map*.

Because this *Toolbar* control is added to handle the actions on the *Map* control, the *BuddyControlType* property was set to *Map*. If it were the

PageLaytout control, which will be discussed later, the property would be set to *PageLayout*.

7 From the Properties window, click on the *BuddyControl* property and an Ellipsis button displays. Click on the button, and then the BuddyControl Collection Editor dialog box appears (Figure 6–11). Click on Add to add *Buddy 'Map1'* as a member. Set the *Name* property to *Map1*and click on OK.

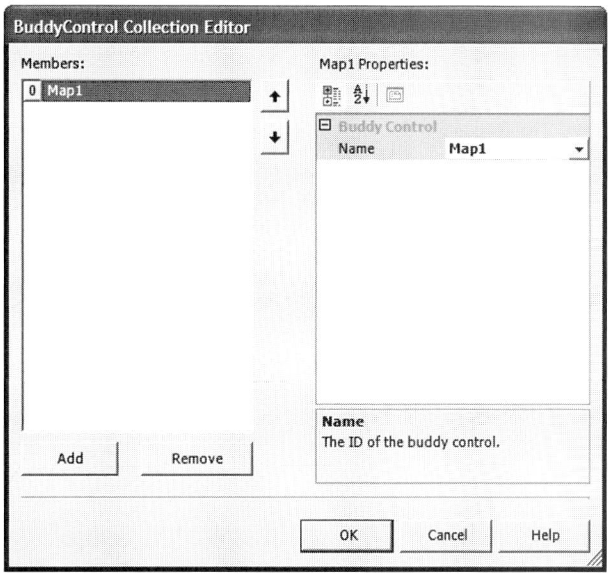

Figure 6–11. BuddyControl Collection Editor dialog.

8 From the Properties window, click on the *ToolbarItems* property and an Ellipsis button displays. Click on the button and the ToolbarItem Collection Editor dialog box appears (Figure 6–12). Click on Add to add the following as three members:

- Zoom In (tool)
- Zoom Out (tool)
- Full Extent (command)

9 Click on Add to add the "zoomin" tool. Set the *Text* property to "Zoom In" and the *ToolTip* property to "Press down left-mouse button to draw a box on the map to zoom in". Click on *ClientToolAction* and select *DragRectangle*. Set the *Name* property "zoomin", set the *ServerToolActionAssembly* to *ESRI.ArcGIS.Server.WebControls*, and set the *ServerToolActionClass* to *ESRI,ArcGIS.Server.WebControls.Tools.MapZoomIn*.

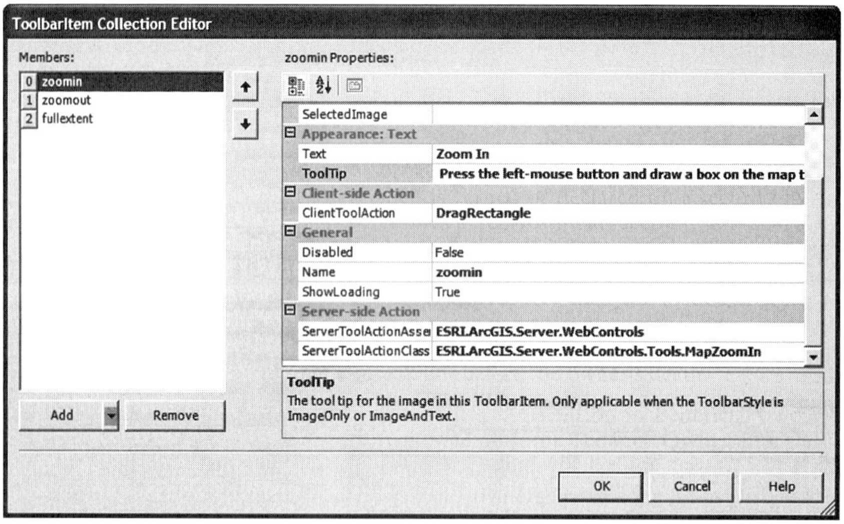

Figure 6–12. ToolbarItem Collection Editor dialog.

10 Click on Add to add the "zoomout" tool. Set the *Text* property to "Zoom Out" and the *ToolTip* property to "Press down left-mouse button to draw a box on the map to zoom out". Click on *ClientToolAction* and select *DragRectangle*. Set the *Name* property "zoomout", set the *ServerToolActionAssembly* to *ESRI.ArcGIS. Server.WebControls*, and set the *ServerToolActionClass* to *ESRI. ArcGIS.Server.WebControls. Tools.MapZoomOut*.

11 Click on the down arrow next to Add and pull down the menu and select *Command* to add a command. Set the *Text* property to "Full Extent", the *ToolTip* property to "Click this button to zoom out and see the entire USA", and the *Name* property to "fullextent".

12 Click on OK to close out the Toolbar Collection Editor.

13 Double click on the *Toolbar* control and then add the following code to *Toolbar1_CommandClick*.

```
Dim webMap As WebMap = Map1.CreateWebMap()
 Try
  webMap.DrawFullExtent()
  Toolbar1.CurrentTool = "zoomin"
 Catch ex As Exception
  webMap.Dispose()
 End Try
```

The procedure *Toolbar1_CommandClick* resets the map extent to the full extent. Moreover, it sets the *CurrentTool* property always to "zoomin", which is the default. If you do not set the *CurrentTool* property back to "zoomin", it will stay at the action right before the full extent command was clicked, which could be "zoomout".

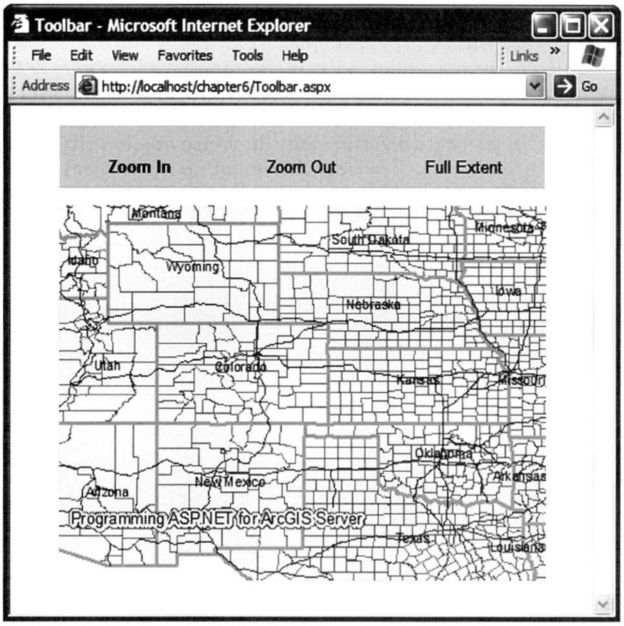

Figure 6–13. An example of using the Toolbar *control and the* Map *control.*

14 Click on the *CurrentTool* property. A drop-down list appears. Select the Zoom-In tool "zoomin" from the list.

The Zoom In tool becomes the default tool. In other words, it will be highlighted when the application is running unless the user selects the Zoom Out tool or Full Extent command.

15 Click on the Start button or press F5 to run the application.

A web page displays with the map (Figure 6–13). Click on the Zoom In tool and draw a box on the map to zoom in (Figure 6–14a). Click on the Zoom Out tool and draw a box on the map to zoom out (Figure 6–14b). Finally, click on the Full Extent command to zoom out and see the entire USA (Figure 6–15).

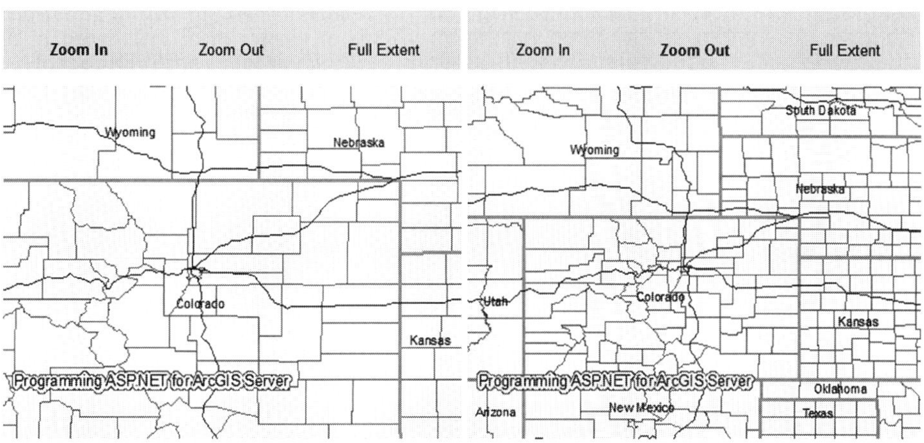

Figure 6–14. A Toolbar *control with the "Zoom In" tool highlighted (left) and "Zoom Out" tool highlighted (right).*

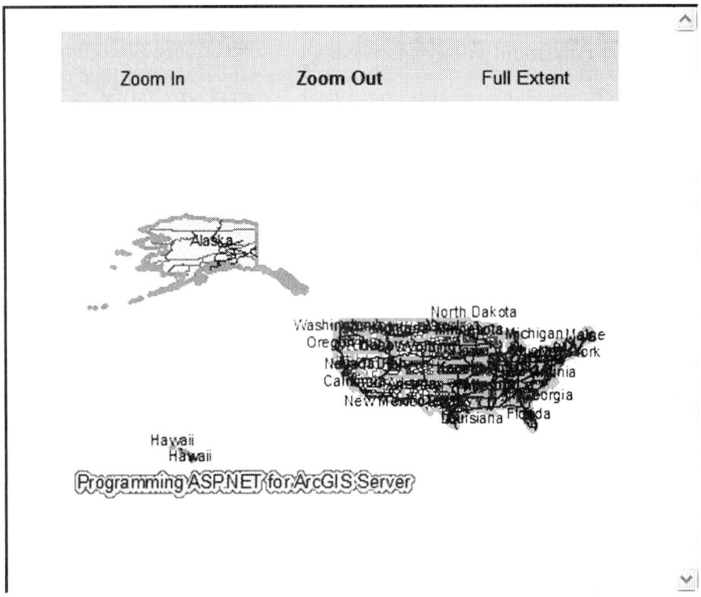

Figure 6–15. The web page after the "Full Extent" command was executed.

Listing 6–2 is the *.aspx* file for the web form with the *Toolbar* control.

Listing 6–2. *.aspx* file for the web form with the *Toolbar* control.

```
<form id="Form1" method="post" runat="server">
<esri:Map id="Map1" style="Z-INDEX: 101; LEFT: 40px;
   POSITION: absolute; TOP: 80px" runat="server"
   Height="300px" Width="400px" Host="localhost"
   ServerObject="usa" CopyrightText-Font="Arial, 10pt"
   CopyrightText-RightToLeft="False"
   CopyrightText-ForeColor="Black"
   CopyrightText-TextDirection="Horizontal"
   CopyrightText-TextAnchorPosition="BottomLeft"
   CopyrightText-TextAlignment="Left"
   CopyrightText-FillColor="Transparent"
   CopyrightText-Shadow="False" CopyrightText-Halo="True"
   CopyrightText-Text="Programming ASP.NET for ArcGIS Server">
</esri:Map>
<esri:Toolbar id="Toolbar1" style="Z-INDEX: 102; LEFT: 40px;
   POSITION: absolute; TOP: 16px" runat="server"
   CurrentTool="zoomin">
<ToolbarItems>
<esri:Tool ToolTip=" Drag a box on the map to zoom in."
   Text="Zoom In"
   ServerToolActionAssembly="ESRI.ArcGIS.Server.WebControls"
   ClientToolAction="DragRectangle"
   ServerToolActionClass="ESRI.ArcGIS.Server.WebControls.Tools.MapZoomIn"
   Name="zoomin">
```

```
</esri:Tool>
<esri:Tool ToolTip="Drag a box on the map to zoom out"
   Text="Zoom Out"
   ServerToolActionAssembly="ESRI.ArcGIS.Server.WebControls"
   ClientToolAction="DragRectangle" ServerToolAction-
   Class="ESRI.ArcGIS.Server.WebControls.Tools.MapZoomOut"
   Name="zoomout">
</esri:Tool>
<esri:Command
   ToolTip="Click this button to zoom to the full extent"
   Text="Full Extent"
   Name="fullextent">
</esri:Command>
</ToolbarItems>
<BuddyControls>
   <esri:BuddyControl Name="Map1"></esri:BuddyControl>
</BuddyControls>
</esri:Toolbar>
<esri:Impersonation id="Impersonation1"
   style="Z-INDEX: 103; LEFT: 32px;
   POSITION: absolute; TOP: 384px"
   runat="server"
   Identity="ENGMEMHMUMPMFMUNOMOMENUMPZGCWIYNIIKPKWUQWVECPZ">
</esri:Impersonation>
</form>
```

Overview Map Control

Often we would like to have a reference of where the main map displays. Such a reference can be provided by the *Overview Map* control. The overview map is also called a locator map. The *Overview Map* control is also used to render a frame of a map, which is similar to the *Map* control. However, the purpose of adding an *Overview Map* control is to provide a point of reference for the area displayed on the main *Map* control. A small box is drawn on the overview map to represent the area currently displayed on the main map. Instead of panning in the main *Map* control, you can simply move that box around inside the *Overview Map* control to pan the main map displayed in the *Map* control. The *Overview Map* control has the same convenience class, *WebMap*, as the *Map* control. In the following example, we will add an *Overview Map* control to the web form created based on the example for the *Toolbar* control.

To add a Map *control to an ASP.NET web form:*

1 Repeat steps 1 through 14 in the example of the *Toolbar* control to add an *Impersonation* control, a *Map* control, and a *Toolbar* control

with the Zoom In and Zoom Out tools and the Full Extent command, and then configure them accordingly.

2 Drag an *Overview Map* control from the ESRI tab of the toolbox onto the web form and set the *ID* property to "OverviewMap1", the *Height* property to "150px", the *Width* property to "200px, and the *Host* property to "localhost".

3 Click on the *ServerObject* property. A drop-down list displays. Select the "usa_locator" server for the *ServerObject* property.

4 Set the *BuddyControl* property to "Map1", the *BorderStyle* property to *Solid*, the *BorderWidth* property to "1px", and the *DragBoxColor* property to *Red*.

5 Click on the Start button or press F5 to run the application.

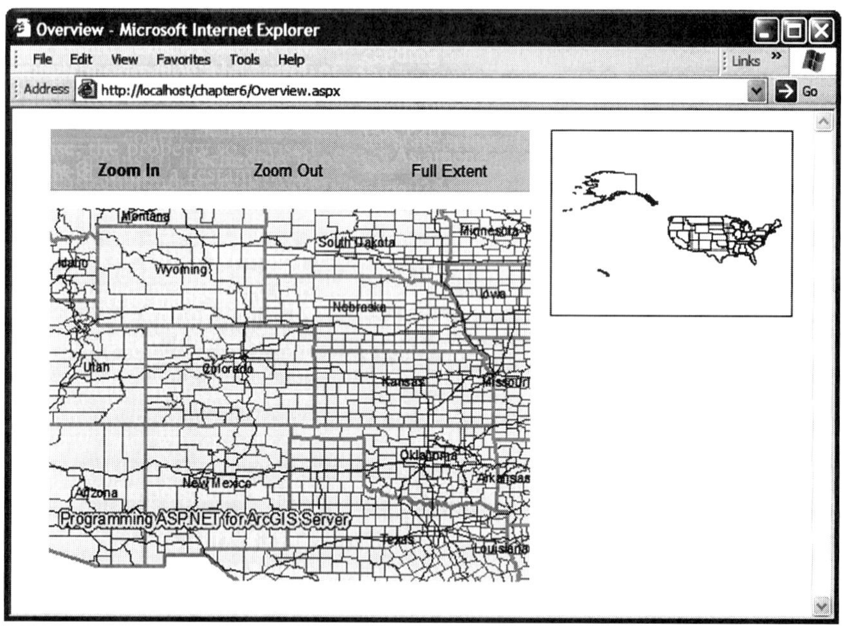

Figure 6–16. Example of using the Overview Map *control together with the* Map *control and the* Toolbar *control.*

A web page displays with the map (Figure 6–16). In addition to the main map and the toolbar, the locator map displays to the right of the map. A red box on the overview map shows the area the *Map* control displays. You can move that red box around and observe how the main map changes. They synchronize with each other. When you move the box to cover Florida, the main map shows the area with Florida in the center. When you move the box to the state of Washington, the main map shows the northwest part of the USA. The *Overview Map* control is very straightforward. Listing 6–3 is the *.aspx* file for the web form with the *Overview Map* control.

Listing 6–3. *.aspx* file for the web form with the *Overview Map* control.

```
<form id="Form1" method="post" runat="server">
<esri:Map id="Map1" style="Z-INDEX: 101; LEFT: 40px;
   POSITION: absolute; TOP: 80px" runat="server"
   Height="300px" Width="400px" Host="localhost"
   ServerObject="usa" CopyrightText-Font="Arial, 10pt"
   CopyrightText-RightToLeft="False"
   CopyrightText-ForeColor="Black"
   CopyrightText-TextDirection="Horizontal"
   CopyrightText-TextAnchorPosition="BottomLeft"
   CopyrightText-TextAlignment="Left"
   CopyrightText-FillColor="Transparent"
   CopyrightText-Shadow="False" CopyrightText-Halo="True"
   CopyrightText-Text="Programming ASP.NET for ArcGIS Server">
</esri:Map>
<esri:Toolbar id="Toolbar1" style="Z-INDEX: 102; LEFT: 40px;
   POSITION: absolute; TOP: 16px" runat="server"
   CurrentTool="zoomin">
<ToolbarItems>
<esri:Tool ToolTip=" Drag a box on the map to zoom in."
   Text="Zoom In"
   ServerToolActionAssembly="ESRI.ArcGIS.Server.WebControls"
   ClientToolAction="DragRectangle"
   ServerToolActionClass="ESRI.ArcGIS.Server.WebControls.Tools.MapZoomIn"
   Name="zoomin">
</esri:Tool>
<esri:Tool ToolTip="Drag a box on the map to zoom out"
   Text="Zoom Out"
   ServerToolActionAssembly="ESRI.ArcGIS.Server.WebControls"
   ClientToolAction="DragRectangle" ServerToolAction-
   Class="ESRI.ArcGIS.Server.WebControls.Tools.MapZoomOut"
   Name="zoomout">
</esri:Tool>
<esri:Command
   ToolTip="Click this button to zoom to the full extent"
   Text="Full Extent"
   Name="fullextent">
</esri:Command>
</ToolbarItems>
<BuddyControls>
   <esri:BuddyControl Name="Map1"></esri:BuddyControl>
</BuddyControls>
</esri:Toolbar>
```

```
<esri:Impersonation id="Impersonation1"
   style="Z-INDEX: 103; LEFT: 32px;
   POSITION: absolute; TOP: 384px"
   runat="server"
   Identity="ENGMEMHMUMPMFMUNOMOMENUMPZGCWIYNIIKPKWUQWVECPZ">
</esri:Impersonation>
<esri:OverviewMap id="OverviewMap1" style="Z-INDEX: 104;
   LEFT: 448px; POSITION: absolute; TOP: 16px"
   runat="server" Height="150px"
   Host="localhost" ServerObject="usa_locator"
   BuddyControl="Map1"
   BorderStyle="Solid" BorderWidth="1px"
   DragBoxColor="Red">
</esri:OverviewMap>
</form>
```

North Arrow Control and Scale Bar Control

Every map needs a North arrow and a scale bar. ArcGIS provides the *North Arrow* control and the *Scale Bar* control. However, these are two *Web* controls and can only be placed on the web form. Thus, if you would like a North arrow and a scale bar on your map you need to do it inside the ArcMap before publishing the service through ArcGIS Server. When you add a *North Arrow* control to the web form, you associate it with the *Map* control placed on the same web form. The *North Arrow* control will point in the north direction for the data frame in the *Map* control. The *Scale Bar* control displays the scale bar of a *Map* control. You can customize the appearance of the *Scale Bar* control, such as the font, color, or size. Both the *NorthArrow* class and the *ScaleBar* class implement all the methods they need, so there are no associated convenience classes for them.

In the following example, we will add six *North Arrow* controls and a *Scale Bar* control to the web form created based on the example for the *Toolbar* control. Each *North Arrow* control uses a different *BarStyle*.

To add six Scale Bar *controls and a* North Arrow *control to an ASP.NET web form:*

1 Repeat steps 1 through 14 in the example of the *Toolbar* control to add an *Impersonation* control, a *Map* control, and a *Toolbar* control with the Zoom In and Zoom Out tools and the Full Extent command, and then configure them accordingly.

2 Drag six *Scale Bar* controls from the ESRI tab of the toolbox onto the web form and place them on top of each other to right side of the *Map* control. Set each of the *ID* properties to "ScaleBar1", "ScaleBar2", "ScaleBar3", "ScaleBar4", "ScaleBar5", and "ScaleBar6", respectively; each of the *BarStyle* properties to *Alternating, DoubleAlternating, Hollow, ScaleLine, SingleDivision*, and *SteppedScaleLine*, respectively; each of the *BuddyControl* properties to "Map1"; each of the *Height* and *Width* properties to 25px and 250px, respectively; and each of the *BarFont/Size* and *BarHeight* properties to 20pt and 15, respectively.

3 Drag a *North Arrow* control from the ESRI tab of the toolbox onto the web form and place it under the *Scale Bar* controls. Set the *ID* property to "NorthArrow1" and the *BuddyControl* property to "Map1".

4 Click on the Start button or press F5 to run the application.

✔ **NOTE:** *The* Hollow *style for the* Scale Bar *control does not show properly at the time of writing. A bug report has been filed according to ESRI.*

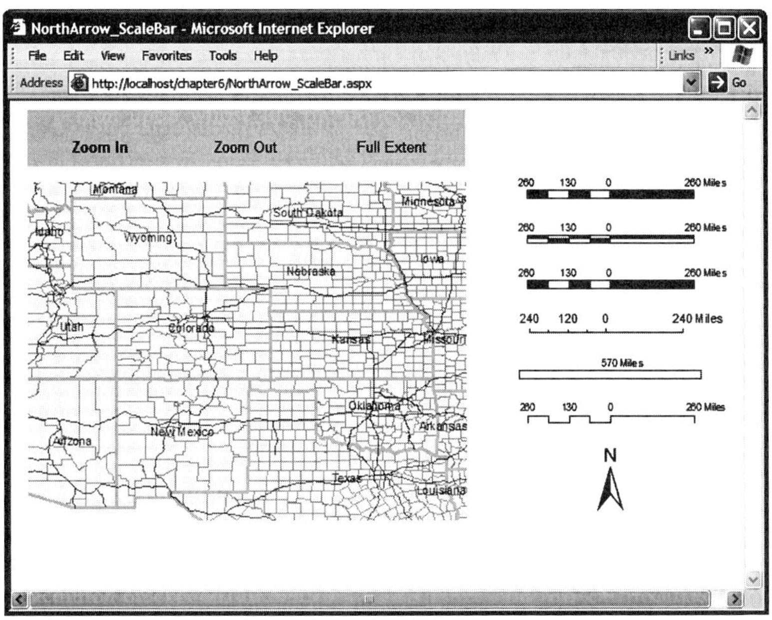

Figure 6–17. Example of using the North Arrow *control and the* Scale Bar *control.*

A web page that is similar to that shown in Figure 6–13 displays with the map with a North arrow and six different styles of scale bars (Figure 6–17). A scale bar can be one segment or multiple segments. You can control the number of segments by setting the *Divisions* property.

Within each division, you can have subdivisions via the *SubDivisions* property. You can also specify whether there is any division before the zero value by setting the *DivisionsBeforeZero* property. For example, the first *Scale Bar* control was set to 2 divisions, 4 subdivisions, and 1 division before the zero value. Listing 6–4 is the *.aspx* file for the web form with six *Scale Bar* controls and a *North Arrow* control.

Listing 6-4. *.aspx* file for the web form with six *Scale Bar* controls and a *North Arrow* control.

```
<form id="Form1" method="post" runat="server">
<esri:Map id="Map1" style="Z-INDEX: 101; LEFT: 16px;
   POSITION: absolute; TOP: 72px" runat="server"
   Height="300px" Width="400px" Host="localhost"
   ServerObject="usa"></esri:Map>
<esri:Toolbar id="Toolbar1" style="Z-INDEX: 102; LEFT: 16px;
   POSITION: absolute; TOP: 8px" runat="server"
   CurrentTool="zoomin">
<ToolbarItems>
<esri:Tool ToolTip="Drag a box to zoom in" Text="Zoom In"
   ServerToolActionAssembly="ESRI.ArcGIS.Server.WebControls"
   ClientToolAction="DragRectangle"
   ServerToolActionClass="ESRI.ArcGIS.Server.WebControls.Tools.MapZoomIn"
   Name="zoomin"></esri:Tool>
<esri:Tool ToolTip="Drag a box to zoom out" Text="Zoom Out"
   ServerToolActionAssembly="ESRI.ArcGIS.Server.WebControls"
   ClientToolAction="DragRectangle" ServerToolAction-
   Class="ESRI.ArcGIS.Server.WebControls.Tools.MapZoomOut"
   Name="zoomout"></esri:Tool>
<esri:Command ToolTip="Zoom out to see USA"
   Text="Full Extent" Name="fullextent">
</esri:Command>
</ToolbarItems>
<BuddyControls>
   <esri:BuddyControl Name="Map1"></esri:BuddyControl>
</BuddyControls>
</esri:Toolbar>
<esri:Impersonation id="Impersonation1"
   style="Z-INDEX: 103; LEFT: 8px; POSITION: absolute; TOP: 376px"
   runat="server" Identity="ENGMEMHMUMPMFMUNOMOMENUMPZGCWIYNIIKPKWUQWVECPZ"
   Width="96px" Height="12px"></esri:Impersonation>
<esri:NorthArrow id="NorthArrow1"
   style="Z-INDEX: 104; LEFT: 496px; POSITION: absolute; TOP: 304px"
   runat="server" BuddyControl="Map1" Height="64px">
</esri:NorthArrow>
<esri:ScaleBar id="ScaleBar1" style="Z-INDEX: 105; LEFT: 432px;
   POSITION: absolute; TOP: 64px"
   runat="server" BuddyControl="Map1"
   DivisionsBeforeZero="1" Width="250px" Height="25px"
```

```
   BarStyle="Alternating"
   BarFont="Arial, 20.25pt" BarHeight="15"
   SubDivisions="4" Divisions="2"></esri:ScaleBar>
<esri:ScaleBar id="ScaleBar2" style="Z-INDEX: 106; LEFT: 432px;
   POSITION: absolute; TOP: 104px"
   runat="server" Width="250px" Height="25px"
   BuddyControl="Map1" BarStyle="DoubleAlternating"
   BarFont="Arial, 20.25pt" BarHeight="15"
   SubDivisions="4" Divisions="2">
</esri:ScaleBar>
<esri:ScaleBar id="ScaleBar3" style="Z-INDEX: 107; LEFT: 432px;
   POSITION: absolute; TOP: 144px"
   runat="server" Width="250px" Height="25px"
   BuddyControl="Map1" DivisionsBeforeZero="1"
   BarStyle="Hollow"
   BarFont="Arial, 20.25pt" BarHeight="15"
   SubDivisions="4" Divisions="2"></esri:ScaleBar>
<esri:ScaleBar id="ScaleBar4" style="Z-INDEX: 108; LEFT: 432px;
   POSITION: absolute; TOP: 184px"
   runat="server" Width="250px" Height="25px"
   BuddyControl="Map1" BarStyle="ScaleLine"
   BarFont="Arial, 20.25pt"
   BarHeight="15" SubDivisions="4"
   Divisions="2"></esri:ScaleBar>
<esri:ScaleBar id="ScaleBar5" style="Z-INDEX: 109; LEFT: 432px;
   POSITION: absolute; TOP: 224px"
   runat="server" Width="250px" Height="25px"
   BuddyControl="Map1" DivisionsBeforeZero="1"
   BarStyle="SingleDivision"
   BarFont="Arial, 20.25pt" BarHeight="15"
   SubDivisions="0" Divisions="1"></esri:ScaleBar>
<esri:ScaleBar id="ScaleBar6" style="Z-INDEX: 110; LEFT: 432px;
   POSITION: absolute; TOP: 264px"
   runat="server" Width="250px" Height="25px"
   BuddyControl="Map1" BarStyle="SteppedScaleLine"
   BarFont="Arial, 20.25pt" BarHeight="15"
   SubDivisions="4" Divisions="2"></esri:ScaleBar>
</form>
```

Table of Contents Control

The *Table of Contents* control is basically a layer selection control. With the control, you can turn on/off layers interactively. It is equivalent to the table of contents you see in ArcView and ArcMap. The *Table of Contents* control lists each layer on the map with a checkbox, a layer name, and the symbology that depicts the layer. The control also has an *AutoPostBack* property, which is set to *True* by default. Both the Map Viewer template and Page Layout template set that property to *True*. There is a drawback with that setting. Every time a layer in the table of contents is checked, the map will be refreshed, which is okay with desktop implementation. But it is not so economical in terms of network traffic with the web implementation. We suggest that you set the *AutoPostBack* property to *False* by default and change the default value (by programming) as needed. You can choose to show all data frames in the map server object within a *Table of Contents* control or just the data frame being displayed in the main *Map* control. A table of contents is linked to a particular *Map* control or *Page Layout* control. The convenience class for the *Table of Contents* control is *WebToc*.

In the following example, we will add a *Table of Contents* control based on the example for the *Toolbar* control. We will disable the *AutoPostBack* and add a Redraw button to control the *PostBack*.

To add a Table of Contents *control to an ASP.NET web form:*

1 Repeat steps 1 through 14 in the example of the *Toolbar* control to add an *Impersonation* control, a *Map* control, and a *Toolbar* control with the Zoom In and Zoom Out tools and the Full Extent command, and then configure them accordingly.

2 Drag a *Table of Contents* control from the ESRI tab of the toolbox onto the web form, and place it to right side of the *Map* control. Set the *ID* property to "Toc1", the *Height* property to "270px", the *Width* property to "200px, the *BuddyControl* property to "Map1", the *AutoPostBack* property to *False*, the *BorderColor* property to *SteelBlue*, the *BorderStyle* property to *Solid*, and the *BorderWidth* property to 1.

3 Drag a Button control from the Web Forms tab of the toolbox onto the web form, and set the ID property to "Button1" and the Text property to "Redraw Map".

4 Double click on the button and add the following code to the `Button1_Click` event to control the *PostBack* for the *Table of Contents* control.

```
Dim webMap As WebMap = Map1.CreateWebMap()
webMap.Refresh()
```

5 Click on the Start button or press F5 to run the application.

A web page (Figure 6–18) that is similar to one shown in Figure 6–13 displays, but with a table of contents to the right. Activate the *Highways* layer by clicking on the checkbox, and you will notice that nothing happens to the map because we disabled the *AutoPostBack* property.

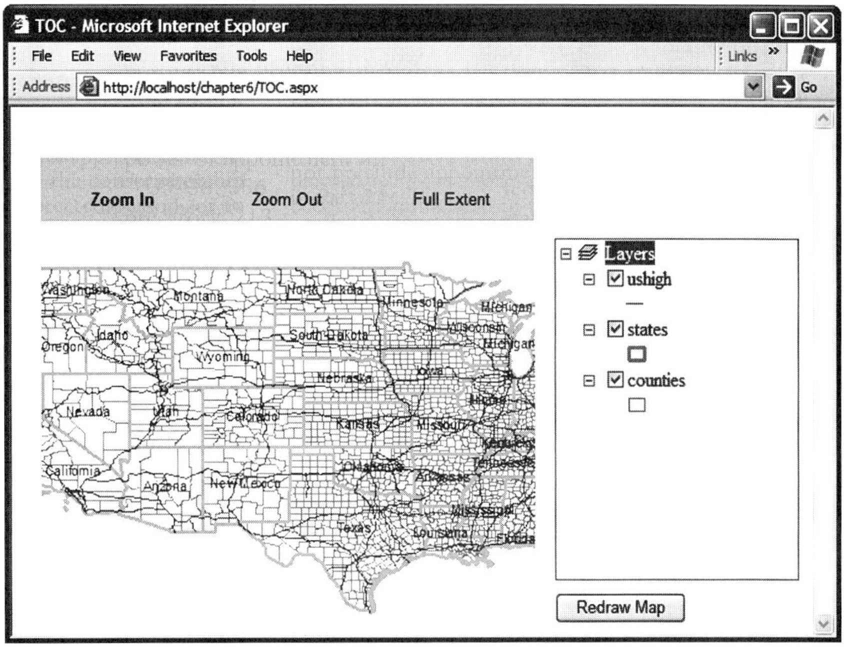

Figure 6–18. Example of using the Table of Contents *control with* AutoPostBack *set to False.*

If you click on the Redraw Map button, the map will be refreshed with only the *States* and *Counties* layers shown. Thus, the code added to the `Button1_Click` event manually enabled the *PostBack* for the *Table of Contents* control, which was exactly what we wanted. With that implementation, if you happen to turn on or off a layer by a mistake you will have a second chance to correct yourself before it takes effect. Also, if you need to turn on and off multiple layers you will have a freedom to do so. Listing 6–5 is the *.aspx* file for the web form with a *Table of Contents* control.

Listing 6–5. *.aspx* file for the web form with a *Table of Contents* control.

```
<form id="Form1" method="post" runat="server">
<esri:Map id="Map1" style="Z-INDEX: 101; LEFT: 40px;
   POSITION: absolute; TOP: 80px" runat="server"
   Height="300px" Width="400px" Host="localhost"
   ServerObject="usa" CopyrightText-Font="Arial, 10pt"
   CopyrightText-RightToLeft="False"
```

```
   CopyrightText-ForeColor="Black"
   CopyrightText-TextDirection="Horizontal"
   CopyrightText-TextAnchorPosition="BottomLeft"
   CopyrightText-TextAlignment="Left"
   CopyrightText-FillColor="Transparent"
   CopyrightText-Shadow="False" CopyrightText-Halo="True"
   CopyrightText-Text="Programming ASP.NET for ArcGIS Server">
</esri:Map>
<esri:Toolbar id="Toolbar1" style="Z-INDEX: 102; LEFT: 40px;
   POSITION: absolute; TOP: 16px" runat="server"
   CurrentTool="zoomin">
<ToolbarItems>
<esri:Tool ToolTip=" Drag a box on the map to zoom in."
   Text="Zoom In"
   ServerToolActionAssembly="ESRI.ArcGIS.Server.WebControls"
   ClientToolAction="DragRectangle"
   ServerToolActionClass="ESRI.ArcGIS.Server.WebControls.Tools.MapZoomIn"
   Name="zoomin">
</esri:Tool>
<esri:Tool ToolTip="Drag a box on the map to zoom out" \
   Text="Zoom Out"
   ServerToolActionAssembly="ESRI.ArcGIS.Server.WebControls"
   ClientToolAction="DragRectangle" ServerToolAction-
   Class="ESRI.ArcGIS.Server.WebControls.Tools.MapZoomOut"
   Name="zoomout">
</esri:Tool>
<esri:Command
   ToolTip="Click this button to zoom to the full extent"
   Text="Full Extent"
   Name="fullextent">
</esri:Command>
</ToolbarItems>
<BuddyControls>
   <esri:BuddyControl Name="Map1"></esri:BuddyControl>
</BuddyControls>
</esri:Toolbar>
<esri:Impersonation id="Impersonation1"
   style="Z-INDEX: 103; LEFT: 32px;
   POSITION: absolute; TOP: 384px"
   runat="server"
   Identity="ENGMEMHMUMPMFMUNOMOMENUMPZGCWIYNIIKPKWUQWVECPZ">
</esri:Impersonation>
<esri:Toc id="Toc1" style="Z-INDEX: 104; LEFT: 440px;
   POSITION: absolute; TOP: 104px" runat="server"
   Width="200px" Height="270px" BuddyControl="Map1"
   BorderColor="SteelBlue" BorderStyle="Solid"
   BorderWidth="1px" AutoPostBack="False"
```

```
   ShowAllDataFrames="False"></esri:Toc>
<asp:Button id="Button1" style="Z-INDEX: 105; LEFT: 440px;
   POSITION: absolute; TOP: 384px" runat="server"
   Text="Redraw Map"></asp:Button>
</form>
```

Page Layout Control

The *Page Layout* control is used to display the layout view of a map. If there is no layout view set through the map server, the control displays the data view of the map. The page layout has a similar concept as the page layout in any word processor. In many cases, we found out the map we picked up from the printer was not exactly what we had in mind. We used to go back and try to adjust it in a few iterations. With the *Page Layout* control, we are provided a WYSIWYG capability to have a map printout to show the exact content we have in mind. The *Page Layout* control is similar to the *Map* control. There are the methods that are offered to zoom and pan on the page layout. There are also different methods that are provided to just zoom and pan the map portion of the layout. The convenience class for the *Page Layout* control is *WebPageLayout*. The methods available in this class are similar to those in the *WebMap* class.

In the following example, we will add a *Page Layout* control to the web form. In addition, we will add a *Toolbar* control with two Zoom In and Zoom Out tools and one command of the Full Extent of the page to the web form.

To add a Page Layout *control to an ASP.NET web form:*

1. Drag a *Page Layout* control from the ESRI tab of the toolbox onto the web form and set the *ID* property to "PageLaout1", the *Height* property to "400px", the *Width* property to "400px, and the *Host* property to "localhost".
2. Click on the *ServerObject* property. A drop-down list displays. Select the "usa" server for the *ServerObject* property.
3. Drag and drop an *Impersonation* control from the ESRI tab of the toolbox onto the web form.
4. From the Properties window, click on the *Identity* property and an Ellipsis button displays. Click on the button and the Impersonation1 Identity dialog box appears. Enter a valid user name and password and click on OK. You can leave the "Domain or Machine Name" box blank.

5 Drag a *Toolbar* control from the ESRI tab of the toolbox onto the web form and place it on top of the *Page Layout* control. Set the *ID* property to "Toolbar1" and the *BuddyControlType* property to *PageLayout*.

6 From the Properties window, click on the *BuddyControls* property and an Ellipsis button displays. Click on the button and then the BuddyControl Collection Editor dialog box appears (Figure 6–19). Click on Add to add Buddy 'PageLayout1' as a member, set the *Name* property to *PageLayout1*, and then click on OK.

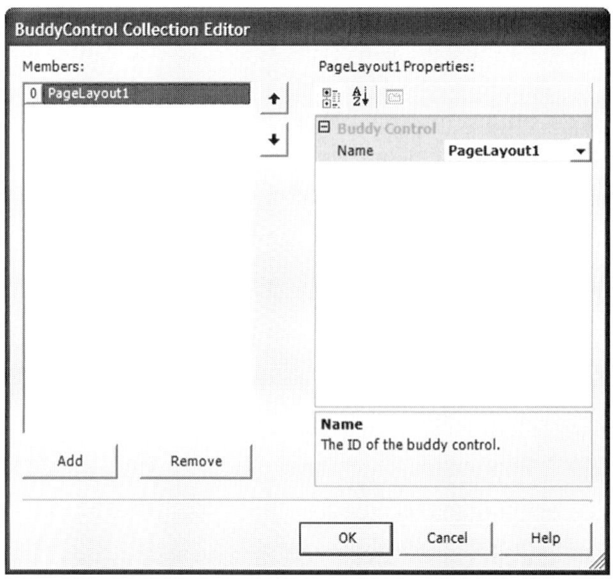

Figure 6–19. BuddyControl Collection Editor dialog.

7 From the Properties window, click on the *ToolbarItems* property and an Ellipsis button displays. Click on the button and then the ToolbarItem Collection Editor dialog box appears. Click on Add to add the following as three members:

- Zoom In (tool)
- Zoom Out (tool)
- Full Extent (command)

8 Click on Add to add the "pagein" tool. Set the *Text* property to "Zoom In" and set the *ToolTip* properties to "Press down left mouse button to draw a box on the map to zoom in". Click on *ClientToolAction* and select *DragRectangle*, set the *Name* property "pagein", set the *ServerToolActionAssembly* to *ESRI.ArcGIS.Server.WebControls*, and set the *ServerToolActionClass* to *ESRI,ArcGIS.Server.WebControls.Tools PageZoomIn'*.

9 Click on Add to add the "pageout" tool. Set the *Text* property to "Zoom Out" and set the *ToolTip* properties to "Press down left mouse button to draw a rectangle and then release it". Click on *ClientToolAction* and select *DragRectangle*, set the *Name* property "zoomout", set the *ServerToolActionAssembly* to *ESRI.ArcGIS.Server.WebControls*, and set the *ServerToolActionClass* to *ESRI.ArcGIS.Server.WebControls.Tools.PageZoomOut*.

10 Click on the down arrow next to Add and pull down the menu and select Command to add a command. Set the *Text* property to "Full Extent", the *ToolTip* property to "Click this button to see the entire page layout", and the *Name* property to "pagefullextent".

11 Click on OK to close out the Toolbar Collection Editor.

12 Double click on the *Toolbar* control and then add the following code to *Toolbar1_CommandClick*.

```
Dim pageLayout As WebPageLayout = PageLayout1.CreateWebPageLayout()
  Try
   pageLayout.PageDrawFullExtent()
   Toolbar1.CurrentTool = "pagein"
  Catch ex As Exception
   pageLayout.Dispose()
  End Try
```

The procedure *Toolbar1_CommandClick* reset the page layout extent to the full extent. Moreover, it sets the *CurrentTool* property always to the "pagein", which is the default. If you do not set the *CurrentTool* property back to "pagein", it will stay at the action right before the full page layout extent command was clicked, which could be "pageout".

13 Click on the *CurrentTool* property. A drop-down list appears. Select the page Zoom In tool "pagein" from the list.

The page Zoom In tool becomes the default tool. In other words, it will be highlighted when the application is running unless the user selects the page Zoom Out tool or page Full Extent command.

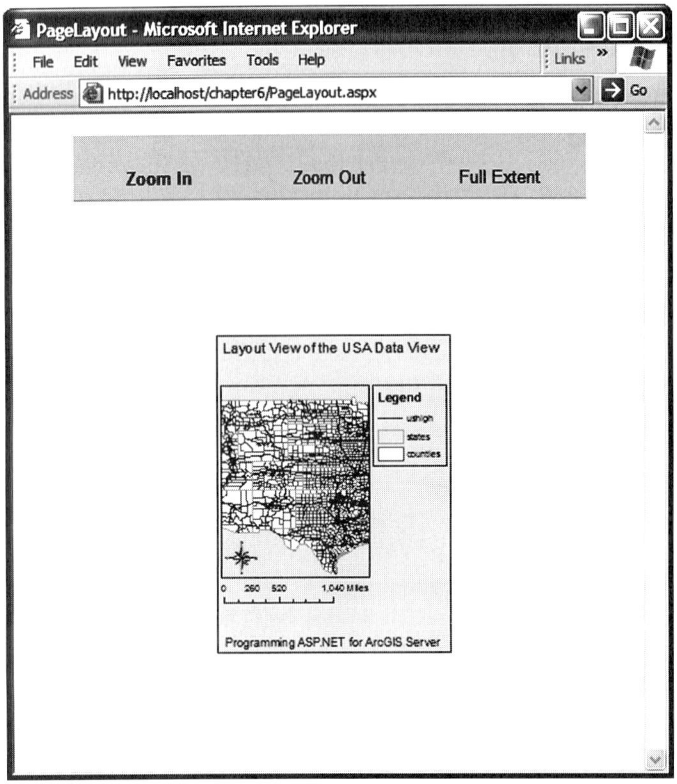

Figure 6–20. Example of using the Page Layout *control.*

14 Click on the Start button or press F5 to run the application.

A web page displays with the map (Figure 6–20). Click on the Zoom In tool and draw a box on the page layout to zoom in (Figure 6–21). Click on the Zoom Out tool and draw a box on the page layout to zoom out. Finally, click on the Full Extent command to zoom out and see the page layout in the full extent of the page (Figure 6–22). The full extent of the page is different from the full extent of the map. The extent of the page deals with the page layout, possibly including the map, overview map, legend, scale bar, North arrow, title, logo, and other text. When you zoom and pan of the page, the map extent does not change at all. From steps 8 through 12, the *ServerToolActionClass* and *WebPageLayout* methods that were used were *PageZoomIn*, *PageZoomOut*, and *PageDrawFullExtent*. If you need to change or adjust the map extent, you need to use those methods that are specifically targeted at the map on the page layout, such as *PageMapZoomIn* and *PageMapZoomOut* out of the *ServerToolActionClass* and *MapDrawFullExtent* out of *WebPageLayout*.

In general, if you use the tools from the *Toolbar* control you use the *ServerToolActionClass*. If you use the commands, you need to add your own code that is most likely from the *WebPageLayout* convenience class. Luckily enough there is minimum chance you would be confused by which one you should use because Visual Studio .NET offers you Intellisense on all classes and the ArcGIS Server ADF for .NET provides you the editor dialogs. Listing 6–6 is the *.aspx* file for the web form with a *Page Layout* control.

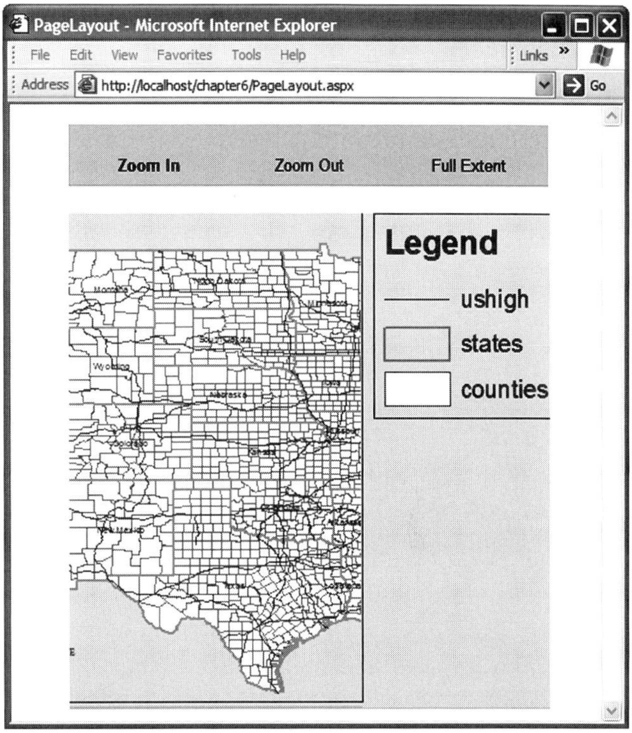

Figure 6–21. Example of using the Page Layout *control after the "Zoom In" tool was applied.*

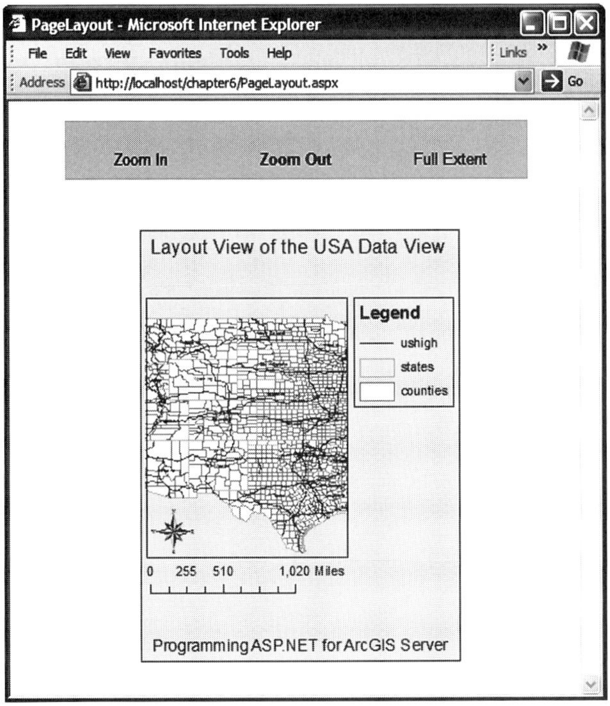

Figure 6–22. Example of using the Page Layout *control after the "Zoom Out" tool was applied.*

Listing 6–6. *.aspx* file for the web form with a *Page Layout* control.

```
<form id="Form1" method="post" runat="server">
<esri:pagelayout id="PageLayout1" style="Z-INDEX: 101; LEFT: 48px;
   POSITION: absolute; TOP: 88px"
   runat="server" CopyrightText-Halo="False"
   CopyrightText-Shadow="False"
   CopyrightText-FillColor="Transparent"
   CopyrightText-TextAlignment="Left"
   CopyrightText-TextAnchorPosition="BottomLeft"
   CopyrightText-TextDirection="Horizontal"
   CopyrightText-ForeColor="Black"
   CopyrightText-RightToLeft="False"
   CopyrightText-Font="Arial, 10pt"
   ServerObject="usa" Host="localhost"
   Width="400px" Height="400px"></esri:pagelayout>
<esri:impersonation id="Impersonation1"
   style="Z-INDEX: 102; LEFT: 488px;
   POSITION: absolute; TOP: 24px"
   runat="server"
   Identity="ENGMEMHMUMPMFMUNOMOMENUMPZGCWIYNIIKPKWUQWVECPZ">
</esri:impersonation>
<esri:toolbar id="Toolbar1" style="Z-INDEX: 103; LEFT: 48px;
   POSITION: absolute; TOP: 16px" runat="server"
   CurrentTool="pagein" BuddyControlType="PageLayout">
<ToolbarItems>
<esri:Tool ToolTip="Drag a box to zoom in " Text="Zoom In"
   ServerToolActionAssembly="ESRI.ArcGIS.Server.WebControls"
   ClientToolAction="DragRectangle"
   ServerToolActionClass="ESRI.ArcGIS.Server.WebControls.Tools.PageZoomIn"
   Name="pagein"></esri:Tool>
<esri:Tool ToolTip="Drag a box to zoom out" Text="Zoom Out"
   ServerToolActionAssembly="ESRI.ArcGIS.Server.WebControls"
   ClientToolAction="DragRectangle"
   ServerToolActionClass="ESRI.ArcGIS.Server.WebControls.Tools.PageZoomOut"
   Name="pageout"></esri:Tool>
<esri:Command ToolTip="Click" Text="Full Extent"
   Name="pagefullextent"></esri:Command>
</ToolbarItems>
<BuddyControls>
   <esri:BuddyControl Name="PageLayout1"></esri:BuddyControl>
</BuddyControls>
</esri:toolbar>
</form>
```

Creating Web Applications Directly with the ArcGIS Server Web Controls

Now we will build a geocoding application using the web controls discussed in Chapter 5 and the ArcGIS Server controls discussed earlier in this chapter. For the simplicity of discussion, we will use a road data set for Arlington County, Virginia. It is the TIGER Line streets from the 2002 Census. ESRI has made all TIGER Lines available from their web site for download. If you need other county data, you can download them from the ESRI web site. We need to do two things before we start developing this web application. First, we need to build an address locator. It is essentially a geocode server object you can build with ArcCatalog. Second, we need to build an ArcGIS project for Arlington County, Virginia. We will illustrate how to create an address locator using ArcCatalog.

In this application, we will also use the *GeocodeConnection* component provided in ArcGIS Server. The *GeocodeConnection* component is used to create geocoding (also called address matching) applications. This component simplifies access to geocode server objects. The *GeocodeConnection* component is not a *Web* control and thus does not have a visual display on a web form. It is derived from *System.ComponentModel.Component* instead of *System.Web.UI* from which the *Web* controls come. You specify what GIS server you want to connect to and what geocode server object you want to access through this component. The convenience class for the *GeocodeConnection* component is *WebGeocode*, which is used to provide geocoding-related properties and methods.

Create an Address Locator—Geocode Server Object

You need to set up a geocoding server object before you can start building a geocoding ASP.NET application. A geocode server object is essentially an address locator. You create an address locator in ArcCatalog. See Chapter 4 for details on how to create an address locator and set up a geocode server object. We will use the geocode server object created in Chapter 4, "ArlingtonAddress", for the application.

Build an ASP.NET Application with GeocodeConnection Component

In the following example, we will add a *GeocodeConnection* component to the web form. In addition, we will add an *Impersonation* control, a *Map* control, a *TextBox* control for the street address, two *DataGrid* controls, and a *Button* control.

To add a GecodeConnection *component to an ASP.NET web form:*

1 Drag a *GeocodeConnection* component from the ESRI tab of the toolbox onto the web form and set the *Host* property to "localhost" and the *ShowAllCandidates* property to *True*.

2 Click on the *ServerObject* property. A drop-down list displays. Select the "ArlingtonLocator" server for the *ServerObject* property.

3 Drag a *Label* control from the ESRI tab of the toolbox onto the web form and set the *Text* property to "Streets".

4 Drag a *TextBox* control from the ESRI tab of the toolbox onto the web form and place it next to the *Label* control. Set the *ID* property to "txtAddress".

5 Drag a *Button* control from the ESRI tab of the toolbox onto the web form and place it next to the *TextBox* control. Set the *Text* property to "Find It".

6 Drag a *DataGrid* control from the ESRI tab of the toolbox onto the web form and place it under the *Label* control. Set the *Font/Name* property to *Arial*, the *AlternatingItemStyle/BackColor* property to "#C0C0FF", the *HeaderStyle/BackColor* property to "#E0E0E0", and the *HeaderStyle/Font/Name* property to *Arial*.

7 Drag another *DataGrid* control from the ESRI tab of the toolbox onto the web form and place it beneath the first *DataGrid* control. Set the *Font/Name* property to *Arial*, the *AlternatingItemStyle/BackColor* property to "#C0C0FF", the *HeaderStyle/BackColor* to "#E0E0E0", and the *HeaderStyle/Font/Name* to *Arial*.

8 Add the following code to the *Button1_Click* event.

```
Dim geocodeconn As WebGeocode = GeocodeConnection1.CreateWebGeocode
Try
  Dim context As IServerContext = geocodeconn.ServerContext
```

```
  Dim pset As IPropertySet = _
            context.CreateObject("esriSystem.PropertySet")
  pset.SetProperty("Street", TxtAddress.Text)
  Dim dataset1 As System.Data.DataSet = _
  geocodeconn.FindAddressCandidates(pset, Nothing, False, False)
  If Not (dataset1 Is Nothing) Then
   DataGrid1.DataSource = dataset1
   DataGrid1.DataBind()
  End If
  Dim dataset2 As System.Data.DataSet = geocodeconn.GetCandidateFields
  If Not (dataset2 Is Nothing) Then
   DataGrid2.DataSource = dataset2
   DataGrid2.DataBind()
  End If
Finally
  CType(geocodeconn, IDisposable).Dispose()
End Try
```

9 Click on the Start button or press F5 to run the application.

A web page with the *GeocodeConnection* component and other controls displays. Enter an address in the Address textbox and click on the Find It button. The web page will be refreshed to show the geocoding results (Figure 6–23).

The *GeocodeConnection* component was used to allow you to match addresses using a geocode server object. Two methods of *FindAddress-Candidates* and *GetCandidateFields* of the *WebGeocode* object were used to obtain candidate addresses and fields. The first *DataGrid* control was used to display candidate addresses, and the other *DataGrid* control was used to list candidate fields. The component also has two properties of *MininumCandidateScore,* and *MinimumMatchScore* to control how the address will be matched. *MininumCandidateScore* sets the minimum candidate score set on the *GeocodeServer* results, whereas *Minimum-MatchScore* defines the minimum match score set on the *GeocodeServer* results.

So far, so good. But how can we display the geocoding result on the map? We will add a *Map* control to the application. To display an address location on a map, we need to use a map server object. We need to pass the coordinates of the geocoded location from the geocoding server object to the map server object. We will use a marker symbol on the map to show the geocoded address location. The map server is called “Arlington_rds”.

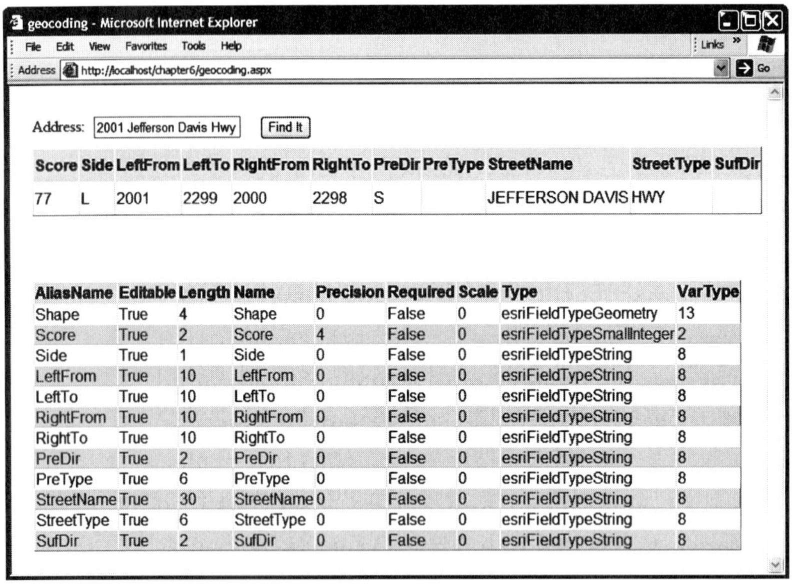

Figure 6–23. Example of using the GecodeConnection *component for a web page.*

10 Drag a *Map* control from the ESRI tab of the toolbox onto the web form and place it under the first *DataGrid* control. Set the *ID* property to "map1", the *Height* property to "200px", the *Width* property to "300px, the *Host* property to "localhost", and the *Visible* property to *False*. The *Map* control will be turned visible when a match is found.

11 Replace the *Button1_Click* event with the following.

```
Dim geocodeconn As WebGeocode = GeocodeConnection1.CreateWebGeocode
Try
  Dim context As IServerContext = geocodeconn.ServerContext
  Dim pset As IPropertySet = _
           context.CreateObject("esriSystem.PropertySet")
  pset.SetProperty("Street", TxtAddress.Text)
  Dim dataset1 As System.Data.DataSet = _
  geocodeconn.FindAddressCandidates(pset, Nothing, False, False)
  If Not (dataset1 Is Nothing) Then
   DataGrid1.DataSource = dataset1
   DataGrid1.DataBind()
  End If
  Dim dataset2 As System.Data.DataSet = geocodeconn.GetCandidateFields
  If Not (dataset2 Is Nothing) Then
   DataGrid2.DataSource = dataset2
```

```
    DataGrid2.DataBind()
   End If
   Dim pt As IPoint = context.CreateObject("esriGeometry.Point")
   pt = geocodeconn.GeocodeAddress(pset, Nothing)
   If (pt.IsEmpty = False) Then
    Map1.Visible = True
    Dim webMap As WebMap = Map1.CreateWebMap()
    Try
     ShowAddress(webMap, pt)
     webMap.Zoom(0.05, pt)
     webMap.Refresh()
    Finally
     CType(webMap, IDisposable).Dispose()
    End Try
   End If
Finally
   CType(geocodeconn, IDisposable).Dispose()
End Try
```

12 Create a new subroutine and call it `ShowAddress`.

```
Public Sub DrawAddress(ByVal webmap As WebMap, ByVal pt As IPoint)
 Dim svrContext As IServerContext = webmap.ServerContext
 Dim mapDes As IMapDescription = webmap.MapDescription
 Dim point As IPoint = svrContext.CreateObject("esriGeometry.Point")
 Dim geometry As IGeometry = point
 geometry.SpatialReference = webmap.SpatialReference
 point.PutCoords(pt.X, pt.Y)
 point.Project(mapDes.SpatialReference)
 If point Is Nothing Then
   Return
 End If
 Dim graphicElements As IGraphicElements = _
   svrContext.CreateObject("esriCarto.GraphicElements")
 Dim rgbColor As IRgbColor = Converter.ToRGBColor(svrContext, _
   System.Drawing.Color.Red)
 Dim simpleMarkerSymbol1 As ISimpleMarkerSymbol = _
   svrContext.CreateObject("esriDisplay.SimpleMarkerSymbol")
 simpleMarkerSymbol1.Color = rgbColor
 simpleMarkerSymbol1.Style = esriSimpleMarkerStyle.esriSMSCircle
 simpleMarkerSymbol1.Size = 9
 Dim markerElement1 As IMarkerElement = _
   svrContext.CreateObject("esriCarto.MarkerElement")
 markerElement1.Symbol = simpleMarkerSymbol1
 Dim pointElement1 As IElement = markerElement1
 pointElement1.Geometry = point
 Dim markerGraphicElement1 As IGraphicElement = pointElement1
 graphicElements.Add(markerGraphicElement1)
```

```
  mapDes.CustomGraphics = graphicElements
End Sub
```

A simple marker symbol was used to show the geocoded address on the map and it was set to red.

13 Click on the Start button or press F5 to run the application.

A web page with an address textbox displays. For example, enter *"2001 Jefferson Davis Hwy"* and then click the Find It button. A match will be found, as shown in Figure 6–24. A red circle shows the location of the address entered. This application can be easily modified to have a similar look-and-feel to one built based on the Geocoding template. Instead, we went step by step to show how a geocoding application can be built from scratch using the *GeocodeConnection* component and *Web* controls.

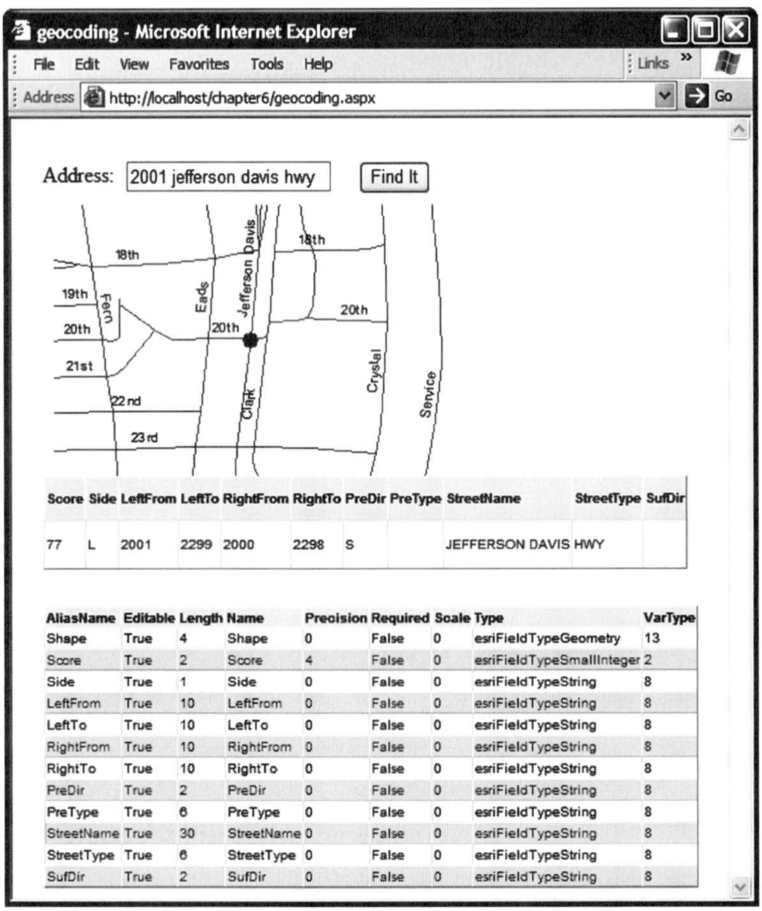

Figure 6-24. Example of using the Map *control and the* GeocodeConnection *component.*

Summary

In this chapter, we have introduced the ArcGIS Server templates that can be used within Visual Studio .NET. We built a simple, straightforward web application based on the Thematic Map template. We then discussed in detail the ArcGIS Server web controls. The first ArcGIS Server web control we discussed was the *Impersonation* control. It is simply a web control, but a certain amount attention is needed otherwise you could be very frustrated. Some of potential errors you would get could drive you totally crazy. We went on to discuss other web controls one by one and created a web application based on each of them. We want to emphasize again that you should disable the *AutoPostBack* on the *Table of Contents* control. Instead, you could add another button to trigger the *PostBack* on that control. Finally, we built a web application using the *GecodeConnection* component. It is not a web control but it is convenient for building web applications with geocoding functionality. Geocoding runs through a geocode server object on an ArcGIS Server instead of a regular map server object. An address locator has to be set up in ArcCatalog before a geocode server object can be created. A geocoded address (a pair of coordinates) has to be transferred from the *GeocodeConnection* component to a *Map* control before it can be mapped.

Considering the nature of additions, changes, and updates to the ArcGIS Server ADF for .NET, the ADF .NET Object Model was not included in this chapter. You can go online and visit the ArcGIS Developer Online web site at *http://arcgisdeveloperonline.esri.com* to obtain the latest edition of the ADF .NET Object Model diagram. A poster to hang up in your office would come in handy.

Like regular web controls, the ArcGIS Server web controls can also be extended and customized. For example, you may want to create a web control that can convert a pair of latitude and longitude coordinates from decimal degrees to degrees, minutes, and seconds, or vice versa. Also, in situations in which you want to enable others to use your page you would need to convert that page into a user control. In the next chapter, we address how to extend and customize web controls, including the ArcGIS Server web controls.

CHAPTER 7

Extending and Customizing Web Controls and Web Templates

THE WEB CONTROLS DISCUSSED IN CHAPTER 5 and the ArcGIS Server web controls and templates discussed in Chapter 6 are sufficient to help you build sophisticated web-based GIS applications and web services. You can drag and drop controls onto a web form and write handler events all without leaving Visual Studio .NET. After you develop a few web applications, you will start thinking whether you can reuse some of the development you did in the previous development. Luckily enough, ASP.NET does let you encapsulate and reuse collections or portions of UI functionality, as well as business logic, in user-defined controls and therefore lets you trim development time and effort even more. There are two types of reusable controls you can create in ASP.NET: *custom* controls and *user* controls. Custom controls also consist of composite controls that can be created by combining existing controls and custom controls that are newly built based on the *System.Web.UI.Control* class. If you just want to have collective UI functionality, you create composite controls. If you want to make a web page into a control, you take the user control approach.

Combining Existing Web Controls—Creating Composite Controls

We appreciate that ASP.NET offers web controls, but sometimes we think it would be nice if they offered a little bit more. For example, every web page needs a *TextBox* control, and the control is always preceded by a *Label* control. It is nice that we can drag and drop a control onto a web form, but it could be very tedious to set up the properties for them and to line them up. It would be nice if there were an enhanced version of the *TextBox* control that had a label/caption property? Well, nothing is available out of the box of ASP.NET. However, ASP.NET provides you a way of combining controls such as the *TextBox* and *Label* controls—creating composite controls. You can create new controls by combining two or more existing controls via class composition. Those controls that are combined are called child controls. A composite control reuses the functionality of its child controls, synthesize properties of its child controls, and handle events raised by its child controls. It can also have its own custom properties and events.

A composite control must first implement the *System.Web.UI.INamingContainer* interface. *INamingContainer* is a marker interface that has no methods. When a control implements *INamingContainer*, ASP.NET creates a new naming scope under that control to guarantee its child controls having unique names in the control hierarchy. Then, the composite control needs to override the protected *CreateChildControls* method inherited from the *Control* class to create instances of the child controls and add them to its *Controls* collection.

In the following, we are going to build a composite control that combines the *TextBox* and *Label* controls. We are going to call it "TextBox2". The new control will have two properties: label and text. The label property is used for the caption of the new textbox control. The *Text* property is used for real inputs. We are going to raise the *TextChanged* event from the child control—*TextBox*.

To create a composite control:

1 Launch Visual Studio .NET and create a project by using the Web Control Library template for the Visual Basic .NET. Name the solution "TextBox2" (Figure 7-1). You will be presented *WebCustomControl1.vb*.

Figure 7–1. New Project dialog box with the Web Control Library template highlighted.

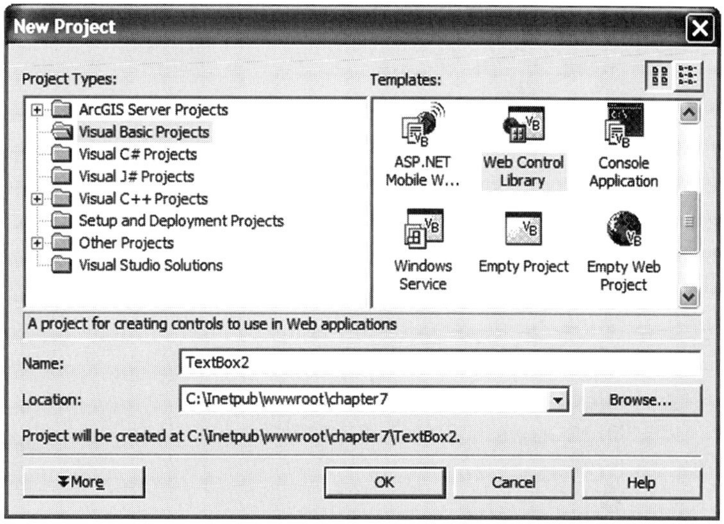

2 Replace the auto-generated code with the following.

```
Imports System.ComponentModel
Imports System.Web.UI
Imports System.Web.UI.WebControls
Namespace CompositeControl
<DefaultProperty("Label"), _
  ToolboxData("<{0}:TextBox2 runat=server></{0}:TextBox2>")> _
  Public Class TextBox2
   Inherits Control
   Implements INamingContainer
  Private _label As Label
  Private _textbox As TextBox
  Public Property Label() As String
    Get
     EnsureChildControls()
     Return _label.Text
    End Get
    Set(ByVal Value As String)
     EnsureChildControls()
     _label.Text = Value
    End Set
  End Property
  Public Property Text() As String
    Get
     EnsureChildControls()
     Return _textbox.Text
    End Get
```

```
    Set(ByVal Value As String)
     EnsureChildControls()
     _textbox.Text = Value
    End Set
  End Property
  Public Event TextChanged As EventHandler
  Protected Overrides Sub CreateChildControls()
    ' add the label
    _label = New Label
    _label.EnableViewState = False
    _label.Font.Name = "Verdana"
    _label.Font.Bold = True
    _label.Font.Size = FontUnit.Point(8)
    _label.Text = Label
    Controls.Add(_label)
    ' add a space
    Controls.Add(New LiteralControl(" "))
    ' add the textbox
    _textbox = New TextBox
    _textbox.EnableViewState = False
    _textbox.Text = Text '
    AddHandler _textbox.TextChanged, AddressOf Text2Changed
    Controls.Add(_textbox)
  End Sub
     ' handler for the TextChanged event
  Protected Sub Text2Changed(ByVal sender As Object, _
               ByVal e As EventArgs)
    Text = _textbox.Text
    RaiseEvent TextChanged(Me, e)
  End Sub
End Class
End Namespace
```

Because we are building a composite control based on the existing controls, we need to include the *System.Web.UI.WebControls* namespace. We want the *TextBox2* control to have the same syntax as that for the *TextBox* control, as follows.

```
<asp:TextBox2 id="TextBox21" runat="server"></asp:TextBox2>
```

✔ **NOTE:** *If you need to customize the GUI of any child control, you need to include the* System.Drawing *namespace.*

Thus, we define the *TextBox2* class and modify the *ToolboxData* as follows.

```
< DefaultProperty("Label"), _
 ToolboxData("<{0}:TextBox2 runat=server></{0}:TextBox2>")> _
  Public Class TextBox2
```

We set the *DefaultProperty* to the *Label* property. When you drag and drop a *TextBox2* control at design time, you will notice that the *Label* property is set to the default (i.e., the property is highlighted). We set two private child controls: a *Label* control and a *Text* control. Then we defined two properties for the *TextBox2* control: `Label` and `Text`. We also defined an event *TextChanged* for the control. We used the `EnsureChildControls` method to ensure that the *TextBox2* control has the *Label* control and the *TextBox* control as its two child controls. The `CreateChildControls` method was overridden with the predefined label and text properties and the predefined `TextChanged` handler. The font for the label child control was set to *Verdana* and *Bold* with a size of 8 points. The `TextChanged` event was set to call a protected `Text2Changed` event, which was essentially the *TextChanged* event raised from the *TextBox* control.

3 Save the project, access the Build menu, and then select Build Solution.

We have built a composite control, *TextBox2*. It should work just like a regular *TextBox* control by design. But it is enhanced with a *Label* property so that you just need to set that property at design time and you are all set. You no longer need to drag and drop a *Label* control. We will create a simple ASP.NET to test the *TextBox2* control.

To create an ASP.NET project to test the TextBox2 *control:*

1 Launch Visual Studio .NET and create a project by using the ASP.NET Web Application template for the Visual Basic .NET. Name the solution "TestTextBox2".

2 Open the toolbox and scroll all the way down and right-mouse click on the empty area to bring up the Toolbox menu (Figure 7-2). Select Add Tab, and a new tab will be appended at the bottom. Enter *My Control* for the tab caption and double click on it to open the tab.

3 Right-click the mouse button to access the menu again, but select Add/Remove Items. A Customize Toolbox dialog box appears (Figure 7–3).

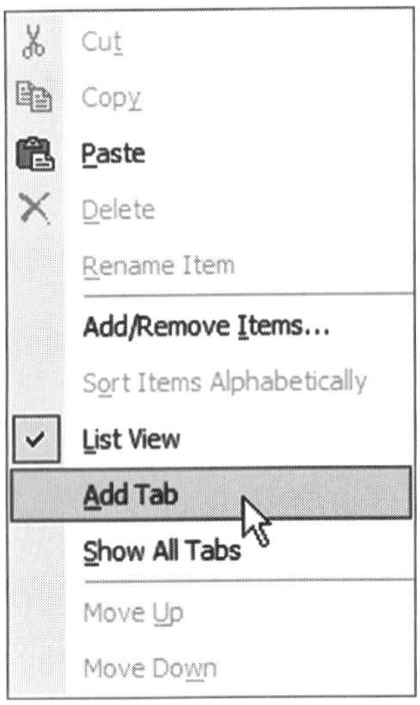

Figure 7–2. Toolbox menu with Add Tab highlighted.

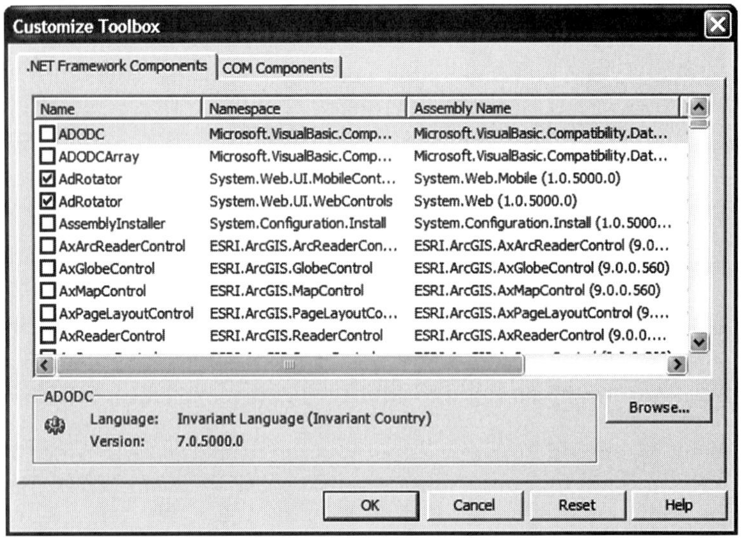

Figure 7–3. Customize Toolbox dialog box of Visual Studio .NET.

4 Click on Browse to navigate to the *bin* directory, where the compiled *TextBox2* control is stored. Select *TextBox2.dll.* A new item called *TextBox2* will be added to the Customize toolbox.

5 Click on OK on the Customize Toolbox dialog box.

The *TextBox2* control will be added to the toolbox "TextBox2" under the "My Control" tab (Figure 7–4). You now have a new control you can drag and drop onto a web form.

6 Drag a *TextBox2* control onto the web form, and set the *Label* property to "ZIP".

The *Label* property should be highlighted, as shown in Figure 7–5. Also, you will notice that the control at design time is just like a regular *TextBox* control (Figure 7–6).

7 Drag a *Label* control onto the same web form.

8 Double click on the *TextBox2* control to display *WebForm1.aspx.vb.*

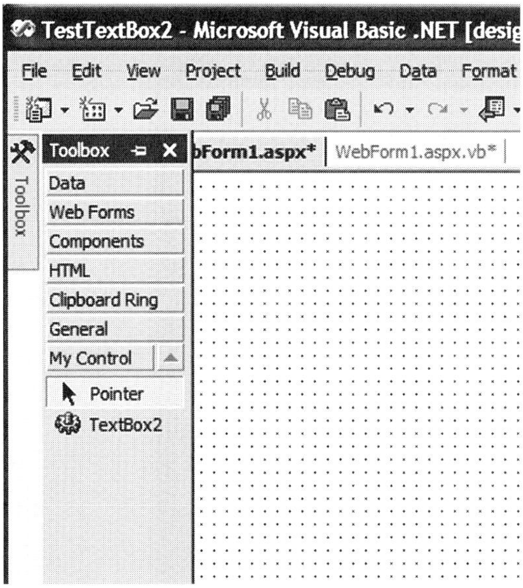

Figure 7–4. Toolbox showing the newly added TextBox2 control under the My Control tab.

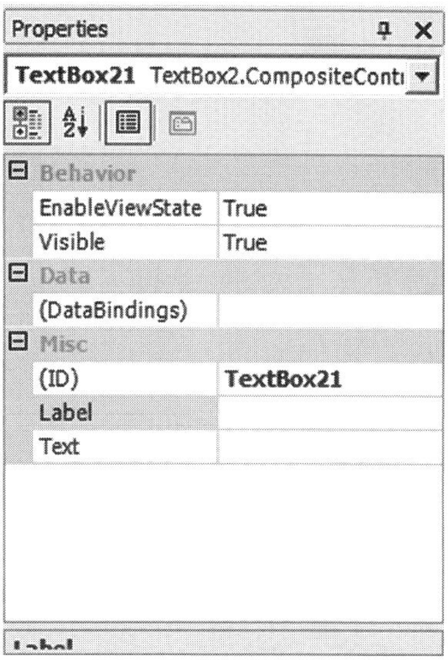

Figure 7–5. The Properties window showing the properties of the TextBox2 control. The Label property was highlighted by design.

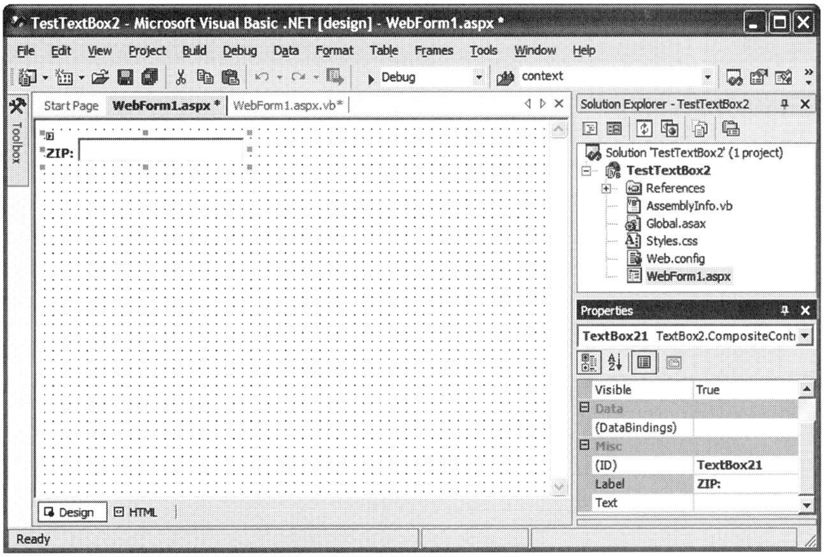

Figure 7–6. TextBox2 control at design time.

9 Replace `TextBox21_TextChanged` with the following code.

```
Private Sub TextBox21_TextChanged(ByVal sender As Object, _
 ByVal e As System.EventArgs) Handles TextBox21.TextChanged
    Label1.Text = "You entered: " & TextBox21.Text
End Sub
```

We will show what a user enters using the *Label* control *Label1*.

10 Click on the Start button or press F5 to run the web page and test it. A web page appears. Type *22202* in the ZIP text box and then press Enter. The web page will be refreshed, saying "You entered: 22202" (Figure 7–7).

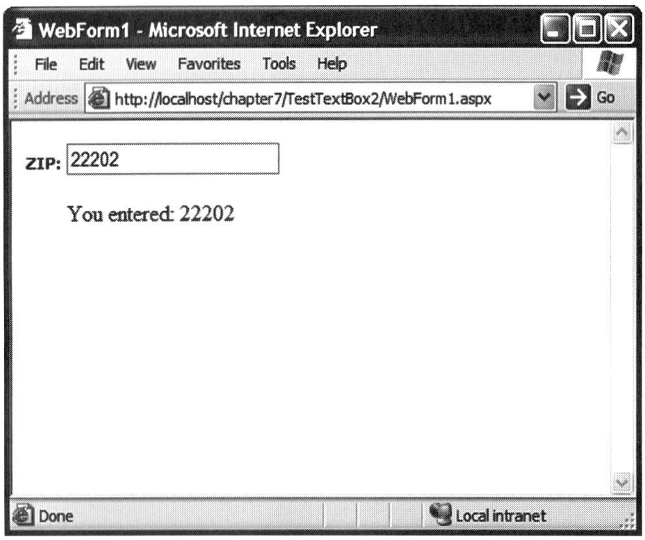

Figure 7–7. A web application that uses the TextBox2 *composite control.*

Congratulations. You have successfully built a composite control. You can follow the same approach to build more. For example, you may want to try to combine the *ScaleBar* and *NorthArrow* controls so that you can drop them together onto a web form. Basically, a composite control is a DLL. Once you reference the DLL, you can simply use the composite control in the same way as you do with other web controls. However, you may want to try the *User* control, discussed in the next section.

Converting a Page to a User Control

The ASP.NET server controls have made web development easier than ever, and provide a great deal of functionality. However, we often run into the situation that we would like to have a control that encapsulates the entire web page or portion of the web page we have just built. You can accomplish this using user controls. A user control is a server control you build. It has both a user interface page and a code-behind file.

Web user controls enable you to easily build controls to meet your specific needs for your applications. You build a user control using the same programming techniques you use to write a web page but save it as a text file with an *.ascx* extension. You can even convert a web page to a user control with a few modifications.

Creating a Web Page

The first step in building a user control is to create a web form page. You do this using Visual Studio .NET and selecting the ASP.NET Web Application template. In the following, you will build a ZIP code entry box with validation. A *TextBox* control and a *RegularExpressionControl* will be used.

To create a web form page:

1. Launch Visual Studio .NET and create a project by using the ASP.NET Web Application template for Visual Basic .NET. Name the solution "ZipUserControl".
2. Drag a *TextBox* control from the Web Forms tab of the toolbox onto the web form and set the *ID* property to "txtZip" and the *Text* property to "ZIP Code".
3. Drag a *RegularExpressionValidator* control from the Web Forms tab of the toolbox onto the web form and place it next to the *TextBox* control. Set the *ID* property to "vldtrRegexZip", the *ErrorMessage* property to "Invalid ZIP Code", and the *ControlToValidate* property to "txtZip".
4. From the Properties window, click on the *ValidationExpression* property and an Ellipsis button displays. Click on the button, and then the Regular Expression Editor dialog box appears with a list of predefined regular expressions. Select "U.S. ZIP code" from the list and then click on OK.
5. Click on the Start button or press F5 to run the web page and test it. A web page displays. If you press the Enter, you will get the error message under the text box saying "Invalid ZIP Code". Enter *22202* and then the error message will disappear. After testing, close the browser that opened.

Listing 7-1 is the *ZipControl.aspx* file for the web page.

Listing 7–1. *ZipControl.aspx* file for the web page.

```
<%@ Page Language="vb" AutoEventWireup="false"
Codebehind="ZipControl.aspx.vb" Inherits="userControl.WebForm1"%>
<!DOCTYPE HTML PUBLIC "-//W3C//DTD HTML 4.0 Transitional//EN">
<HTML>
<HEAD>
   <title>WebForm1</title>
   <meta name="GENERATOR" content="Microsoft Visual Studio .NET 7.1">
   <meta name="CODE_LANGUAGE" content="Visual Basic .NET 7.1">
   <meta name="vs_defaultClientScript" content="JavaScript">
   <meta name="vs_targetSchema"
      content="http://schemas.microsoft.com/intellisense/ie5">
</HEAD>
<body MS_POSITIONING="GridLayout">
<form id="Form1" method="post" runat="server">
<asp:TextBox id="txtZip" style="Z-INDEX: 101; LEFT: 32px;
   POSITION: absolute; TOP: 32px"
   runat="server">ZIP Code</asp:TextBox>
<asp:RegularExpressionValidator id="vldtrRegex"
   style="Z-INDEX: 102; LEFT: 32px;
   POSITION: absolute; TOP: 56px"
   runat="server" ErrorMessage="Invalid ZIP Code"
   ControlToValidate="txtZip"
   ValidationExpression="\d{5}(-\d{4})?">
</asp:RegularExpressionValidator>
</form>
</body>
</HTML>
```

Now you have a web page you can convert to a user control. Converting a web page to a user control is relatively straightforward.

Converting a Page to a User Control

To convert a web page to a user control, you need to remove all HTML tags (such as `<html>`, `<head>`, `<meta>`, `<body>`, and `<form>`) from the web form page. Once the user control is ready, you can drag and drop it onto a web form in the same way as you do with other web controls. You must also change the ASP.NET directive type in the web page from `@ Page` to `@ Control`. The `@ Control` directive specifies the settings used by the page and user control compilers.

To convert a web page to a user control:

1 Return to *WebForm1.aspx* in the designer, and switch to HTML view by clicking on the HTML tab at the bottom of the designer window.

2 Delete the *<html>* tags, *<!doctype>* tag, *<head>* tags and inner content, *<body>* tags, and *<form>* tags from the *.aspx* file (shown previously). The *.aspx* file looks as follows.

```
<%@ Page Language="vb" AutoEventWireup="false"
  Codebehind="ZipControl.aspx.vb" Inherits="userControl.WebForm1"%>
  <asp:TextBox id="txtZip" style="Z-INDEX: 101; LEFT: 32px;
         POSITION: absolute; TOP: 32px"
         runat="server">ZIP Code</asp:TextBox>
  <asp:RegularExpressionValidator id="vldtrRegex"
         style="Z-INDEX: 102; LEFT: 32px;
         POSITION: absolute; TOP: 56px"
          runat="server" ErrorMessage="Invalid ZIP Code"
         ControlToValidate="txtZip"
         ValidationExpression="\d{5}(-\d{4})?">
   </asp:RegularExpressionValidator>
```

3 Change the ASP.NET directive type from *@ page* to *@ Control* as follows.

```
<%@ Control Language="vb" AutoEventWireup="false"
  Codebehind="ZipControl.asmx.vb" Inherits="userControl.ZipControl"%>
```

4 Change the `Codebehind` attribute reference of the directive from *ZipControl.aspx.vb* to *ZipControl.ascx.vb*. User controls are identified by the file name extension *.ascx*.

5 Change the `Inherits` attribute of the directive to *userControl.ZipControl*.

6 Return to the *ZipControl.aspx* file and change the base class from *System.Web.UI.Page* to *System.Web.UI.UserControl* and the Public Class declaration from `WebForm1` to `ZipControl`.

```
Public Class ZipControl
  Inherits System.Web.UI.UserControl
```

7 Save and close the code-behind file and the *.aspx* file.

8 Highlight *ZipControl.aspx* in Solution Explorer and click on it again to rename its extension from *.aspx* to *.ascx*.

✔ **NOTE:** *All user controls must have* .ascx *as their file extension.*

9 The file is now a user control and ready to be consumed by web pages in your project.

Listing 7–2 is the *ZipControl.ascx* file for the user control.

Listing 7–2. *ZipControl.ascx* file for the user control.

```
<%@ Control Language="vb" AutoEventWireup="false"
Codebehind="ZipControl.ascx.vb" Inherits="userControl.ZipControl"%>
<asp:textbox id="txtZip" style="Z-INDEX: 101; LEFT: 32px;
   POSITION: absolute; TOP: 32px" runat="server">ZIP Code
</asp:textbox>
<asp:regularexpressionvalidator id="vldtrRegex"
   style="Z-INDEX: 102; LEFT: 32px; POSITION: absolute; TOP: 56px"
   runat="server" ValidationExpression="\d{5}(-\d{4})?"
   ControlToValidate="txtZip" ErrorMessage="Invalid ZIP Code">
</asp:regularexpressionvalidator>
```

Testing the User Control

You can test the user control you have just created by creating a new web form page and adding the control to it. In the following, you will create a new web form and then drag and drop the *ZipControl* user control onto the form.

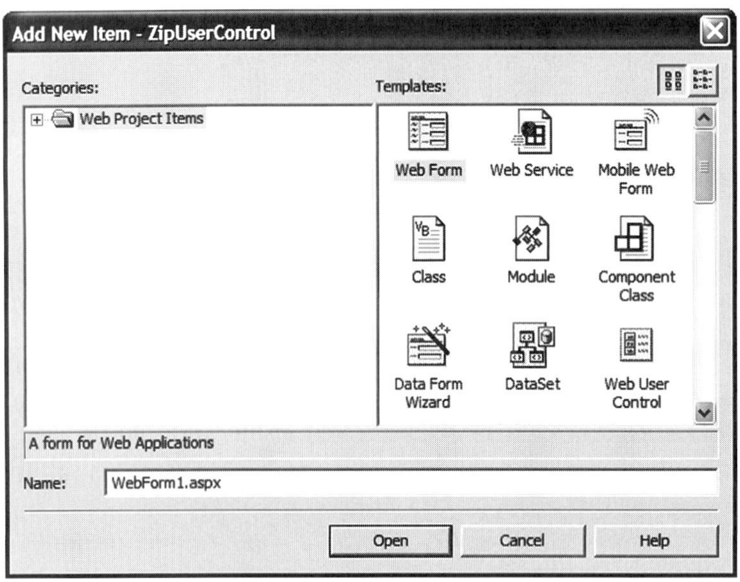

Figure 7–8. Add New Item dialog box of Visual Studio .NET.

To test the user control with a new web form:

1 On the Project menu, click on Add Web Form. The Add New Item dialog box appears with Web Form selected (Figure 7–8).

2 Leave the default name *WebForm1.aspx* and click on Open. The new web form page opens in the designer.

3 Drag *ZipControl.ascx* from Solution Explorer onto the web form (Figure 7–9).

4 Right-click on *WebForm1.aspx* in Solution Explorer and click on Set As Start Page on the shortcut menu.

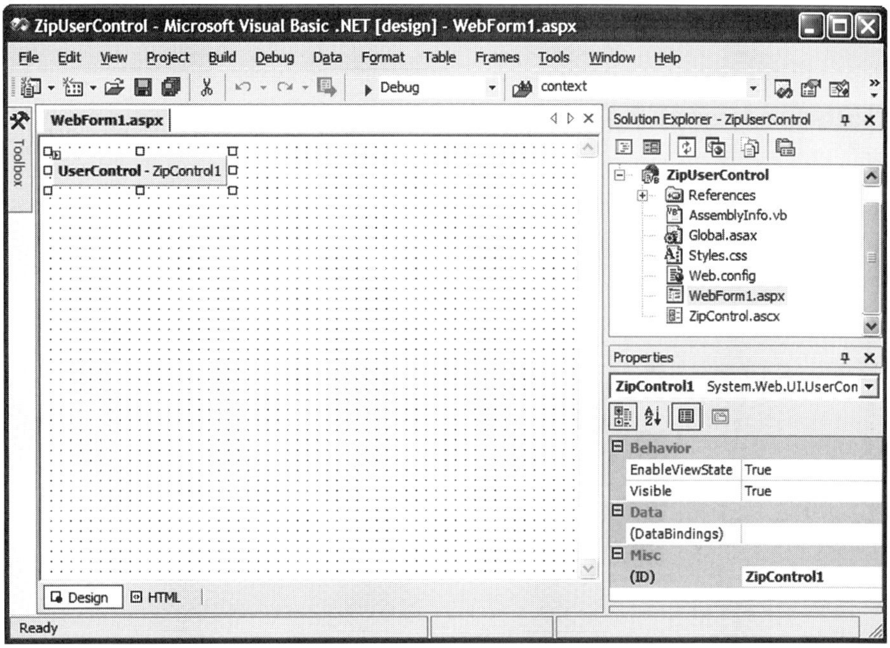

Figure 7–9. ZipControl user control at design time.

5 Click on the Start button or press F5 to run the web page and test it. You will notice that the new web page works in the same way as *ZipControl.aspx* does. Press the Enter key and you will get the error message under the text box saying "Invalid ZIP Code" (Figure 7–10). Enter *22202* and then the error message will disappear. It makes no difference at all to the user who browses this web page. However, the project file *WebForm1.aspx* is quite different from *ZipControl.aspx*. It only has one server control, `uc1:ZipControl`, which is the user control you just built.

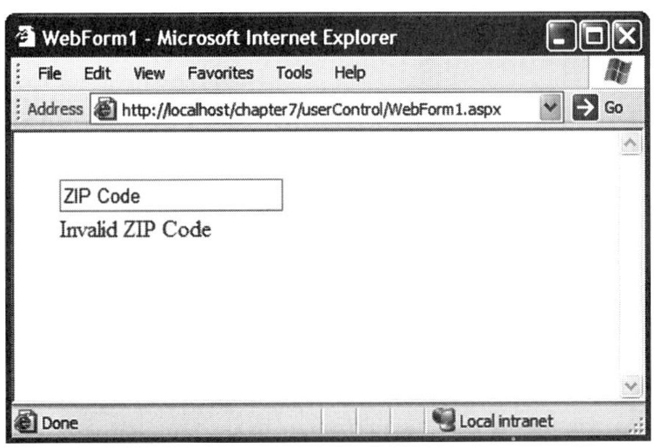

Figure 7–10. Web application that uses the ZipControl user control.

Listing 7–3 is the *WebForm1.aspx* file for the new web form page (*WebForm1.aspx*).

Listing 7–3. *WebForm1.aspx* file for the new web form page (*WebForm1.aspx*).

```
<%@ Page Language="vb" AutoEventWireup="false"
   Codebehind="WebForm1.aspx.vb" Inherits="userControl.WebForm1"%>
<%@ Register TagPrefix="uc1" TagName="ZipControl"
   Src="ZipControl.ascx" %>
<!DOCTYPE HTML PUBLIC "-//W3C//DTD HTML 4.0 Transitional//EN">
<HTML>
<HEAD>
   <title>WebForm1</title>
   <meta name="GENERATOR" content="Microsoft Visual Studio .NET 7.1">
   <meta name="CODE_LANGUAGE" content="Visual Basic .NET 7.1">
   <meta name="vs_defaultClientScript" content="JavaScript">
   <meta name="vs_targetSchema"
      content="http://schemas.microsoft.com/intellisense/ie5">
</HEAD>
<body MS_POSITIONING="GridLayout">
<form id="Form1" method="post" runat="server">
   <uc1:ZipControl id="ZipControl1" runat="server"></uc1:ZipControl>
</form>
</body>
</HTML>
```

Congratulations. You have successfully built a user control. You can follow the same approach to build more. For example, you can create a user control with the *Map* control and the *Toolbar* control for your team so that the entire team can use that user control for development. All applications to be built by your team will have the same look-and-feel. You can technically create your own ArcGIS Server template with the user-control approach. One thing to keep in mind is that a user control is a text file and has *.ascx* as the file extension.

Customizing ArcGIS Web Controls

As discussed in Chapter 6, ArcGIS Server web templates are a good starting point for creating web applications. But often you have a special requirement for your application for which you cannot use templates. You start building your application using ArcGIS web controls. The *Toolbar* control is used in nearly every application. In this section, we will discuss how to customize the toolbar by adding a new item to the toolbar.

Using Icon Buttons on the Toolbar Control

In Chapter 6, we created an example using the *Toolbar* control. However, the tools on the toolbar were all text buttons. We can replace those text buttons on the toolbar with icon buttons. Before we do that, we need to create icon images. Luckily enough, the project created based on the ArcGIS Server Map Viewer template has all of the images we need. We just need to copy those icon image files from *C:\inetpub\wwwroot\chapter6\templates\agsMap\images* to *C:\inetpub\wwwroot\chapter7\iconToolbar\images*. Now we can add icons to the toolbar.

To add image icons to the Toolbar *control:*

1. Recreate a web application based on the steps discussed in the "Toolbar Control" section in Chapter 6. You can also copy and rename that project and place it in *C:\inetpub\wwwroot\chapter7\iconToolbar.*
2. Click on the Start button or press F5 to run the web page and test it. A web page with a text-based toolbar displays (Figure 7–11).
3. Close the application and switch back to the *iconToolbar* project. We are going to add icon image buttons to the toolbar.

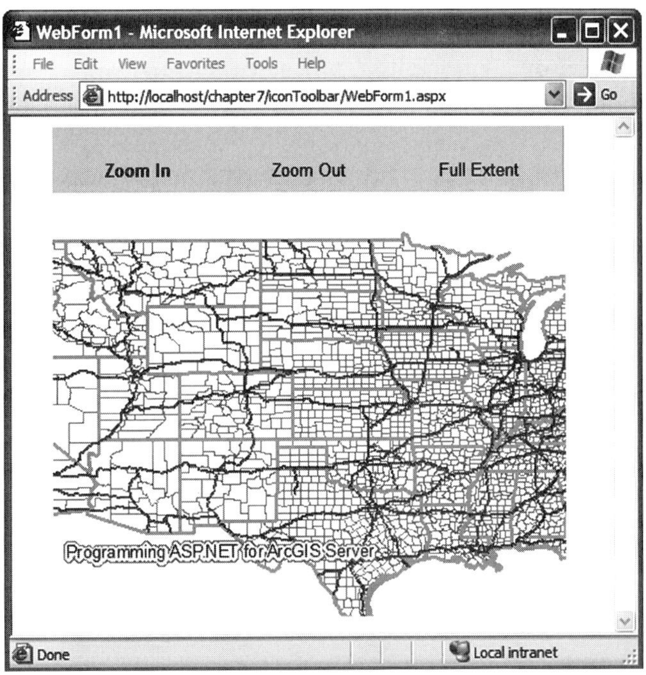

Figure 7–11. A web application that only uses text buttons for the toolbar.

4 From the Properties window, click on the *ToolbarItems* property and an Ellipsis button displays. Click on the button, and then the ToolbarItem Collection Editor dialog box appears. The three tools on the toolbar are *zoomin*, *zoomout*, and *fullextent* (listed in the editor). The *DefaultImage* property is the default property and is highlighted, and there is an Ellipsis button in the *DefaultImage* text box.

5 Click on the Ellipsis button, and then the Select Image dialog box appears (Figure 7–12). Click on Browse to navigate to the image directory and select *zoomin.gif*. Click on OK to close the dialog box.

6 Repeat step 5 to add *zoominU.gif* and *zoominD.gif* for the *HoverImage* and *SelectedImage* properties, respectively (Figure 7–13).

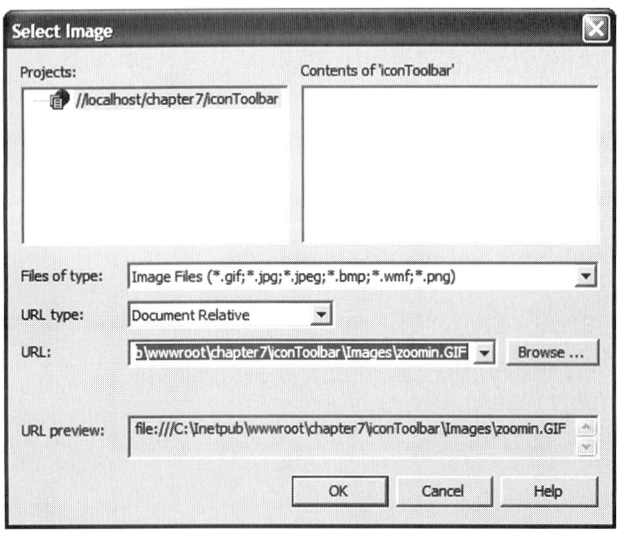

Figure 7–12. Select Image dialog box for the ToolbarItem Collection editor.

✔ **NOTE:** *The* DefaultImage *icon image is used to define the general appearance of a tool on the toolbar. The* HoverImage *icon image is used define the appearance of a tool with a mouseover action. The* SelectedImage *icon image is used to define the appearance of a tool with a selection or mouse press action. The* DisabledImage *icon image is used to define the appearance of a tool when the tool is disabled.*

7 Click on the Start button or press F5 to run the web page and test it.

A web page similar to that shown in Figure 7–1 displays (Figure 7–14). Because the *zoomin* tool is the default tool (it is selected), the *SelectedImage* icon image file *zoominD.gif* is shown on the web page. Both the *zoomout* and *fullextent* tools are shown in the *DefaultImage* icon image files, *zoomout.gif* and *fullextent.gif*. If you move the mouse over the *zoomout* tool, you will see that the appearance of the *zoomout* tool is replaced by the *HoverImage* icon image file *zoomoutU.gif*. If you click on the *zoomout* tool, its appearance will be replaced by the *SelectedImage* icon image file *zoomoutD.gif*.

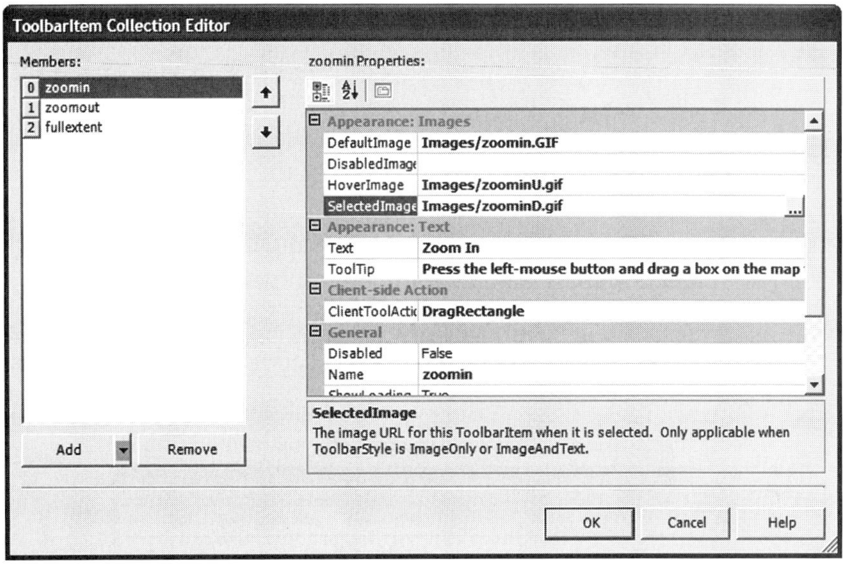

Figure 7–13. The ToolbarItem Collection Editor dialog box shows that the DefaultImage, HoverImage, *and* SelectedImage *properties have been set.*

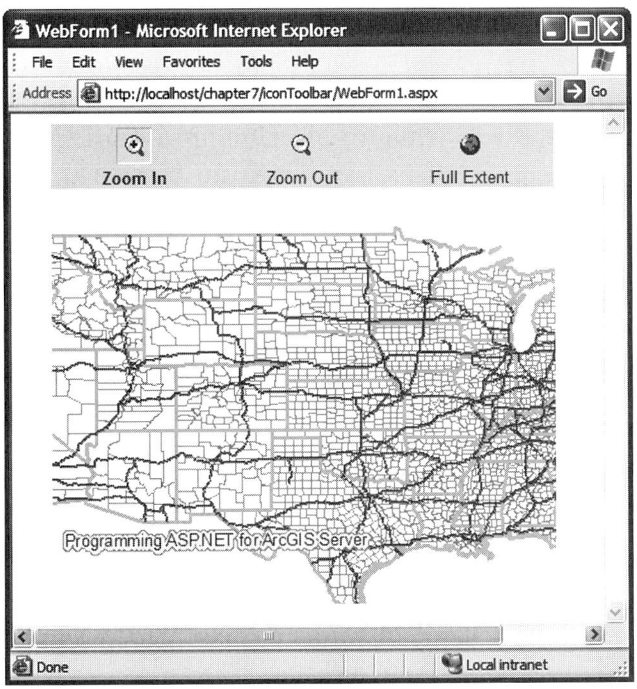

Figure 7–14. A web application that uses both text and image-icon buttons for the toolbar.

✔ **NOTE:** *To make the toolbar that only shows the image icons, set the* ToolbarStyle *property* ImageOnly.

8 Close the application and switch back to the *iconToolbar* project.

9 From the Properties window, set the *ToolbarStyle* to *ImageOnly* (*ImageAndText*, *ImageOnly*, and *TextOnly* are three options; the default is *ImageAndText*.), the *Height* property to "32px", and the *Width* property to "100px" (each icon image is 28 x 28 pixels).

10 Click on the Start button or press F5 to run the web page and test it. A web page with image icons only displays (Figure 7–15).

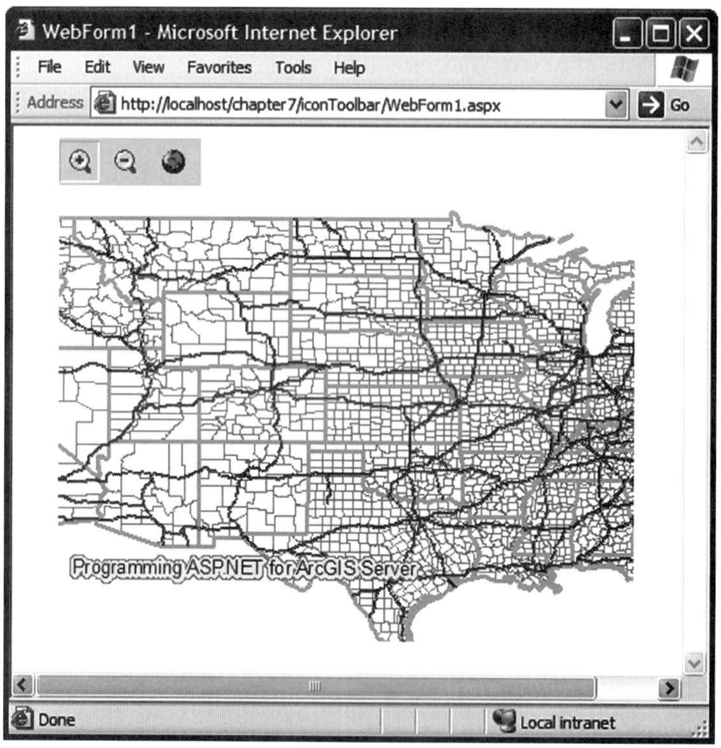

Figure 7–15. A web application that only uses image-icon buttons for the toolbar.

Adding a New Item to a Toolbar Control

Four types of items can be added to the *Toolbar* control: *Tool*, *Command*, *Separator*, and *Space*. A tool interacts with the *Map* control or *Page Layout* control and has a client-side action or a server-side action, or both. For example, Zoom In, Zoom Out, and Identify are tools. A command does not interact with the *Map* control or *Page Layout* control. Instead, it simply executes the code-behind, such as zooming to the full extent or previous extent. A separator can be used to separate or group tools and commands. It simply draws a vertical bar on the toolbar between the two items it separates. A space can be used to increase the space between two items and therefore to enhance the visibility of items on the toolbar.

To define a new item for a *Toolbar* control, you use the ToolbarItem Collection Editor (shown in Figure 7–13). The default property for a tool or command is the *Name* property. You need to set the *Text* and *Tooltip* properties in general. The *Tooltip* property is one of best user-friendly features implemented in ArcGIS Server web controls. The tip message you enter at design time will show up at runtime when a mouse is over the tool or command. To make the GUI more attractive, you would use image icons. You need to define three of fours images for each icon: *DefaultImage*, *HoverImage*, and *SelectedImage*. *DisabledImage* is used when a tool or command has a disable status. The *ClientToolAction* property is for you to specify the JavaScript name for the client-side action. ArcGIS Server comes with a suite of JavaScript libraries and can be found at *<ArcGIS Install Location>\DotNet\VirtualRootDir\aspnet_client\esri_arcgis_server_ webcontrols\9_0\JavaScript*. The *ServerToolActionAssembly* and *ServerToolActionClass* are only available for tools, not commands. Table 7–1 summarizes those common properties that need to be defined for a tool or command.

Table 7–1: Common Tool/Command Properties Requiring Definition

Property	Description
DefaultImage	General icon image for a tool or command
DisabledImage	Dimmed icon image for a tool or command
HoverImage	Icon image for mouse-over action
SelectedImage	Icon image for selection or mouse press
Text	Short text for a tool or command
Tooltip	Short description of a tool or command
ClientToolAction	Client action JavaScript name
Name	ID for a tool or command

Property	Description
ServerToolActionAssembly*	Name of an assembly containing the server tool action
ServerToolActionClass*	Server tool action class

* Available for tools only.

In the following, we will modify the "iconToolbar" project by adding a pan tool, a separator, and space to the toolbar.

To add new items to the Toolbar *control:*

1 Open the "iconTool" project in Visual Studio .NET.

2 Click on the *Toolbar* control.

From the Properties window, click on the *ToolbarItems* property and an Ellipsis button displays. Click on the button, and then the ToolbarItem Collection Editor dialog box appears. The three tools on the toolbar are *zoomin*, *zoomout*, and *fullextent* (listed in the editor).

3 Click on Add to add the "pan" tool. Set the *ID* property to "pan", and the remaining properties to the values listed in Table 7–2.

Table 7–2: Property Values for Pan Tool

Property	Value
DefaultImage	Images/pan.gif
DisabledImage	—
HoverImage	Images/panU.gif
SelectedImage	Images/panD.gif
Text	Pan
Tooltip	Drag the map image to pan
ClientToolAction	DragImage
Name	pan
ServerToolActionAssembly	ESRI.ArcGIS.Server.WebControls
ServerToolActionClass	ESRI.ArcGIS.Server.WebControls.MapPan

✔ **NOTE:** *After you add an icon image file using the Select Image dialog box, you need to change its absolute path to a relative path. For example, the absolute path for the pan tool is:*

files:///C:/Inetpub/wwwroot/chapter7/iconToolbar/Image/pan.gif

You need to change it to *Image/pan.gif*. If you define your own JavaScript, select the *custom* JavaScript option for the *ClientToolAction* property and the Custom ClientToolAction Editor dialog box pops up to allow you to enter your JavaScript code. If you define your own server action class, enter the assembly name for *ServerToolActionAssembly* and the class name for *ServerToolActionClass*.

4 Click on the Up arrow to move the pan tool before the *fullextent* tool.

5 Click on the Down arrow next to Add and pull down the menu and select *Separator* to add a separator. A separator does not have a property.

6 Click on the Up arrow to move the separator between the *zoomout* tool and the pan tool.

7 Click on the Down arrow next to Add and pull down the menu and select *Space* to add a space. Set the Size property to "3". Size 1 is 10 pixels wide. So, size 3 is 30 pixels wide.

8 Click on the Start button or press F5 to run the web page and test it. A web page with a toolbar containing *tool*, *command*, *separator*, and *space* displays (Figure 7–16).

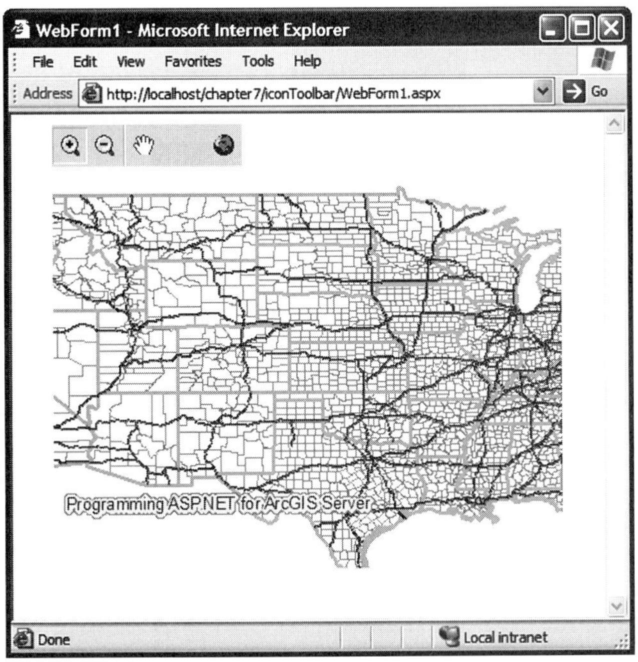

Figure 7–16. A web application with a toolbar that contains tool, command, separator, and space displays.

A separator was used to group the *zoomin* and *zoomout* tools together and separate it from the pan tool. Three spaces each with a width of 10 pixels were used to space out between the pan tool and the *fullextent* tool.

Customizing ArcGIS Server Web Templates

As discussed in Chapter 6, there are seven templates shipped with ArcGIS Server: *Map Viewer, Search, Page Layout, Geocoding, Thematic, Buffer Selection*, and *Web Service Catalog*. Each provides rich functionality. They can be customized to fit your specific needs. You can customize them in three ways.

- Change the GUI
- Modify the functionality for a certain component
- Add new component

Changing the GUI can be relatively easily done. Most likely you only need to modify the HTML portion of the web form. For example, the following is the HTML code of *Default.aspx* from a map viewer that generates the title and logo for a web page. As you can see, the title and logo section is enclosed with an HTML table. You just need to replace the entire table to display a proper title and logo. The default shows the ESRI logo and has a title of "Map Viewer". You can replace the ESRI logo with the logo of your organization and create a fancy banner bar for your application.

```
<!-- Title and Logo -->
<TABLE id="Table1" style="Z-INDEX: 101; LEFT: 10px; WIDTH: 100%;
    POSITION: absolute; TOP: 6px; HEIGHT: 65px"
    cellSpacing="1" cellPadding="1" width="300"
    bgColor="#000099" border="0">
<TR>
 <TD style="WIDTH: 43px"><IMG alt="" src="Images/globe.gif"></TD>
 <TD style="FONT-WEIGHT: bold; FONT-SIZE: x-large; COLOR: white;
   FONT-FAMILY: Arial, Helvetica, Sans-Serif"> Map Viewer</TD>
</TR>
</TABLE>
```

You can also modify the functionality for a component. Before you do that, make sure the component has client-side action, server-side action, or both. Client-side actions are controlled by the JavaScript functions that only execute on the client. You can replace the existing JavaScript function with another prewritten ArcGIS Server JavaScript function. A prewritten JavaScript function requires an ID, the name of the tool, and a Boolean value indicating where to refresh the map image between postback and page refresh. Server-side function modification may require a certain amount of ArcGIS Server and ArcObjects programming. Make sure you have a .NET version of object diagrams for ArcGIS Server object models and ArcObjects object models.

Adding a new component to an existing template can be very challenging. It will require the same amount of effort as you would write a component just based on ArcGIS Server web controls. Again, it may also require addition of both client-side and server-side actions. For the client-side actions, you should make use of the prewritten JavaScript functions in ArcGIS Server. What you need may be just a simple modification of an existing JavaScript function. For the server-side actions, make sure you retrieve the server context and create an instance of the *WebMap* or *WebPageLayout* object if the new component includes the map or page layout. The discussions on how to create a composite control and a user control may also be useful for adding a new component to an existing template.

Summary

In this chapter, we have introduced how to combine existing web controls to create a new composite control. We create a new hybrid *TextBox* control based on the *Label* control and the *TextBox* control. We covered how to convert a web page into a user control. A similar hybrid *TextBox* control was built for a ZIP code entry with validation. We then discussed how to customize ArcGIS Server web controls and web templates.

To combine existing web controls to make a new control is common in .NET programming. We created a new *TextBox* control called *TextBox2* by adding a label/caption to the *TextBox* control. There are a few web controls that need to have a label/caption property. For example, whenever we use a list control, we always drag and drop a *Label* control to the web form before the list control. The label/caption is used to identify what that list is for or about and should be part of the list control. On the other hand, a user control appears to be relatively easy to create. You create a web page first and then convert it to a control by a few simple configurations in *.aspx* file and save that file into *.ascx*.

Composite controls (also called custom controls) are compiled code, which makes them easier to use but more difficult to create. You can drag and drop a composite control in the same way as you do with a regular web control. A composite control can be added to the toolbox and displayed in Visual Studio .NET with full Properties window support and all the other design-time features of ASP.NET server controls. In addition, you can share a single copy of the composite control in the global assembly cache with the rest of your organization. You may consider using a composite control if there is a lot dynamic content and data binding in the control.

User controls are text files and easy to create. You develop a user control in a manner similar to development of a web form page. User controls can be written with code-behind. However, they cannot be added to the toolbox because they are compiled dynamically at runtime. They are simply placeholder signs when added to a page. Also, the only way to share the user control between applications is to put a separate copy in each application, which may not be that easy to maintain if the control needs to be changed or updated. You may consider using a user control if there is a special requirement to the control's look-and-feel.

Customizing ArcGIS Server web controls and templates can be very challenging. It may be involved in both client-side and server-side actions. Try to use the prewritten JavaScript code for the client-side actions. Before you start server-side coding, you need to find a .NET version of object models for both ArcGIS Server and ArcObjects.

So far we have been talking about how to create, change, and customize GUI for web applications. In Chapter 8 we will cover the subjects that are GUI-less. We will discuss in detail how to create web services using the ArcGIS Server Web Service Catalog template (covered in Chapter 6) and how to build web services based on ArcGIS-Server server objects.

CHAPTER 8

Building Web Services for ArcGIS Server

WEB SERVICES ARE LIKE "POKER CHIPS." That is, you can trade them with other parties like you do with ActiveX controls. The difference is that these trades are made over the web, and the services themselves never leave your web site—only the methods and data exposed by the web service get traded. In the GIS arena, data are the primary commodities of web services that are exchanged, shared, or subscribed to. For example, you can visit the Geography Network web site to get the image and vector data in ArcXML format. You can also subscribe to the GDT street maps from ESRI's ArcWeb Services, which come in genuine web service format. You can even call a query, a place finder, or a proximity analysis from the ArcWeb Services. Increasing numbers of web sites are offering GIS web services.

In this chapter, we discuss how to build web services for the ArcGIS Server. You can create two types of GIS web services: *ArcGIS Server web services* and *application web services*. You can create an ArcGIS Server web service using the ArcGIS Server Web Service Catalog template. You can also publish ArcGIS-Server server objects, such as map server objects and geocode server objects, through ArcGIS Server web services. You can use them in ArcGIS clients, such as ArcMap or other ArcGIS Server application. An application web service is an ASP.NET web service that is built based on ArcGIS Server objects. You can publish that service for other IT and GIS professionals to use.

We are going to introduce the definition of web services and standards first. Then, we will explore how to create and consume web services.

We will also examine web service security, managing the state of web services, and publishing and deploying web services.

Overview of Web Services

Web services are referred to as XML web services, meaning that business logic or data that reside remotely can be consumed locally. If we expand the concept a bit, static images, hyperlinks to web applications, XML data streams, and true web services that are exchanged in HTTP on the web can be called web services. But genuine web services have their description language, communication protocol, and XML-based registry so that they can be found easily.

What Is a Web Service?

Web services are software components that reside remotely, but can be used over the Web to provide functionality and data. Think of a web service as a platform-independent remote dynamic-link library. You can use web services in Windows or UNIX environments.

According to Microsoft's definition, web services provide "the ability to exchange messages in a loosely coupled environment using standard protocols such as HTTP, XML, XSD, SOAP, and WSDL." The .NET framework (Figure 8–1) simplifies the web service technology into three major components: web service, web service consumer, and web service registry. The web service can be registered in a Universal Description, Discovery and Integration (UDDI), or created with a DISCO (Discovery Document) file. A web service client or consumer will discover a web service from UDDI or get the DISCO file at design time to obtain a reference to the web service. The communication between a web service and a web service client is through HTTP, XML, and SOAP, which send requests to the web service and receive responses back from the web service.

One major benefit of web services is the ease of integration with external data sources and functionality. Web services also address interoperability issues the software industry has been struggling with for years. Web services are built based on common protocols that are platform independent. You can build a web application by utilizing existing web services. For example, you can retrieve base-map data layers via ESRI's ArcWeb Services to get the data you need to rapidly develop a proto-

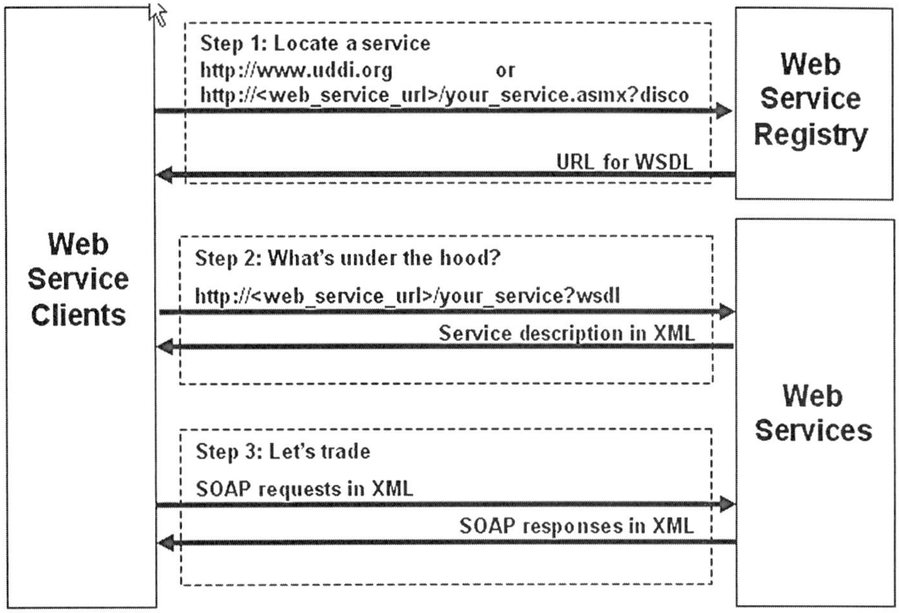

Figure 8–1. The .NET web service architecture.

type. Furthermore, web services themselves are reusable components and can be used as building blocks for enterprise applications.

The ASP.NET Web Service File

Web services are part of ASP.NET. A web service is defined in an *.asmx* file in ASP.NET. You can also develop a web service in any .NET-compatible language, such as Visual Basic .NET or C#. The following is an example of an *.asmx* file.

```
<%@ WebService Language="VB" Class="HelloWorld" %>
```

This example contains a special page directive `WebService` that identifies the file to be compiled at runtime as a web service. It was created with Visual Basic .NET, and the web-service class name was called "HelloWorld".

✔ **NOTE:** *An ASP.NET web service requires the use of the* Sytem.Web.Services *namespace.*

To create a web service, Visual Studio .NET provides an ASP.NET Web Service template. It does all the plumbing required to build an ASP.NET web service. You can even build a new web service by referencing existing web services. We highly recommend that you use the ASP.NET Web Service template to build your web services. Visual Studio .NET will also automatically generate the WSDL (Web Service Description Language) to describe your web service.

The Web Service Description Language (WSDL)

WSDL (Web Service Description Language) is an XML-based language that describes the structure and various features of web services. IBM and Microsoft developed the WSDL technology, which is recommended by W3C. With WSDL, a potential client can learn how to call the methods exposed by a web service, what the parameters and types are offered by the web service, and what type of return values can be expected from those methods. A web service is basically a WSDL document. A typical WSDL document has the root element called *<definitions>*, making the basic structure of a WSDL document as follows.

```
<definitions xmlns="http://schemas.xmlsoap.org/wsdl/">
  … WSDL document elements
</definitions>
```

The *<definitions>* element can contain a single *<types>* element defining the types used in data transfer, which usually contains a schema.

```
<types>
 <schema
  targetNamespace="http://localhost/chapter8/WebServiceTest/Service1">
   … Type definitions
 </schema>
</types>
```

The *<definitions>* element can also contain any number of the following elements to configure the various pieces of information required for web service description.

- *<message>* (contains *<part>* elements)
- *<portType>* (contains *<operation>* elements)
- *<binding>* (contains *<operation>* elements)
- *<service>* (contains *<port>* elements)

An example of a WSDL document follows.

```
<?xml version="1.0" encoding="utf-8" ?>
<definitions xmlns:http="http://schemas.xmlsoap.org/wsdl/http/"
   xmlns:soap="http://schemas.xmlsoap.org/wsdl/soap/"
   xmlns:s="http://www.w3.org/2001/XMLSchema"
   xmlns:s0="http://localhost/chapter8/SampleWebService/Service1"
   xmlns:soapenc="http://schemas.xmlsoap.org/soap/encoding/"
   xmlns:tm="http://microsoft.com/wsdl/mime/textMatching/"
   xmlns:mime="http://schemas.xmlsoap.org/wsdl/mime/"
   targetNamespace="http://localhost/chapter8/SampleWebService/Service1"
   xmlns="http://schemas.xmlsoap.org/wsdl/">
<types>
   <s:schema elementFormDefault="qualified"
   targetNamespace="http://localhost/chapter8/SampleWebService/Service1">
   <s:element name="DD2DMS">
   <s:complexType>
   <s:sequence>
      <s:element minOccurs="1" maxOccurs="1" name="DD"type="s:double" />
   </s:sequence>
   </s:complexType>
   </s:element>
   <s:element name="DD2DMSResponse">
   <s:complexType>
   <s:sequence>
      <s:element minOccurs="0" maxOccurs="1" name="DD2DMSResult"
         type="s:string" />
   </s:sequence>
   </s:complexType>
   </s:element>
   </s:schema>
</types>
<message name="DD2DMSSoapIn">
   <part name="parameters" element="s0:DD2DMS" />
</message>
<message name="DD2DMSSoapOut">
   <part name="parameters" element="s0:DD2DMSResponse" />
</message>
<portType name="Service1Soap">
<operation name="DD2DMS">
   <input message="s0:DD2DMSSoapIn" />
   <output message="s0:DD2DMSSoapOut" />
</operation>
</portType>
<binding name="Service1Soap" type="s0:Service1Soap">
   <soap:binding
      transport="http://schemas.xmlsoap.org/soap/http"
      style="document" />
```

```
    <operation name="DD2DMS">
    <soap:operation
    soapAction="http://localhost/chanpter8/SampleWebService/Service1/DD2DMS"
       style="document" />
    <input>
      <soap:body use="literal" />
    </input>
    <output>
       <soap:body use="literal" />
    </output>
    </operation>
    </binding>
    <service name="Service1">
       <port name="Service1Soap" binding="s0:Service1Soap">
          <soap:address location="http://localhost/chapter8/
             SampleWebService/Service1.asmx" />
    </port>
    </service>
</definitions>
```

You use the WSDL to find out information about a target web service, such as the methods that are available to call. If you are creating a web service, you need to make the WSDL of the web service available for consumption. There is no need to worry about manually creating the WSDL. Visual Studio .NET automatically generates a WSDL for you when you build the web service using the IDE. To access the web service mentioned previously using Visual Studio .NET, simply type the following URL in a web browser:

http://localhost/chapter8/SampleWebService/Service1.asmx?wsdl

✔ **NOTE:** *To view a web service in a browser, you need to enter the URL for the web service (an* .asmx *file) by appending* "?wsdl" *as a query string.*

The Simple Object Access Protocol (SOAP)

SOAP is an XML-based protocol used to invoke remote methods. It is a set of rules that allows programs of various operating systems to communicate with each other. SOAP uses HTTP and XML as mechanisms for information exchange. A typical SOAP message consists of:

- *SOAP envelope:* Defines an overall framework for the message, what is in the message, who the message recipient is, and where the message is optional or mandatory.
- *SOAP encoding rules:* Define a set of rules for exchanging instances of application specific-defined data types.
- *SOAP RPC representation:* Defines a convention for representing remote procedure calls and responses.

When you invoke a web service, a SOAP message is embedded in an HTTP POST and sent to the web service for processing. An example of such a message follows.

```
POST /SampleWebService/Service1.asmx HTTP/1.1
Host: localhost
Content-Type: text/xml; charset=utf-8
Content-Length: length
SOAPAction: "http://localhost/chapter8/SampleWebService/Service1/DD2DMS"
<?xml version="1.0" encoding="utf-8"?>
<soap:Envelope xmlns:xsi="http://www.w3.org/2001/XMLSchema-instance"
   xmlns:xsd="http://www.w3.org/2001/XMLSchema"
   xmlns:soap="http://schemas.xmlsoap.org/soap/envelope/">
   <soap:Body>
      <DD2DMS xmlns="http://localhost/chapter8/SampleWebService/Service1">
         <DD>double</DD>
      </DD2DMS>
   </soap:Body>
</soap:Envelope>
```

The web service processes the message, executes the requested operation, and returns the result to you as another SOAP message, as follows.

```
HTTP/1.1 200 OK
Content-Type: text/xml; charset=utf-8
Content-Length: length
<?xml version="1.0" encoding="utf-8"?>
<soap:Envelope xmlns:xsi="http://www.w3.org/2001/XMLSchema-instance"
   xmlns:xsd="http://www.w3.org/2001/XMLSchema"
   xmlns:soap="http://schemas.xmlsoap.org/soap/envelope/">
   <soap:Body>
      <DD2DMSResponse
         xmlns="http://localhost/chapter8/SampleWebService/Service1">
         <DD2DMSResult>string</DD2DMSResult>
      </DD2DMSResponse>
   </soap:Body>
</soap:Envelope>
```

The message is returned to you as an HTTP response with an HTTP block at the top of the message. When you create a web service using Visual Studio .NET, the IDE does everything for you, including configuring the web service and the client to use SOAP for communication.

Web Services Discovery

When a web service is created, how to advertise and make it available to consumers is a legitimate question. In other words, how does a consumer know what service is available and where it is available? One feasible solution is to register the web service in a distributed and replicated directory. Registering web services means that potential consumers can search or browse service entries and make use of those services that fit their needs. You can register web services through a UDDI service that is maintained by *UDDI.org*.

UDDI is an XML-based registry that allows businesses worldwide to list their web services on the Internet. Its ultimate goal is to streamline online transactions by enabling companies to find one another on the web and to make their systems interoperable for e-commerce. Initially, the idea was to have a few very large public repositories containing UDDI information that could be used by anyone to locate web services. There are online registries operated by IBM, Microsoft, HP, and SAP. You can use the following URLs to search the registered web services.

- Microsoft: *http://uddi.microsoft.com.com/*
- IBM: *http://uddi.ibm.com/*
- HP: *http://uddi.hp.com/uddi/*
- SAP: *http://udditest.sap.com/*

Use UDDI registries to search for businesses that provide web services, select the appropriate WSDL, and build your proxies. The following is an example of the Place Finder Sample web service from ESRI listed in the Microsoft UDDI repository (Figure 8–2).

The trend seems to have been that individual organizations use their own UDDI repository. The UDDI specification uses W3C and Internet Engineering Task Force (IETF) standards, such as XML, HTTP, and DNS (Domain Name System) protocols. UDDI has also adopted early ver-

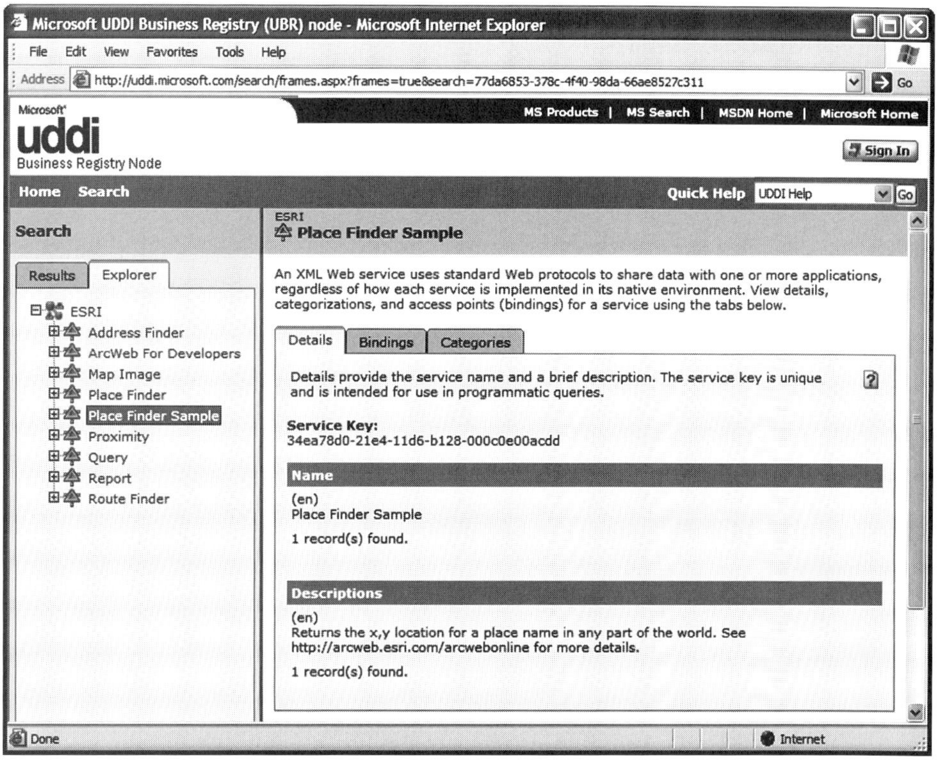

Figure 8–2. The Place Finder Sample web service offered by ESRI is listed in the Microsoft UDDI Repository at http://uddi.microsoft.com.

sions of the proposed SOAP messaging guidelines for cross-platform programming. Microsoft .NET provides a utility to register web services in the UDDI repository.

As an alternative, web services created using ASP.NET can be published and discovered by using DISCO, a discovery document associated with the web service. The *.disco* document is an XML document that contains links to the resources that describe the web service, such as a WSDL file containing the interface contract. The following example is a *.disco* file from the previously cited web service.

```
<?xml version="1.0" encoding="utf-8" ?>
<discovery xmlns:xsd="http://www.w3.org/2001/XMLSchema"
   xmlns:xsi="http://www.w3.org/2001/XMLSchema-instance"
   xmlns="http://schemas.xmlsoap.org/disco/">
```

```
  <contractRef
    ref="http://localhost/chapter8/SampleWebService/Service1.asmx?wsdl"
    docRef="http://localhost/chapter8/SampleWebService/Service1.asmx"
    xmlns="http://schemas.xmlsoap.org/disco/scl/" />
<soap address="http://localhost/chapter8/SampleWebService/Service1.asmx"
  xmlns:q1="http://localhost/chapter8/SampleWebService/Service1"
  binding="q1:Service1Soap"
  xmlns="http://schemas.xmlsoap.org/disco/soap/" />
</discovery>
```

To discover the web service you just created using Visual Studio .NET, simply type the following URL into a web browser:

http://localhost/chapter8/SampleWebService/Service1.asmx?disco

Creating and Consuming a Simple Web Service

In this section, we will create a simple web service that converts a coordinate pair from decimal degrees to degree/minute/second format. Decimal degrees are used internally for computation, but we generally prefer to read a pair of latitude and longitude coordinates in degrees, minutes, and seconds instead of decimal degrees. We know that one degree is 60 minutes and one minute is 60 seconds. So, to convert 38.2345 degrees to degrees/minutes/seconds, we take the decimal degrees of 0.2345 and multiply the value by 60 to get 14.07 minutes. Then we continue with the decimal minutes of 0.07 and multiply the value with 60 to get 4.2 seconds. This sequence of calculations results in 38.2345 degrees, which is equal to 38 degrees, 14 minutes, and 4.2 seconds.

Now that we have verified the calculation, we will build a simple web service that converts coordinate pairs from decimal degrees to degrees/minutes/seconds. Then we will create a web application to consume the web service.

Creating a Simple Web Service

Now we are going to create a web service called Converter. This web service has a method called DD2DMS (stands for “decimal degrees to degrees, minutes, seconds”) that takes a decimal degree as an input and returns a string value in degree/minute/second format.

To create a simple web service:

1 Launch Visual Studio .NET and create a project by using the ASP.NET Web Service template for the Visual Basic .NET. Name the solution "DD2DMS" (Figure 8–3).

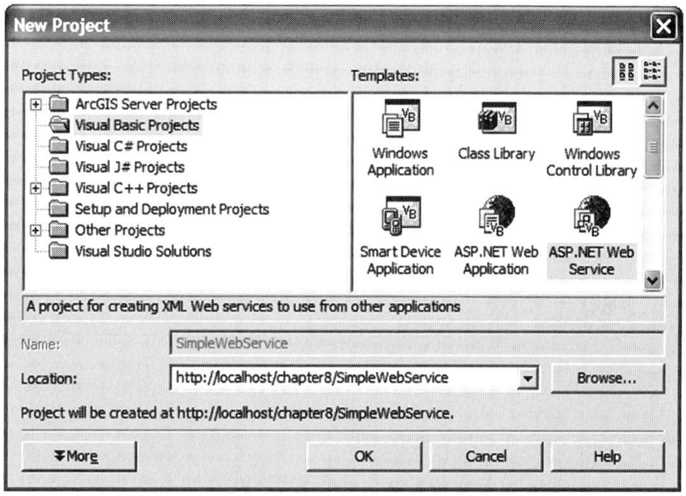

Figure 8–3. Web page generated by an ASP.NET Web Service file Converter.asmx.

A web service project has been created. Instead of creating *WebForm1.aspx*, it creates *Service1.asmx*. The *Service1.asmx* file contains the code for the Converter web service. Again, all .NET web services have the *.asmx* file extension.

2 Rename *Service1.asmx* to *Converter.asmx*.

3 Click on "click here to switch to code view" in the designer environment to switch to code view.

4 Change the name of the class from `Public Class Service1` to `Public Class Converter`.

5 Change

```
System.Web.Services.WebService(Namespace: =http://tempuri.org/DD2DMS/
   Service1).
```

to

```
WebService(Description:="Converting Decimal Degrees to Degrees, Minutes,
   and Seconds", Namespace: =http://localhost/chapter8/DD2DMS/Converter).
```

`Description` and `Namespace` are the two most commonly used web service attributes. You should set them up every time you create a web service.

6 Locate the `<WebMethod()>` section and replace it with the following code.

```
<WebMethod()> _
    Public Function DD2DMS(ByVal DD As Double) As String
    If DD = 0 Then
      Return ""
    Else
      Try
        'declare DMS variable
        Dim valDMS As String
        Dim dg As Int16
        Dim mn As Int16
        Dim sc As Double
        Dim dx As Double
        dg = Int(DD)
        dx = DD - dg
        If dx > 0 Then
          dx = dx * 60
          mn = Int(dx)
          dx = dx - mn
        End If
        If dx > 0 Then
          dx = dx * 60
        End If
        If dx > 0 Then
          sc = Format(dx, "#0.00")
        End If
        valDMS = CStr(dg) + " " + CStr(mn) + " " + CStr(sc)
        Return valDMS
      Catch ex As Exception
        Return ""
      End Try
    End If
  End Function
```

The *WebMethod* encapsulates the functionality of the conversion from decimal degrees to degrees/minutes/seconds, which is defined as `DD2DMS`. A service needs to start with a `WebMethod` attribute. Without this attribute, the method will not be exposed from the service.

7 Click on the Start button or press F5 to run the web service page and test it.

You can also build the solution first and test the web service by using the URL *http://localhost/chapter8/DD2DMS/Converter.asmx* in a browser.

A web service test page appears (Figure 8–4). The page shows the service name "*Converter*", the description of the service, and the conversion method DD2DMS. Click on DD2DMS to bring up the web service testing page (Figure 8–4). Enter *38.2345* in the decimal degree (DD) text box and click on Invoke. The DD2DMS method of the web service then converts 38.2345 to 38:14'4.2" (a colon was used to represent the degree symbol for simplicity), and the resulting value is displayed in XML in a new browser window (Figure 8–5). Close the result window and click on "here" to return to the Converter web service page. Click on Server Description to see the WSDL for the conversion web service (Figure 8–6).

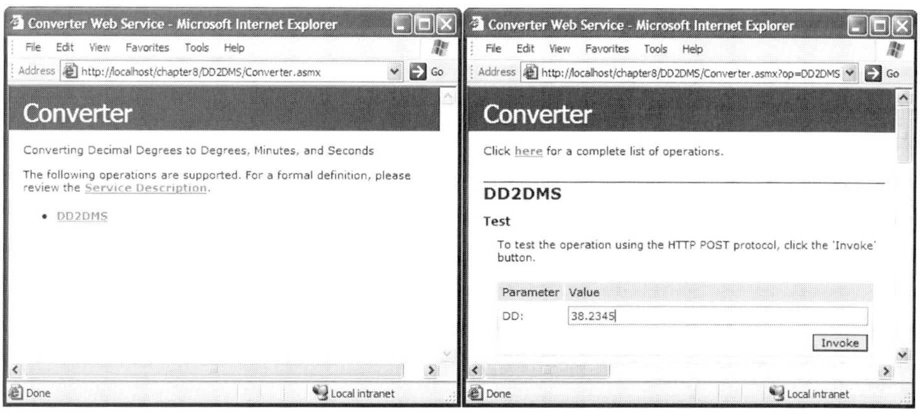

Figure 8–4. Web page generated by an ASP.NET Web Service file Converter.asmx (left) and web service test page for the "DD2DMS" method of the Converter web service (right).

✔ **NOTE:** *A web service URL and service file name plus* ?WSDL *is the protocol to view a WSDL. For example, to view the WSDL for the Converter web service, enter the URL:* http://localhost/chapter8/DD2DMS/Converter.asmx?WSDL.

Congratulations! You have successfully created and tested your first web service. As you can see, Visual Studio .NET creates the WSDL automatically if you build your web services using the ASP.NET Web Service template. You are not exposed to the SOAP protocol, but this communication occurs behind the scenes. You can now create a client application to consume the web service.

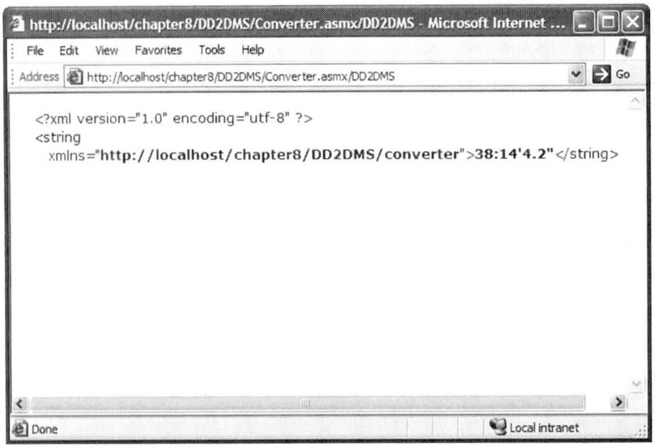

Figure 8–5. Return value in XML from the Converter web service.

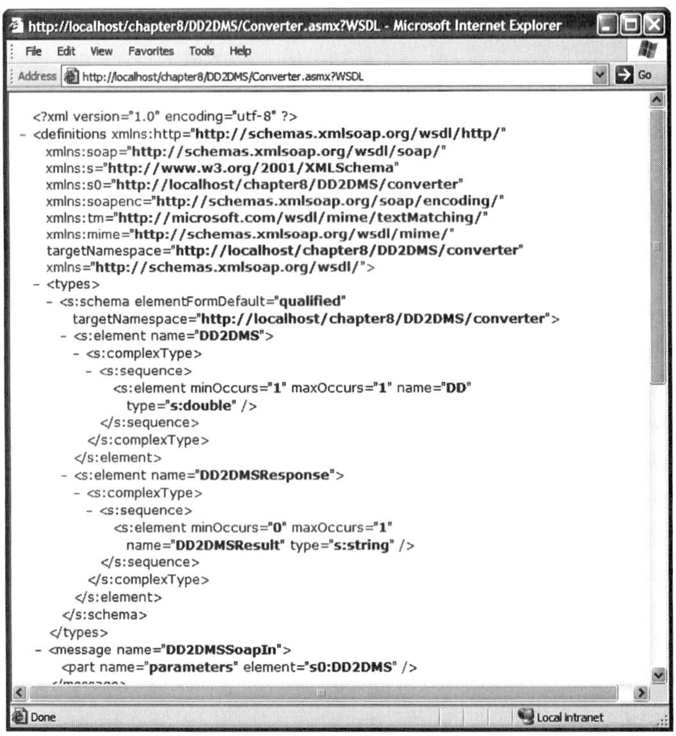

Figure 8–6. WSDL for the Converter web service.

Consuming a Simple Web Service

You can consume a web service in both web and Windows applications. In this section, we will focus on how to consume a web service in a web application. We will build a web form page application to convert latitude and longitude coordinates from decimal degrees to degrees/minutes/seconds. We will use a couple of *TextBox* controls to input a pair of latitude and longitude coordinates and a *Button* control to invoke the conversion. The conversion will be executed based on the Converter web service built in the previous section.

To build an ASP.NET web service client application:

1 In Visual Studio .NET, create a new Visual Basic .NET project using the ASP.NET Web Application template and give the application the name *ConverterWSClient*.

2 Drag a *Label* control from the Web Forms tab of the toolbox onto the web form and set the *Text* property to "Converting Decimal Degrees to Degrees, Minutes, and Seconds", the Font/Bold property to *True*, and the Font/Name to *Arial*.

3 Drag another *Label* control from the Web Forms tab of the toolbox onto the web form and set the *Text* property to "Latitude (Decimal Degrees)" and the Font/Name to *Arial*.

4 Drag a *TextBox* control from the Web Forms tab of the toolbox onto the web form and place it next to the label. Set the *ID* property to "txtLat".

5 Drag a third *Label* control from the Web Forms tab of the toolbox onto the web form and place it under the second label. Set the *Text* property to "Longitude (Decimal Degrees)" and the Font/Name to *Arial*.

6 Drag a *TextBox* control from the Web Forms tab of the toolbox onto the web form and place it next to the longitude label. Set the *ID* property to "txtLong".

7 Drag a *Button* control from the Web Forms tab of the toolbox onto the web form and place it under the third label. Set the *ID* property to "btnResults" and the *Text* property to "Convert".

8 Drag a fourth *Label* control from the Web Forms tab of the toolbox onto the web form and place it under the button. Set the *Text* property to blank, the Font/Bold property to *True*, and the Font/Name to *Arial*.

We use this label to display a pair of converted coordinates in the degree/minute/second format.

9 Before we can consume the Converter service we just built, we have to add a web reference to it. This is done as follows.

- Highlight Reference of the *ConverterWSClient* project inside Solution Explorer.
- Right-mouse click on it and select Add Web Reference.
- The Add Web Reference dialog box appears (Figure 8-7). Enter *http://localhost/chapter8/DD2DMS/Converter.asmx?WSDL* for the URL box and then click on the Go button.
- The Converter service info displays (Figure 8-8). After briefly examining the service, you should see the description and method of the Converter web service. Click on Add Reference.
- The Converter web service will be added to the project as one of the web References. It will be named *localhost*. Rename it *simpleWebService*.

10 Add the following code to the `btnRefults_Click` event.

```
Dim dmsClient As New smipleWebService.Converter
Dim lat, lng As String
If txtLat.Text <> "" And txtLong.Text <> "" Then
  lat = dmsClient.DD2DMS(txtLat.Text)
  lng = dmsClient.DD2DMS(txtLong.Text)
  If lat <> "" And lng <> "" Then
   lblResults.Text = lat & ", " & lng
  End If
End If
```

11 Click on the Start button or press F5 to run the web application.

A web page with two text boxes and a button displays (Figure 8-9). Enter a pair of latitude and longitude coordinates (e.g., 38.2345 and -77.1234) and click on Convert. The web application calls the Converter web service, performs the conversion, concatenates the converted latitude and longitude coordinates, and displays the result using the fourth label.

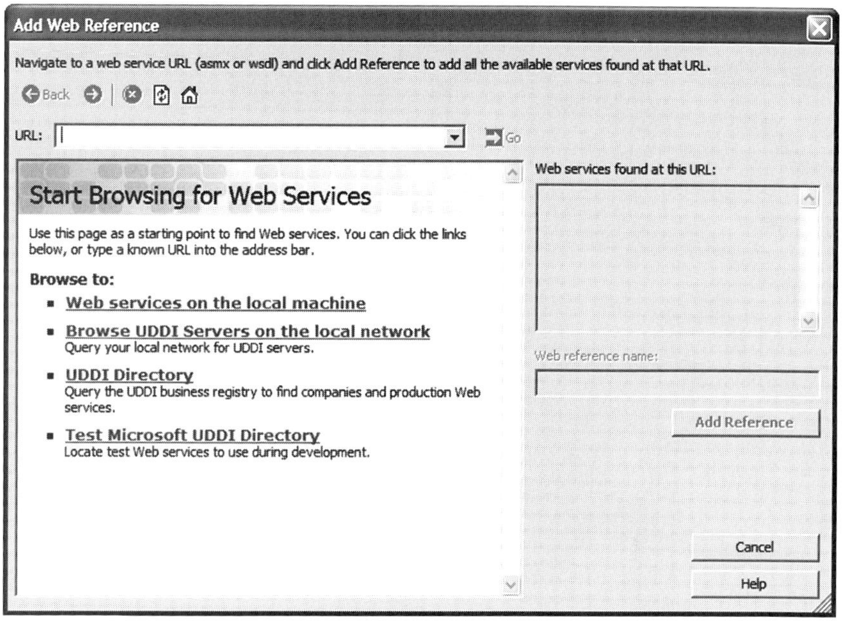

Figure 8–7. Add Web Reference dialog box of Visual Studio .NET.

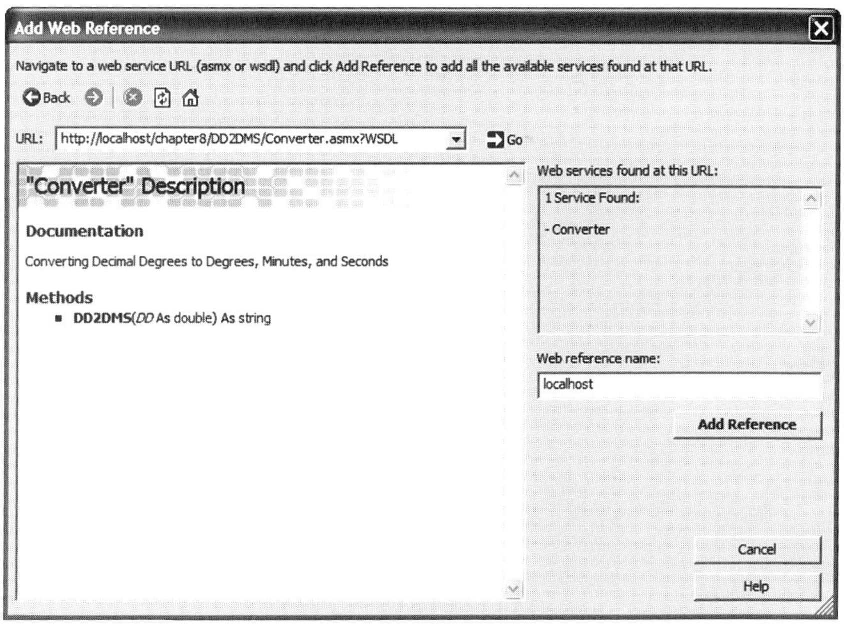

Figure 8–8. Converter web service shown in the Add Web Reference dialog box of Visual Studio .NET.

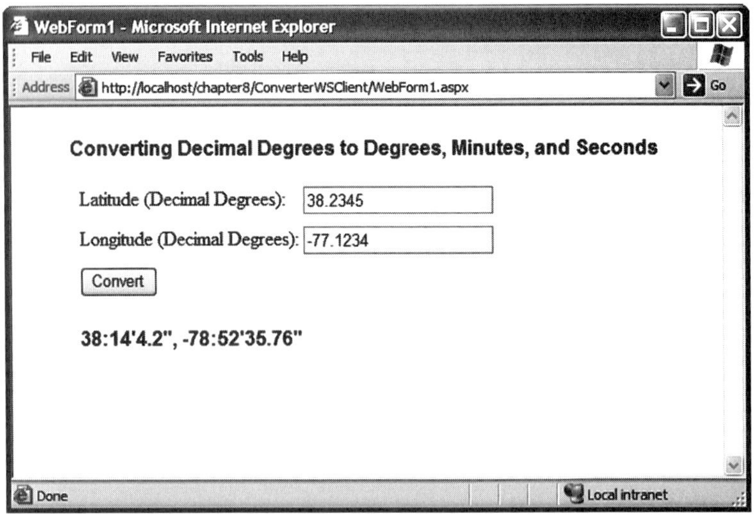

Figure 8–9. The ConverterWSClient web application converts a pair of latitude and longitude coordinates based on the Converter web service.

Listing 8–1 is the *.aspx* file for the web form with four *Label* controls, two *TextBox* controls, and a *Button* control.

Listing 8–1. *.aspx* file for the web form with four *Label* controls, two *TextBox* controls, and *Button* control.

```
<form id="Form1" method="post" runat="server">
<asp:Label id="Label1" style="Z-INDEX: 101; LEFT: 48px;
   POSITION: absolute; TOP: 24px" runat="server"
   Font-Bold="True" Font-Names="Arial">
   Converting Decimal Degrees to Degrees, Minutes, and Seconds
</asp:Label>
<asp:Label id="Label2" style="Z-INDEX: 102; LEFT: 56px;
   POSITION: absolute; TOP: 64px" runat="server">
   Latitude (Decimal Degrees):</asp:Label>
<asp:TextBox id="txtLat" style="Z-INDEX: 103; LEFT: 240px;
   POSITION: absolute; TOP: 64px"
   runat="server"></asp:TextBox>
<asp:Label id="Label3" style="Z-INDEX: 104; LEFT: 56px;
   POSITION: absolute; TOP: 96px" runat="server">
   Longitude (Decimal Degrees):</asp:Label>
<asp:TextBox id="txtLong" style="Z-INDEX: 105; LEFT: 240px;
   POSITION: absolute; TOP: 96px"
   runat="server"></asp:TextBox>
<asp:Button id="btnConvert" style="Z-INDEX: 106; LEFT: 56px;
   POSITION: absolute; TOP: 128px"
   runat="server" Text="Convert"></asp:Button>
<asp:Label id="lblResults" style="Z-INDEX: 107; LEFT: 56px;
   POSITION: absolute; TOP: 176px"
```

```
   runat="server" Font-Names="Arial"
   Font-Bold="True"></asp:Label>
</form>
```

As you can see, you can use a web service in the same way you use a local DLL (a web service is a remote DLL). Once you are granted access to a web service, you can use it. The corresponding WSDL should tell you everything you want to know about that particular web service, including its methods and properties. You can also see the methods and properties after you create a local instance of the web service object.

Now you know how to build a web service. In the next section, we discuss how to build and access an ArcGIS Server web service. The only difference is that you can access ArcObjects through ArcGIS Server web services.

Creating and Consuming ArcGIS Server Web Services

ArcGIS Server implements two types of web Services: ArcGIS Server web services and application web services. The ArcGIS Server web services are web services created using ADF supplied by ArcGIS Server and ArcGIS Server objects. The web services are processed and executed from within ArcGIS Server. The application web services are web services created using native web service template of the web server (e.g., ASP.NET web service) and can be executed from outside ArcGIS Server. When creating an application web service in ArcGIS Server, the web service uses ArcObjects on the server to implement certain GIS functions such as geocoding and spatial searches.

Creating ArcGIS Server Web Services

ArcGIS Server web services are the web services of ArcGIS server objects published using the ArcGIS Server Web Service Catalog template. The Web Server Catalog template publishes server objects in a web service catalog. Server objects, such as the map server object and geocode server object, can be published as ArcGIS web services on a GIS server. You can access most of the ArcObjects types via ArcGIS Web Services.

To create an ArcGIS Server web service, you need to have at least one server object of any type (e.g., a map server object or geocode server object). In Chapter 4 we created four map server objects and one geocode server object. We will create an ArcGIS Server web service to publish those server objects. In the following example, we will use the ArcGIS Server Web Service Catalog template to create a project and publish those server objects.

To create a web service catalog using the ArcGIS Server Web Service Catalog template:

1 In Visual Studio .NET, create a new Visual Basic .NET project using the Web Service Catalog Application template under the *ArcGIS Server Projects* folder and name it "*agsWebServiceCatalog*" for the application (Figure 8–10).

The Create a Web Service Catalog dialog box appears (Figure 8–11).

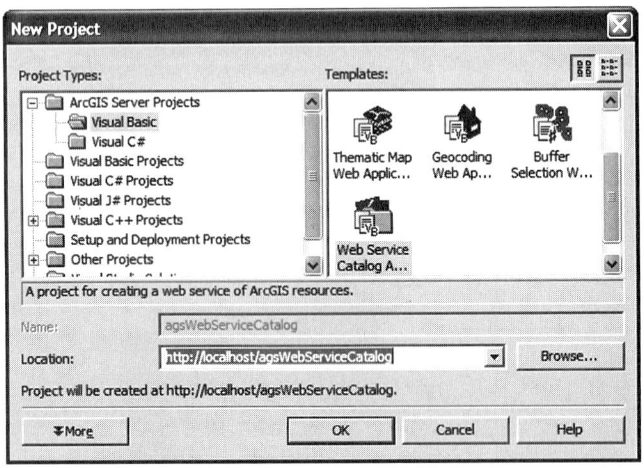

Figure 8–10. New Project dialog box of Visual Studio .NET.

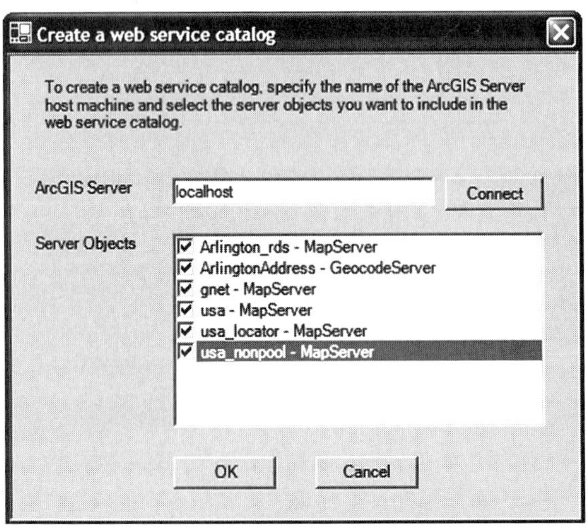

Figure 8–11. The Create a Web Service Catalog dialog box.

2 Enter *localhost* for the ArcGIS Server and click on Connect.

All server objects available on the server will be listed in the Server Objects box (Figure 8–11).

3 Select all of the server objects and click on OK.

4 Select the *Impersonation* control on the web form.

5 From the Properties window, click on the *Identity* property and an Ellipsis button displays. Click on the button and the Impersonation1 Identity dialog box appears. Enter a valid user name and password and click on OK. You can leave the “Domain or Machine Name” box blank.

6 Click on the Build menu and select Build Solution.

You now have a web service catalog of all ArcGIS Server objects on the “localhost” server. Because the catalog is a true SOAP/WSDL service, you need to enter the following URL to view the WSDL (Figure 8–12):

http://localhost/agsWebServiceCatalog/default.aspx?wsdl

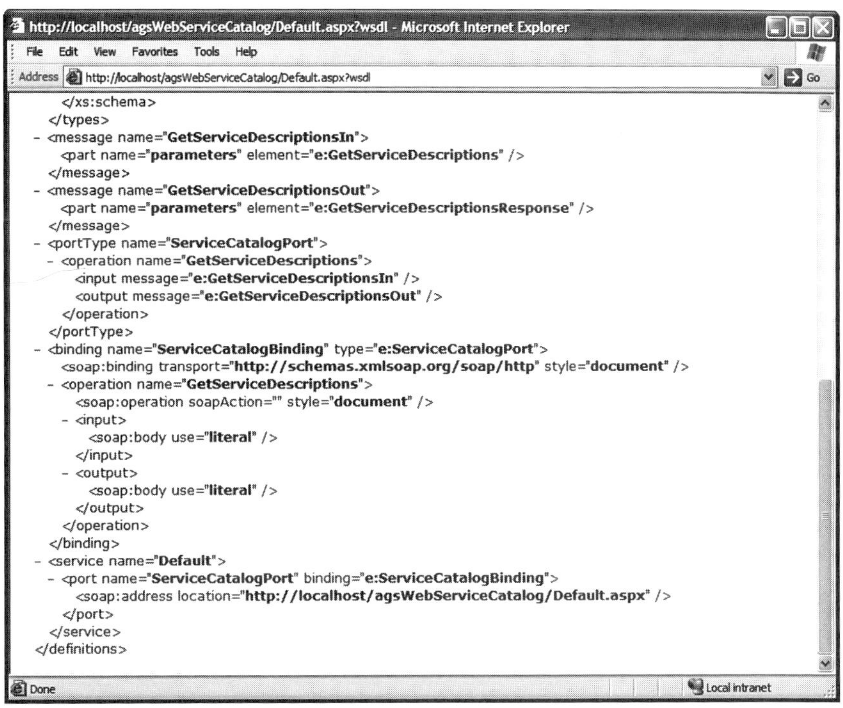

Figure 8–12. WSDL for an ArcGIS Server web service catalog.

Now you are ready to connect to a web service catalog in ArcCatalog.

7 Launch ArcCatalog and click on Add ArcGIS Server. The Add ArcGIS Server dialog box appears (Figure 8–13).

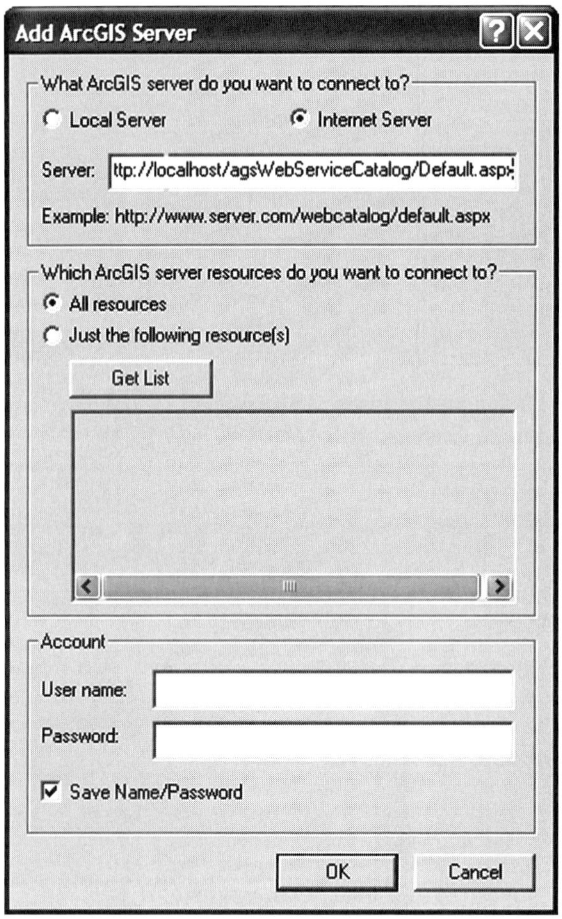

Figure 8–13. Add ArcGIS Server dialog box.

8 Check Internet Server, enter *http://localhost/agsWebServiceCatalog/default.aspx* for the Server box, and click on OK. The "agsWebServiceCatalog" web service catalog will be added to the server under the *GIS Servers* folder (Figure 8–14).

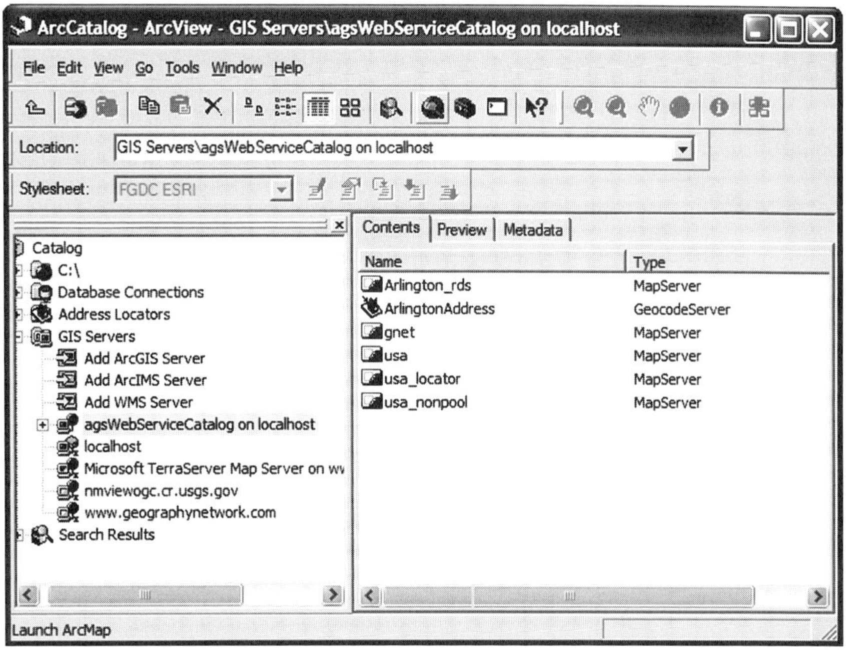

Figure 8–14. ArcGIS Server web service shown in ArcCatalog.

Creating Application Web Services for ArcGIS Server

Application web services can be created using the ASP.NET Web Service template in Visual Studio .NET. You follow the same procedure as you did to create the simple web service example discussed previously. The difference is that you need to use the ArcGIS Server namespaces in addition to regular ASP.NET namespaces. In general, an application web service returns generic data types that may also include ArcObects types. The service can also be used by non-ArcGIS clients.

In the following example, we will create a web service called "SiteService" that has a method called "FindSiteLocations", which can be used to locate sites along the streets identified by a user-defined address. Then we will create a client application to consume the web service and display site locations that are found.

Before we start the web service, we need to create a geocode server object using ArcCatalog. A geocode server object requires an address locator, which you can create using ArcCatalog. For the "SiteService" web service, we used an address locator *ArlingtonLocator.loc* and a geocode object "ArlingtonAddress" created based on the address locator. See "Creating an Address Locator" and "Creating a Geocode Server Object" in Chapter 4 for the details on how to create an address locator and a geocode server, respectively. Once we have the geocode server, we can start creating the "SiteService" application web service.

The "SiteService" web service needs to connect to the ArcGIS Server to create an instance of the geocode server object. It requires a connection to the geocode server via the *ServerConnect* object, which is available in the *ESRI.ArcGIS.Server.WebControls* namespace. Thus, we need to use the *ESRI.ArcGIS.Server.WebControls* namespace. The "FindSiteLocations" method geocodes a given address and finds those sites that are within a given distance. A proximity analysis will be conducted by buffering a geocoded location. The following ArcGIS assemblies will be used in the project.

- *ESRI.ArcGIS.Geodatabase*
- *ESRI.ArcGIS.Geometry*
- *ESRI.ArcGIS.Location*
- *ESRI.ArcGIS.Server*
- *ESRI.ArcGIS.Server.WebControls*
- *ESRI.ArcGIS.esriSystem*

To create an application web service for ArcGIS Server:

1. In Visual Studio .NET, create a new Visual Basic .NET project using the ASP.NET Web Service and name it "SiteLocations" for the application.
2. In the Solution Explorer window, select *Service1.asmx* and rename it *"SiteService.asmx"*.
3. Right-click on References and select Add Reference. Add the following references.
    - *ESRI.ArcGIS.Geodatabase*
    - *ESRI.ArcGIS.Geometry*
    - *ESRI.ArcGIS.Location*
    - *ESRI.ArcGIS.Server*

- *ESRI.ArcGIS.Server.WebControls*
- *ESRI.ArcGIS.esriSystem*

You may not be able to directly locate *ESRI.ArcGIS.Server.WebControls* from the dialog box. In this case, click on the Browse button, navigate to *<Install Location>\DotNet*, and double click on *ESRI.ArcGIS.Server.WebControls.dll*. The assembly should be added to the list.

4 Click on OK to add all references.

5 Click on "click here to switch to code view" in the designer environment to switch to code view.

6 Add `Imports` statements at the beginning, as follows.

```
Imports System
Imports System.Collections
Imports System.ComponentModel
Imports System.Data
Imports System.Diagnostics
Imports System.Web
Imports System.Web.Services
Imports ESRI.ArcGIS.Geodatabase
Imports ESRI.ArcGIS.Location
Imports ESRI.ArcGIS.Server
Imports ESRI.ArcGIS.Geometry
Imports ESRI.ArcGIS.Server.WebControls
Imports ESRI.ArcGIS.esriSystem
```

7 Change the name of the class from `Public Class Service1` to `Public Class SiteService`.

8 Change

```
System.Web.Services.WebService(Namespace: =http://tempuri.org/SiteLocations/
    Service1)
```

to

```
WebService(Namespace: =http://localhost/chapter8/SiteLocations/
    SiteService)
```

9 Replace *<WebMethod>* with the following code.

```
<WebMethod()> _
Public Function FindSiteLocations(ByVal Address As String, ByVal Distance
  As Double) As SitePoints()
   Dim WebObj As New WebObject
   If Address = "" Or Distance = 0.0 Then
```

```
      Return Nothing
    Else
      Dim connection As _
   ESRI.ArcGIS.Server.WebControls.ServerConnection
      connection = New _
   ESRI.ArcGIS.Server.WebControls.ServerConnection
      connection.Host = "localhost"
      connection.Connect()

      Dim som As IServerObjectManager
      som = connection.ServerObjectManager

      Dim sc As IServerContext
      sc = som.CreateServerContext("Sites", "GeocodeServer")
      Try

        Dim so As IServerObject
        so = sc.ServerObject

        Dim gc As IGeocodeServer
        gc = so

        Dim ps As IPropertySet
        ps = sc.CreateObject("esriSystem.PropertySet")
        ps.SetProperty("Street", Address)

        Dim res As IPropertySet
        res = gc.GeocodeAddress(ps, Nothing)

        Dim gcPoint As IPoint
        gcPoint = res.GetProperty("Shape")

        Dim segc As ISegmentCollection
        segc = sc.CreateObject("esriGeometry.Polygon")
        segc.SetCircle(gcPoint, Distance)

        Dim geom As IGeometry
        geom = segc

        Dim gcso As IGeocodeServerObjects
        gcso = gc

        Dim reftabs As IReferenceDataTables
        reftabs = gcso.AddressLocator

        Dim enumreftabs As IEnumReferenceDataTable
        enumreftabs = reftabs.Tables
        enumreftabs.Reset()

        Dim reftab As IReferenceDataTable
        reftab = enumreftabs.Next

        Dim dsname As IDatasetName
        dsname = reftab.Name

        Dim wsnm As IName
```

```
      wsnm = dsname.WorkspaceName

      Dim fws As IFeatureWorkspace
      fws = wsnm.Open

      Dim fc As IFeatureClass
      fc = fws.OpenFeatureClass("Sites")

      Dim sf As ISpatialFilter
      sf = sc.CreateObject("esriGeoDatabase.SpatialFilter")
      sf.Geometry = geom
      sf.GeometryField = fc.ShapeFieldName
      sf.SpatialRel = esriSpatialRelEnum.esriSpatialRelIntersects

      Dim fcursor As IFeatureCursor
      fcursor = fc.Search(sf, True)
      WebObj.ManageLifetime(fcursor)

      Dim lName As Int16
      Dim lType As Int16

      Dim flds As IFields
      flds = fc.Fields
      lName = flds.FindField("FEATURE_NAME")
      lType = flds.FindField("FEATURE_TYPE")

      Dim siteList As New ArrayList
      Dim f As IFeature
      f = fcursor.NextFeature
      Do While Not f Is Nothing
        Dim pt As IPoint = f.Shape
        Dim ts As New SitePoints(f.Value(lName), f.Value(lType),_
          pt.X, pt.Y)
        siteList.Add(ts)
        f = fcursor.NextFeature
      Loop

     Dim siteArray As SitePoints()
      If siteList.Count > 0 Then
        ReDim siteArray(siteList.Count)
        siteList.CopyTo(siteArray)
        Return siteArray
      End If

      sc.ReleaseContext()
    Catch ex As Exception
      sc.ReleaseContext()
    End Try

  End If

End Function
```

10 Create the *SitePoints* class as follows. This class will be used by the *FindSiteLocation* function.

a. In Solution Explorer, right-click on the *SiteLocations* project and select Add/Add New Item.

b. In the Add New Item dialog box, click on Class and set "SitePoints.vb" for the Name.

c. Click on Open. This will add a new blank class (*SitePoints*) to your project.

11 Modify code for the SitePoints class as follows.

a. Add the following *Imports* statements to the beginning of the code.

```
Imports System.Xml.Serialization
Imports System
```

The class will use the *XMLInclude* attribute to serialize the results returned from the web service into XML. The *XMLInclude* attribute is defined in *System.Xml.Serialization*.

b. Add the following code to the class.

```
Imports System.Xml.Serialization
Imports System
<XmlInclude(GetType(SitePoints))> _
    Public Class SitePoints
  Dim Name As String
  Dim Type As String
  Dim X As Double
  Dim Y As Double

  Public Sub New(ByVal sName As String, ByVal sType As _
    String, ByVal dX As Double, ByVal dY As Double)
    Name = sName
    Type = sType
    X = dX
    Y = dY
  End Sub

  Public Sub New()

  End Sub

  Property SiteName() As String
    Get
      Return Name
    End Get
```

```
    Set(ByVal Value As String)
      Name = Value
    End Set
  End Property

  Property SiteType() As String
    Get
      Return Type
    End Get
    Set(ByVal Value As String)
      Type = Value
    End Set
  End Property

  Property SiteX() As String
    Get
      Return X
    End Get
    Set(ByVal Value As String)
      X = Value
    End Set
  End Property

  Property SiteY() As String
    Get
      Return Y
    End Get
    Set(ByVal Value As String)
      Y = Value
    End Set
  End Property
End Class
```

When the *FindSiteLocation* calls the class, the class returns four attributes (in XML format): *SiteName*, *SiteType*, *SiteX*, and *SiteY*.

12 Add an *Impersonation* control as follows.

a. In Solution Explorer, double click on *Web.config* to open the Web.config code window.

b. Insert the following code under `<system.web>`.

```
<identity impersonate="true" userName="yourusername"    password="yourpassword" />
```

13 Click on the Start button or press F5 to run the web page and test it. You can also build the solution first and test the web service by using the URL *http://localhost/chapter8/SiteLocations/SiteService.asmx* in a browser. A web service test page displays (Figure 8-15).

Click on "FindSiteLocations" to bring up the web service test page (Figure 8–16).

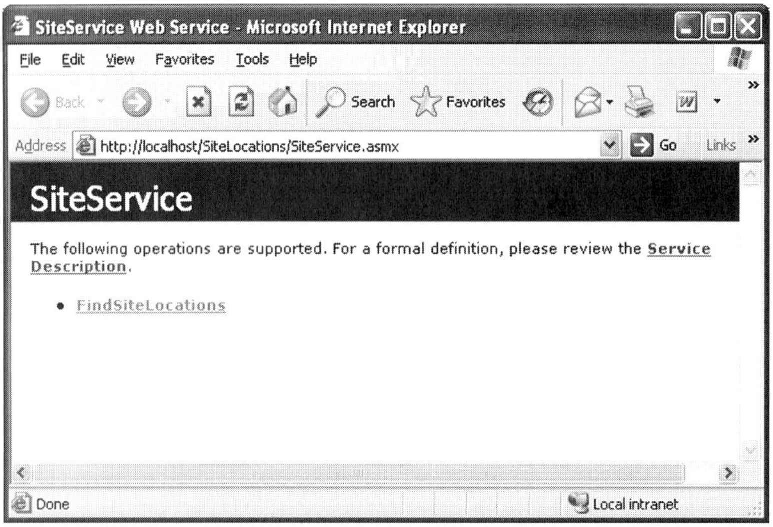

Figure 8–15. The web service test page for "SiteService".

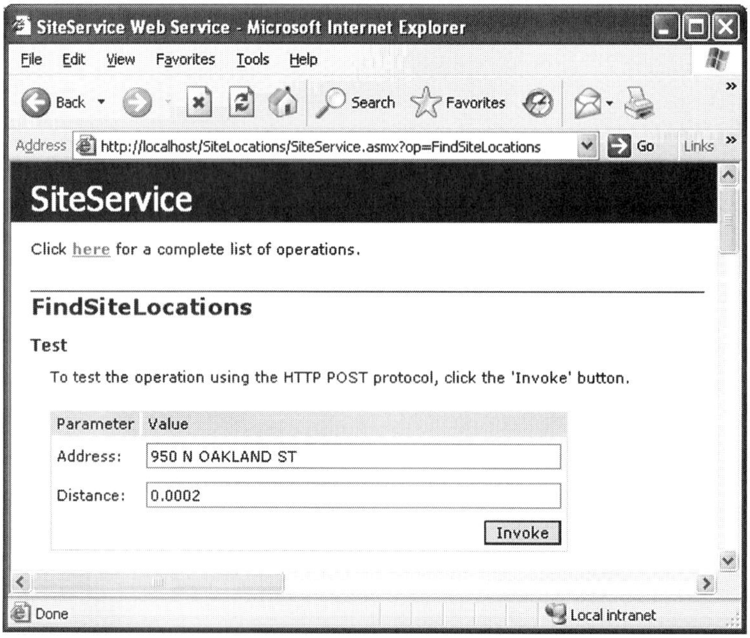

Figure 8–16. The test page for the "FindSiteLocations" method of the "SiteService" web service.

Enter *950 N OAKLAND ST* for the address and *"0.0002"* for the distance. Click on Invoke. A web page appears showing the following XML.

```
 <?xml version="1.0" encoding="utf-8" ?>
<ArrayOfSitePoints  xmlns:xsd="http://www.w3.org/2001/XMLSchema"
   xmlns:xsi="http://www.w3.org/2001/XMLSchema-instance"
   xmlns="http://localhost/chapter8/SiteLocations/SiteService">
  <SitePoints>
   <SiteName>Saint George Episcopal Church</SiteName>
   <SiteType>church</SiteType>
   <SiteX>-77.1047200004873</SiteX>
   <SiteY>38.8822199995135</SiteY>
  </SitePoints>
  <SitePoints xsi:nil="true" />
</ArrayOfSitePoints>
```

The test result shows a site, "Saint George Episcopal Church", which was found within 0.0002 mapping units (decimal degrees) of the given address.

Consuming an Application Web Service in ArcGIS Server

You can consume an application web service in a desktop application, a web application, or another web service. Similar to the procedure discussed under "Consuming a Simple Web Service," you need to add a web reference before you can use the "SiteService" web service. Once the web service is added to your project, you can call the "FindSiteLocation" method.

In the following example, we are going to create a web application using the Map Viewer Web Application template from ArcGIS Server to consume the "SiteService" web service. The concept of this example is that we will create a standard map viewer application and set the server object of the map control to a map server of the web service catalog "agsWebServiceCatalog". When the user enters an address and clicks on the "Find Site" button, the "Find Site" click event will consume the "SiteService" web service by calling two methods: *GetGeocodePoint* and *FindSiteLocation*. The *GetGeocodePoint* method returns a pair of coordinates for the geocoded point for the address. The *FindSiteLocation* method returns the coordinates for the sites of schools, churches, and hospitals found within a given distance of the geocoded point. We then display both geocoded location and site locations on the map.

1 First, we need to add the *GetGeocodePoint* method to "SiteService" web service. In Visual Studio .NET, open the *SiteLocations* project. In Solution Explorer, double click on *SiteService.asmx*. In the code window, scroll to the end of the page and place the following code immediately after the *FindSiteLocation* method.

```
 <WebMethod()> _
Public Function GetGeoCodePoint(ByVal Address As String) As SitePoints()
    Dim WebObj As New WebObject
    If Address = "" Then
      Return Nothing
    Else
      Dim connection As _
              ESRI.ArcGIS.Server.WebControls.ServerConnection
      connection = New _
              ESRI.ArcGIS.Server.WebControls.ServerConnection
      connection.Host = "localhost"
      connection.Connect()
      Dim som As IServerObjectManager
      som = connection.ServerObjectManager
      Dim sc As IServerContext
      sc = som.CreateServerContext("Sites", "GeocodeServer")
      Try
        Dim so As IServerObject
        so = sc.ServerObject
        Dim gc As IGeocodeServer
        gc = so
        Dim ps As IPropertySet
        ps = sc.CreateObject("esriSystem.PropertySet")
        ps.SetProperty("Street", Address)
        'ps.SetProperty("Zone", ZipCode)
        Dim res As IPropertySet
        res = gc.GeocodeAddress(ps, Nothing)
        Dim gcPoint As IPoint
        gcPoint = res.GetProperty("Shape")
        Dim siteList As New ArrayList
        Dim ts As New SitePoints("GeoCoding Point", "GeoCode",
              gcPoint.X, gcPoint.Y)
        siteList.Add(ts)
        Dim siteArray As SitePoints()
        If siteList.Count > 0 Then
          ReDim siteArray(siteList.Count)
          siteList.CopyTo(siteArray)
          Return siteArray
        End If
```

```
      sc.ReleaseContext()
    Catch ex As Exception
      sc.ReleaseContext()
    End Try
  End If
End Function
```

This method returns a pair of X and Y coordinates for the geocoded point. You can preview the modified web service by clicking on Start or pressing F5 (Figure 8-17). You should see that the *GetGeocodePoint* method was inserted into "SiteService" by comparing Figures 8-15 and 8-17.

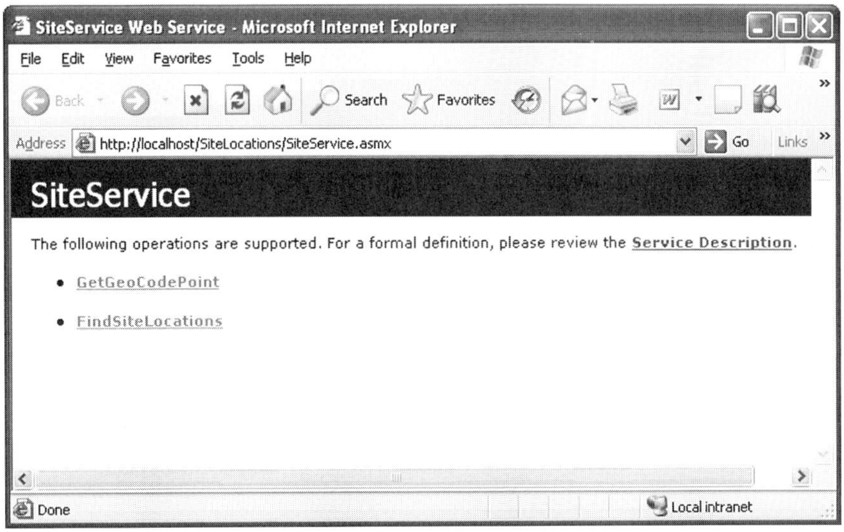

Figure 8-17. The "SiteService" web service containing two web methods.

2 Now let's create a new web application. In Visual Studio .NET, create a new Visual Basic .NET project using the Map Viewer Web Application template under the *ArcGIS Server Projects* folder and name it "WSClientWebMap" for the application. When the designer window opens, you should see the following preloaded ArcGIS Server web controls.

- A toolbar control with preloaded map tools *zoomin*, *zoomout*, *pan*, *fullextent*, *zoomback*, *zoomnext*, and *identify*
- A *Map* control

- A TOC control
- An Overview Map control
- A Scalebar control
- A Northarrow control
- An Impersonation control
- A banner with ESRI logo

3 Add the web service catalog to the project as a reference. Right-mouse click on References and select Add Web Service. The Add Web Reference dialog box appears. Enter the following URL in the URL box and click on the Go button.

http://localhost/chapter8/agsWebServiceCatalog /Default.aspx?WSDL

4 The "agsWebServiceCatalog" information displays. After briefly examining the service, change the web reference name to "WebCatalog" and click on Add Reference. The "agsWebServiceCatalog" web service will be added to the project under the name "WebCatalog".

5 Now we will set the host object and server object for the map control and the map overview control when the page is loaded. Use the following code for the `Page_Load` event.

```
    Dim wc As WebCatalog.Default = New WebCatalog.Default
    Dim wcs As WebCatalog.ServiceDescription()
    wcs = wc.GetServiceDescriptions
    Dim sd As WebCatalog.ServiceDescription
    Dim soname As String
    Dim i As Int16
    For i = 0 To wcs.Length - 1
      sd = wcs(i)
      If sd.Type = "MapServer" Then
        soname = sd.Name
        Exit For
      End If
    Next
' set host and server object for map control and overview map control
    Map1.Host = "localhost"
    Map1.ServerObject = soname
    OverviewMap1.Host = "localhost"
    OverviewMap1.ServerObject = soname
```

6 Click on the *TOC* control and set the *BuddyControl* property to "Map1".

7 Click on the *Impersonation* control. From the Properties window, click on the *Identity* property and an Ellipsis button displays. Click on the button, and then the Impersonation1 Identity dialog box appears. Enter a valid user name and password and click on OK. You can leave the "Domain or Machine Name" box blank.

8 Drag a *Label* control from the Web Forms tab of the toolbox onto the web form and set the *Text* property to "Find Site Location".

9 Drag a *TextBox* control from the Web Forms tab of the toolbox onto the web form and set the *ID* property to "txtLocation".

10 Drag a *Button* Control from the Web Forms tab of the toolbox onto the web form and set the *ID* property to "btnFindSite" and the *Text* property to "Find Site".

11 Click on the HTML Tab in the designer window and change the text for the banner to SAMPLE WEB SERVICE MAP VIEWER.

12 Click on the Start button or press F5 to run the web page and test it.

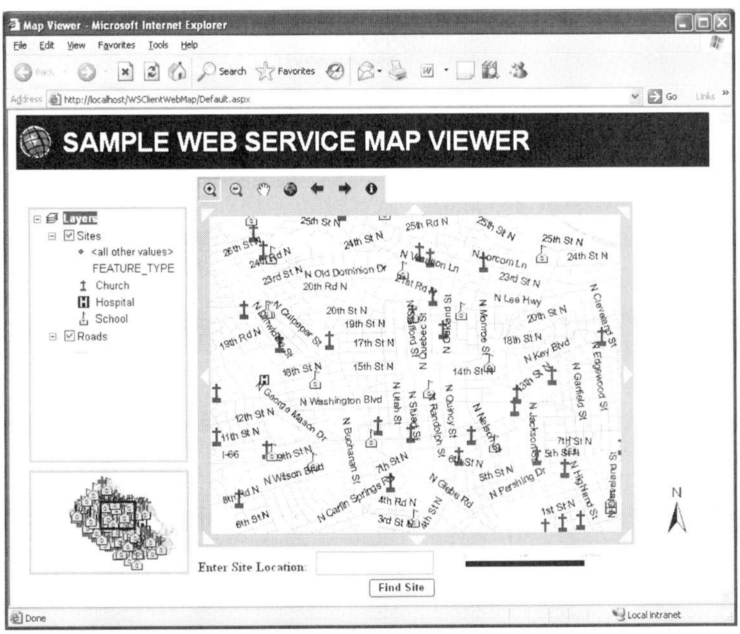

Figure 8–18. A map viewer application that displays the "ArlingtonSites" map service.

A web page displays with an additional site location box and the "Find Site" button. It maps the roads, schools, churches, and hospitals for Arlington County, Virginia (Figure 8–18).

Now we have the application. If you enter an address and click on the "Find Site" button, nothing will happen because we have not linked the application to the "SiteService" web service yet. The following are the steps for adding the "SiteService" web service as a reference.

1 In the Solution Explorer window, click on References and make sure the following ArcGIS Server assemblies are listed.

- *ESRI.ArcGIS.Server.WebControls*
- *ESRI.ArcGIS.Carto*
- *ESRI.ArcGIS.Geometry*
- *ESRI.ArcGIS.Server*
- *ESRI.ArcGIS.System*
- *ESRI.ArcGIS.Display*
- *ESRI.ArcGIS.Location*

See the previous section on how to add references.

2 Now we will add the "SiteService" web service to the project as a reference. Right-mouse click on References and select Add Web Service. The Add Web Reference dialog box appears. Enter the following URL in the URL box and click on the Go button.

http://localhost/chapter8/SiteLocations/SiteService.asmx?WSDL

3 The "SiteService" web service information displays (Figure 8–19). After briefly examining the description and methods of the "SiteServicer" web service, change the web reference name to "SiteLocations" and click on Add Reference.

The "SiteService" Web service will be added to the project as the second web reference under the name "SiteLocations".

4 Double click on the `btnFindSite` button and add the following code to the `btnFindSite_Click` event.

```
If txtLocation.Text = "" Then
  Exit Sub
End If
Dim webmap As WebMap = Map1.CreateWebMap()

Dim sl As SiteLocations.SiteService = New SiteLocations.SiteService
Dim objs As Object() = sl.FindSiteLocations(txtLocation.Text, 0.001)
If objs Is Nothing Then
  txtLocation.Text = "Can't location the site, try a different address"
  Exit Sub
End If
Dim wsServerContext As IServerContext = webmap.ServerContext
Dim graphicElements As IGraphicElements = _
    wsServerContext.CreateObject("esriCarto.GraphicElements")
```

```
Dim wsPoint As IPoint = wsServerContext.CreateObject("esriGeometry.Point")
Dim wsMapDescription As IMapDescription = webmap.MapDescription
Dim gcPoint As IPoint = wsServerContext.CreateObject("esriGeometry.Point")

Dim pts As Object() = sl.GetGeoCodePoint(txtLocation.Text)

' set geocoding point
gcPoint.PutCoords(pts(0).SiteX, pts(0).SiteY)
AddLocation(wsServerContext, wsMapDescription, graphicElements, _
       gcPoint, esriSimpleMarkerStyle.esriSMSCross)

' add site graphic elements
Dim i As Int16
For i = 0 To objs.Length - 1
  If objs(i) Is Nothing Then Exit For
  wsPoint.PutCoords(objs(i).SiteX, objs(i).SiteY)
  AddLocation(wsServerContext, wsMapDescription, graphicElements, _
wsPoint, esriSimpleMarkerStyle.esriSMSDiamond)
Next

'pass array of elements to graphic container of MapDescription
wsMapDescription.CustomGraphics = graphicElements
webmap.Zoom(0.5, gcPoint)
webmap.Refresh()
Map1.Draw()
```

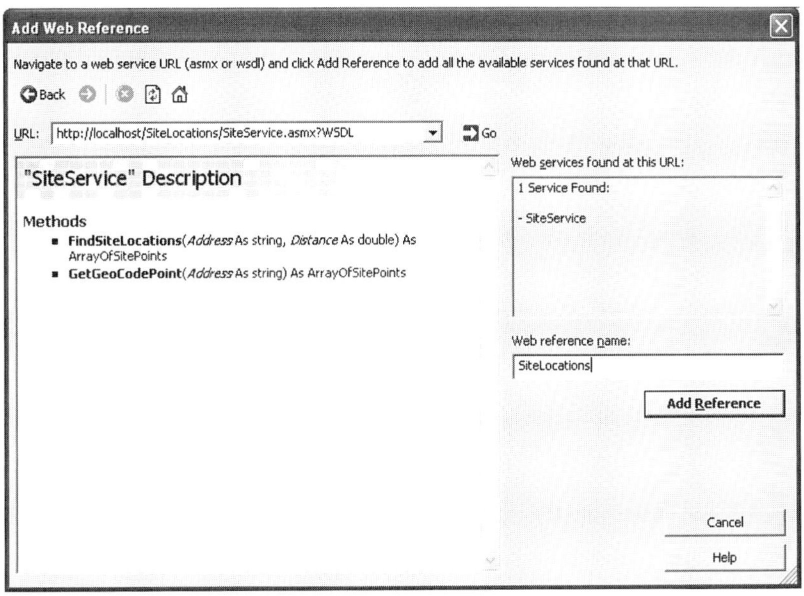

Figure 8–19. The "SiteService" web service is shown in the Add Web Reference dialog box.

Note that a function "AddLocation" was called, so we need to add that function.

✔ **NOTE:** *Although we hard-coded the search distance "0.001" in the code, you can add a search-distance textbox on the web page to allow the user to specify a search distance.*

5 Add the following function immediately after the `btnFindSite_click` event.

```
Private Sub AddLocation(ByVal wsServerContext As IServerContext, _
ByVal wsMapDescription As IMapDescription, _
       ByVal graphicElements As IGraphicElements, _
       ByVal wPoint As IPoint, ByVal ptStyle As Int16)
    wPoint.Project(wsMapDescription.SpatialReference)
    Dim wsGeometry As IGeometry = wPoint
    ' define symbol
    Dim rgbColor As IRgbColor = Converter.ToRGBColor(wsServerContext, _
       System.Drawing.Color.Red)
    Dim simpleMarkerSymbol1 As ISimpleMarkerSymbol = _
       wsServerContext.CreateObject(_
       "esriDisplay.SimpleMarkerSymbol")
    simpleMarkerSymbol1.Color = rgbColor
    simpleMarkerSymbol1.Style = ptStyle
    simpleMarkerSymbol1.Size = 10
    'create marker element
    Dim markerElement1 As IMarkerElement = _
       wsServerContext.CreateObject("esriCarto.MarkerElement")
    markerElement1.Symbol = simpleMarkerSymbol1
    Dim pointElement1 As IElement
    Dim markerGraphicElement1 As IGraphicElement
    'set geometry of point element
    wsGeometry.SpatialReference = wsMapDescription.SpatialReference
    pointElement1 = markerElement1
    pointElement1.Geometry = wPoint
    markerGraphicElement1 = pointElement1
    graphicElements.Add(markerGraphicElement1)
  End Sub
```

6 Click on the Start button or press F5 to run the web application and test it.

A web page displays. Enter *4220 N FAIRFAX DR* in the site location box and click on the "Find Site" button. The web page will refresh with a cross symbol in the middle of the map that shows the geocoded location and with a diamond symbol that shows the resultant site within the search distance (Figure 8–20). In this example, the web application called the "FindSiteLocations" method of the "SiteService" web service to geocode the address and locate a church that was within a given distance (0.001 decimal degrees). A proximity analysis was conducted by buffering the geocoded location at the given distance, and the result was displayed as a diamond symbol on the map. The application also called the "GetGecodePoint" method to retrieve the pair of the geocoded coordinates, and the geocoded point was displayed as a cross symbol on the map.

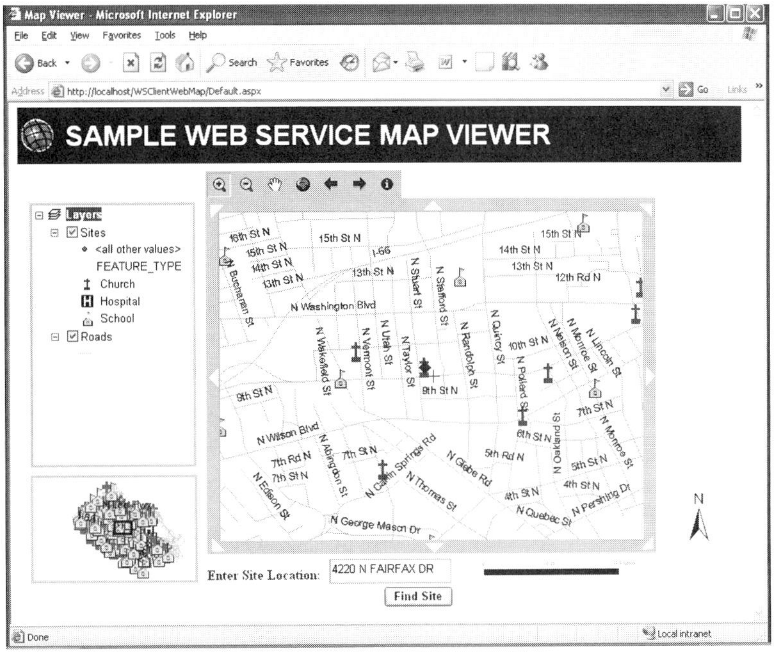

Figure 8–20. The web application that consumes the SiteService web service to locate a site near "4220 N FAIRFAX DR".

Securing Web Services

With the growing popularity of the Web and e-commerce, users are becoming increasingly aware of the vulnerabilities of the information they share with other people over the Web. Security has become a critical issue for many web applications. Using the functionality offered by XML and SOAP, a web service can go right through a firewall, which makes the web service more vulnerable to cracking and other security concerns. In this section, we will first discuss the security issues web services face and techniques you can use to increase security. Finally, we will address how to develop a secured web service.

Web Service Security Issues

In general, security involves controlling users' access to information using techniques such as authentication, permission, and encryption. In an open environment such as the Internet, however, a small loophole in an application structure can cause the application to be cracked. The main security issues relating to web services include:

- Authentication
- Authorization
- Non-repudiation
- Integrity
- Privacy

Authentication

Authentication is the validation of identity, such as identity of a user, a computer, or an application. Authentication represents the physical existence of an identity. Authentication can be implemented in web services using various techniques. The most commonly used technique is the user-name/password system, which can be used at both the client and the server. Other techniques may include cryptography, digital signature, or digital certificates.

Authorization

Authorization depends on the authentication. Once the client is authenticated, it can be authorized to take an action on the data and resources depending on the privileges assigned to it. A standard implementation of this system is the operating system access to a computer or in a network. For example, in Windows environment a user can be assigned to the administration group, power user group, or specific application user group with different privileges for accessing files and data. In the case of web services, there are two fundamental ways to authorize access to a given resource: file authorization and URL authorization.

Non-repudiation

Non-repudiation is a service that ensures that the data sent by the client safely reaches the server. It generates and stores evidence of data transmission. Non-repudiation is implemented by using digital signatures and digital certificates and keeping the private keys secure. It is supported by a certification authority (CA) that issues digital certificates for the generated signatures or keys by the user or any entity.

Integrity

Integrity is the authentication of the data to make sure that the data transmitted has not been tampered with. The most commonly used technique for integrity is the implementation of various encryption algorithms to make the data unintelligible to others during the transmission. When the data is encrypted, nobody can change it without a proper key (to decrypt it). Therefore, encryption ensures data integrity.

Privacy

Similar to integrity, privacy is also about securing data. Whereas integrity is concerned with data tampering, privacy defines mechanisms for keeping data private. Just like integrity, privacy of data is also implemented using encryption mechanisms.

Web Service Security Techniques

Most of these security techniques are based on point-to-point security; that is, security techniques are implemented between sender and receiver. The security techniques are implemented with the combination of encryption algorithms and keys. The encryption and decryption techniques are generally called cryptography. The combinations of encryption algorithms and key settings result in different security technologies. The following are some examples of such security technologies.

Digital Signature

Digital signature is a code that can be attached to an electronically transmitted message. It allows the receiver to check the sent message and see whether it was tampered with before receiving. Digital signatures use public key cryptography for encryption.

Digital Certificate

Digital certificate is the identity of the sender. It contains the sender's details and its public key. The receiver can rely on this certificate and use the listed public key to verify the sender. Digital certificates are issued by a CA. The information stored in the certificate is encoded using two standards: Abstract Syntax Notation 1 (ANS 1) and Distinguished Encoding Rules (DER).

Secure Sockets Layer Protocol (SSL)

The SSL is an important security measure to overcome the limitations of HTTP. URLs that start with *https://* are using the SSL to transfer information over the Internet. The SSL specifies that a document can be transmitted over the SSL connection using a public key to encrypt a session key. The public key acts as a password for the document at the receiver's side. The receiver uses its private key to match it with the public key that is received with the document.

When developing a secured web service, you may choose the appropriate security technology for your application based on business requirements. The next section discusses how to develop a secured web service.

Developing Secured Web Services

When considering security for a web service, there are two key security principles you must consider: *authentication* and *authorization*. Authentication is the process of validating an identity based on credentials (e.g., a user name and password) against an authority. Authorization, on the other hand, determines whether the identity is authorized to access the resource. Web services created using ASP.NET can choose their security options from the authentication and authorization options provided by ASP.NET or customized SOAP-based security. ASP.NET operates in conjunction with IIS to provide several options for authentication and authorization. ASP.NET also offers impersonation capability to execute a request using the credentials of the client.

Windows Authentication

Both IIS and ASP.NET provide support for authenticating web services using security built into Windows. Windows provides three options for authentication: Basic, Digest, and Integrated Windows. Each option can be used with SSL (Securer Sockets Layer). You can use IIS to configure Windows authentication for web services. IIS allows users to specify security at either the directory level or file level. To set the security for a web service on a per-file basis, set permissions for the web service on the *.asmx* file in IIS. The *.asmx* file is the entry point into the web service. You can also modify the ASP.Net configuration file to specify Windows authentication. To do so, you can set the mode attribute of the authentication XML element in a configuration file to “Windows” as follows.

```
<authentication mode="Windows"></authentication>
```

Authorization Options

Windows provides two fundamental ways of authorizing access to a given resource: *file authorization* and *URL authorization*. File authorization can be used whenever Windows authentication is used, as the permissions are set in IIS on a per-file basis. URL authorization can be used with any of the built-in authentication mechanisms supported by ASP.NET. With URL authorization, configuration is done through a configuration file, where users can be selectively granted or denied access to any files associated with ASP.NET, including *.asmx* files. The following is an example of using the *<authorization>* section within the configure file to set grant or deny access.

```
<configuration>
  <system.web>
  <authorization>
    <allow roles="Admin" />
    <deny users="*" />
  </authorization>
  </system.web>
</configuration>
```

This example allows access to all *Admin* roles and denies access to all users.

Managing the State of Web Services

Web services implemented from the *WebService* class have the same state management options as other ASP.NET applications. The *WebService* class contains many of the common ASP.NET objects, including the *Session* and *Application* objects. The *Application* object provides a mechanism for storing data that is accessible to all code running within the web application, whereas the *Session* object allows data to be stored on a per-client session basis. If the client application supports cookies, a cookie can identify the client session. Data stored in the *Session* object is available only when the *EnableSession* property of the *WebMethod* attribute is set to *true* for a class deriving from *WebService*. A class deriving from *WebService* automatically has access to the *Application* object.

The following is an example that accesses and stores state specific to a particular client session.

```
'declare a Web service:
<%@ WebService Language="VB" Class="SiteServiceUsage" %>
'Add a reference to the System.Web.Services namespace
Imports System.Web.Services
'Derive the class implementing the XML Web service from WebService:
Public Class LocationServerUsage: Inherits WebService
'Declare a Web service method, setting the EnableSession property of the
    WebMethod attribute to true:
< WebMethod(EnableSession:=True) > _
    Public Function PerSessionSiteServiceUsage() as Integer
'Store state in the Session object, specifying a name for the state for later
    retrieval
'- increment the usage count:
Session("MyServiceUsage") = 1
'Access the state variable stored in the Session object:
Session("MyServiceUsage") = CInt(Session(MyServiceUsage"))+ 1
```

The following is an example of how to access and store state specific to the web application hosting the web service.

```
'declare a Web service:
<%@ WebService Language="VB" Class="SiteServiceUsage" %>
'Add a reference to the System.Web.Services namespace
Imports System.Web.Services
'Derive the class implementing the XML Web service from WebService:
Public Class LocationServerUsage: Inherits WebService
'Declare a Web service method:
< WebMethod > _
Public Function ApplicationServiceUsage() as Integer
'Store state in Application object, specifying a name for the state for
    later retrieval:
Application("appMyServiceUsage") = 1
'Access the state variable stored in the Application - increment usage count:
Application("pMyServiceUsage" = _
    CInt(Application("appMyServiceUsage")) + 1
```

The difference between the two web methods is if three clients call the second web method consecutively the last one receives returned value of 3, but if the three clients call the first web method consecutively each will receive the same result of 1. The first web method is for a particular client session.

Publishing and Deploying Web Services

Deploying a web service involves copying the *.asmx* file and any assemblies used by the web service to the web server. Suppose you have a web service named "SiteService". To deploy the web service, you create a virtual directory on the web server and place the web service *.asmx* file in that directory. A typical deployment would have the following directory structure.

\Inetpub

\wwwroot

\SiteLocations

\SiteService.asmx

\Bin

Assemblies (dlls) used by your web service that are not part of the .NET framework.

You can also use the Visual Studio .NET built-in tools to deploy your web service to a production server, as follows.

1. In Solution Explorer, click on the *SiteLocations* project.
2. Click on Project and then on Copy Project.
3. In the Copy Project dialog box, specify the location on your production web server.
4. Click on OK.

After copying, you should see the directory structure.

Summary

In this chapter, we provided an overview of the basics of web services technology, including the definition, description language (WSDL), communication protocol (SOAP), and the online registry (UDDI). You can search for GIS-related web services, such as ESRI's ArcWeb Services or Microsoft's MapPoint, from a UDDI web site. We then discussed in detail how to create web services for ArcGIS Server. ArcGIS Server implements two types of web Services: ArcGIS Server web services and application web services. The first includes all of the plumbing an ArcGIS client needs and can be consumed by an ArcGIS client or another ArcGIS Server application.

You use the ArcGIS Server Web Service Catalog template to create an ArcGIS Server web service. But you cannot add functions to the web service. The functionality of the web service depends on the original map documents or address locators. The second targets a client that can consume a SOAP/WSDL web service. Such a client can be any IT component and is not necessarily an ArcGIS client. An application web service can be created based on individual ArcGIS-Server server objects and ArcObjects. You can add extra components to the web service. Before you create a web service for ArcGIS Server, you need to make sure you have at least one GIS server object (whether it is a map server object or geocode server object) on the server.

The techniques on how to access ArcGIS Server web services and application web services were demonstrated in the examples built in this chapter. Before you consume a web service, you need to add that web service to the project you are working on. To add a web service, you need to know the web service URL, whether it is an *.asmx* or WSDL file. The web service catalog created with the ArcGIS Server Web Service Catalog template is retrieved by a web application URL (*.aspx*) by appending "?wsdl" as a query string. If you obtain a web service from a UDDI registry, that web site should provide you the URL of the WSDL for the web service.

At the end of chapter, we briefly reviewed some other issues with web service security, maintaining the state of web services, and publishing and deploying web services. Our intention is to provide you with basic knowledge of those topics. For more information on any of these topics, see the *.NET Framework Developer's Guide*. So far in this book, we have covered web applications and web services for ArcGIS Server. Up to now, we have focused on desktop-based clients that use the web applications and consume the web services. In Chapter 9 we will explore in detail how to build wireless GIS applications for mobile devices.

CHAPTER 9

Developing Mobile Applications

IN PRECEDING CHAPTERS, WE TALKED ABOUT HOW TO DEVELOP ArcGIS Server web applications using ASP.NET. We dealt with web forms and web controls and viewed web pages on PCs. In this chapter, we will talk about how to build mobile GIS applications for personal data assistants (PDAs) and cellular phones. PDAs and cellular phones are much smaller devices than PCs. We cannot simply port all web form applications we have talked about so far to those mobile devices. Luckily, ASP.NET has a sibling called *ASP.NET Mobile*. ASP.NET Mobile offers mobile web forms and mobile web controls for mobile devices. In addition, an application developed using ASP.NET Mobile can also be viewed on PCs. We will start with an introduction to the mobile Internet and talk about mobile emulators, against which developers must have to test their mobile applications. We will then discuss mobile web controls, which are customized web controls for mobile devices. We will also examine mobile web forms. This chapter finishes with how to build mobile GIS applications.

Introduction to Wireless Internet Based GIS Applications

Wireless devices such as PDAs and cellular phones have become part of our daily life, just like personal computers. As a result, GIS over mobile devices is in high demand. Emergency response teams, for example, are well served by mobile GIS. Accessing spatial information on a portable device empowers first responders on the scene. Similarly, emergency response teams can collect information using mobile GIS applications, and feed this information back to the command center for situation assessment and decision making. Survey and environmental monitoring people also benefit from mobile GIS. Recreation users would even love to have one on hiking or fishing trips. Wireless GIS will soon be part of our leisure life.

However, deploying a wireless GIS solution is not as easy as it sounds because we are dealing with devices that are far less powerful than personal computers. Developing wireless Internet applications is quite different from developing regular web applications (applications you access through a web browser on your PC). For regular web development, you develop, test, and use an application for the same environment. We all use web browsers to view these applications. For wireless Internet application development, however, you develop an application on PCs but test and use the application on targeted mobile devices.

Personal Data Assistants (PDAs)

In general, there are two types of PDAs: Windows Mobile based PDAs such as HP's iPaqs and Java-based PDAs such as Palms and BlackBerry. When PDAs were new, Microsoft had a product called Windows CE for compact and mobile devices. Then Microsoft released Pocket PC as its replacement. In 2003, Microsoft renamed the product Windows Mobile 2003, which has two editions: one for PDAs and the other for smart phones. Smart phones are cellular phones powered by Microsoft Pocket PC or Windows Mobile 2003. Just like personal computers, mobile devices have a variety of functionality (see Table 9–1).

Table 9–1: Comparison of Mobile Devices

Device	E-mail	Voice	Web	Bluetooth	Wi-Fi
BlackBerry	Encrypted via BlackBerry Enterprise Server (secure)	Via carriers	WAP, HTTP, and HTTPS	Yes	No
PDA	Web client (insecure)	Via carriers	WAP, HTTP, and HTTPS	Yes	Yes
Cell phones	WAP client (insecure)	Via carriers	WAP, HTTP	Yes	No

Palm and BlackBerry are both based on Java technology. Palm is the product line of PalmOne located in Milpitas, California, and BlackBerry is the product line of Research In Motion located in Waterloo, Ontario, Canada. Palm and BlackBerry are both handheld devices, but BlackBerry offers a unique encrypted e-mail feature. BlackBerry requires a

BlackBerry Enterprise Server to serve its services such as encrypted e-mail. Because BlackBerry devices allow users to easily access e-mail, BlackBerry devices are generally purchased by corporations and organizations for their employees. The latest BlackBerry devices come with an organizer, cell phone, wireless Internet, encrypted e-mail, and even Bluetooth. BlackBerry claims that it is the only PDA that can be operated with one hand (and although it is a feature we do not recommend it!). For example, you can simply roll and push the trackwheel to read e-mails, which can be pushed and scrolled as well.

✔ **NOTE:** *Palm runs an operating system called Palm OS from PalmSource of Sunnyvale, California. PalmOne and PalmSource are two independent companies. They used to be united as Palm. Palm bought its direct competitor Handspring in 2003 and then formed PalmOne and PalmSource. PalmOne produces the handheld devices, whereas PalmSource makes Palm OS.*

Cellular Phones

There are two types of wireless systems prevalent in the world now: Global System for Mobile Communications (GSM) and Code Division Multiple Access (CDMA). GSM is widely used in over 170 countries. If you own one of those quad-band GSM cellular phones, you can basically use your phone when you travel around the world (1900 MHz and 850 MHz used in the United States, and 900 MHz and 1800 MHz used elsewhere). Cingular, AT&T, and T-Mobile are GSM providers, whereas Verizon and Sprint are CDMA providers. Nextel uses an old technology called iDen. At the time of this writing, Cingular has successfully acquired AT&T Wireless, and Nextel is merging with Sprint. Soon we will not have to remember which carrier offers what. In addition, we deliver our wireless GIS solutions through the wireless Internet, and we only need to make sure that the cell phone we are using to test our solutions supports wireless Internet.

✔ **NOTE:** *CDMA is one of several digital wireless transmission methods. CDMA was invented by Qualcomm of San Diego, California. GSM was invented in Europe. TDMA stands for Time Division Multiple Access, which is an old technology invented by Lucent, Murray Hill, New Jersey. iDen stands for Integrated Dispatch Enhanced Network, a Motorola technology based on TDMA.*

Compact HTML and WML

PDAs and cellular phones do not use HTML. Instead, they use Compact HTML (cHTML) and Wireless Markup Language (WML), respectively. cHTML is a subset of HTML designed for small information appliances such as PDAs, whereas WML is specifically designed for mobile phones. Due to the hardware restrictions of small appliances such as small memory, low power CPU, small and even mono-color display, and restricted input method, style-rich HTML cannot be adopted for small devices. Thus, we use cHTML, which gives us basically web elements so that we can still build web pages with bare-bones HTML forms. cHTML still runs via the HTTP protocol, whereas WML communicates through the Wireless Application Protocol (WAP). A WML page is called a deck, which consists of multiple cards (a card equivalent to a screen of information). A WML page is served through the WAP gateway from a carrier. Both PDAs and cell phones have their own browsers: iPaq uses a trimmed-down version of Internet Explorer called Pocket IE. Palm and Balckberry have proprietary browsers. Different cell phones have different versions of wireless browsers.

ASP.NET Mobile

In addition to multiple development environments, web protocols, and markup languages, the major challenge to wireless web development is limited screen size. This challenge is exacerbated in GIS application development because GIS application display maps require a big and sharp screen for optimum results. As mentioned previously, there are hundreds or even thousands of devices, making it impossible to code for the various screen sizes. With ASP.NET Mobile, which is part of ASP.NET 1.1, you have the tools you need to quickly write mobile web applications for a wide variety of mobile browsers. ASP.NET Mobile was called the Microsoft Mobile Internet Toolkit (MMIT), an extension for ASP.NET 1.0. Now it is part of ASP.NET 1.1 and comes with Visual Studio .NET 2003. ASP.NET Mobile relieves developers from the challenge of writing and maintaining numerous web applications, each targeted to a specific wireless browser. It comes with mobile web server controls and can render the appropriate markup (HTML 3.2, WML 1.1, cHTML) while dealing with different screen sizes, orientations, and device capabilities.

Mobile Emulators

As we mentioned previously, mobile devices differ from one to another. Keep in mind that we are talking about hundreds of models of cellular phones and PDAs, if not thousands. Furthermore, the same device renders a wireless Internet page differently from one carrier to another in the United States. You definitely need an emulator/simulator when developing any wireless application. You absolutely must have at least one emulator to develop a wireless GIS application. The reason is that an emulator runs on a PC, which is most likely connected via a broadband Internet or better via HTTP. The emulator will then run much faster than a real device. It would also be much more economical than a real device in terms of service charges (your real device is most likely billed by minutes or bytes). We will primarily focus on three types of emulators.

- Microsoft PDA emulator
- BlackBerry simulator
- Openwave phone simulator

These are well-known emulators/simulators for wireless application development. You can download these emulators/simulators from the Microsoft, BlackBerry, and Openwave web sites.

Installing Microsoft PDA Emulator

The Microsoft PDA emulator can be downloaded from the Microsoft download web site. You need to download the Pocket PC 2003 SDK and the emulator images from the web site. Then you need to download eMbedded Visual C++ 4.0 and its Service Pack 4 if you want to have a standalone emulator. Once you download them, you must perform the following sequence to install Pocket PC SDK 2003 and eMbedded Visual C++ 4.0.

1. eMbedded Visual C++ 4.0
2. eMbedded Visual C++ 4.0 Service Pack 4
3. Packet PC 2003 SDK
4. Pocket PC 2003 Emulator Images

Otherwise, the standalone version of the Pocket PC 2003 emulator will not be available.

✔ **NOTE:** *Pocket PC has been renamed Windows Mobile. At the time of writing, the latest version is Windows Mobile 2003. You can find the Pocket PC SDK and its emulator images at the following web sites:*

- *Pocket PC SDK:* http://www.microsoft.com/downloads/details. aspx?FamilyID = 9996b314-0364-4623-9ede-0b5fbb133652&displayang = en
- *Emulator images:* http://www.microsoft.com/downloads/details. aspx?FamilyID = 57265402-47a8-4ce4-9aa7-5fe85b95de72&DisplayLang = en

  If you cannot find them, search "Pocket PC Emulator" or "Pocket PC SDK" from Google and you should be able to find them. eMbedded Visual C++ 4.0 and eMbedded Visual C++ 4.0 Service Pack 4 are available at the following web site.

- *eMbedded Visual C++ 4.0:* http://www.microsoft.com/downloads/details.aspx?FamilyID = 1dacd3d-50d1-41b2-a107-fa75ae960856&displayang = en
- *eMbedded Visual C++ 4.0 SP 4:* http://www.microsoft.com/downloads/details.aspx?FamilyID = 4a4ed1f4-91d3-4dbe-986e-a812984318e5&displaylang = en

You need to install eMbedded Visual C++ 4.0 before you install its Service Pack 4. You will need to get the CD key from the web site (TRT7H-KD36T-FRH8D-6QH8P-VFJHQ) and install eMbedded Visual C++ 4.0 SP 4. Then you install Pocket PC 2003 SDK and the emulator images. Follow the instructions on the screen to install eMbedded Visual ++ 4.0 and Pocket PC 2003 SDK. After the installation, click on *Start/Programs/Microsoft Pocket PC 2003 SDK/Pocket PC 2003 Emulator* to start the emulator.

To install the emulator for Visual Studio .NET 2003, you must log on your PC with administrator privileges so that you can install or remove the emulator. After you install the two downloads from the Microsoft download web site, you can then launch Visual Studio .NET 2003 and do the following.

1 From the Visual Studio .NET Tools menu, select Connect to Device.

This accesses the Connect to Device dialog box, shown in Figure 9–1. The default platform is Pocket PC. You can also select Windows CE from the Platform option list.

2 From the Devices option list, click on Pocket PC 2003 Emulator.

3 Click on the Connect button.

4 If prompted to accept the emulator driver security certificate, click on OK. If successful, the emulator appears on screen (Figure 9–2).

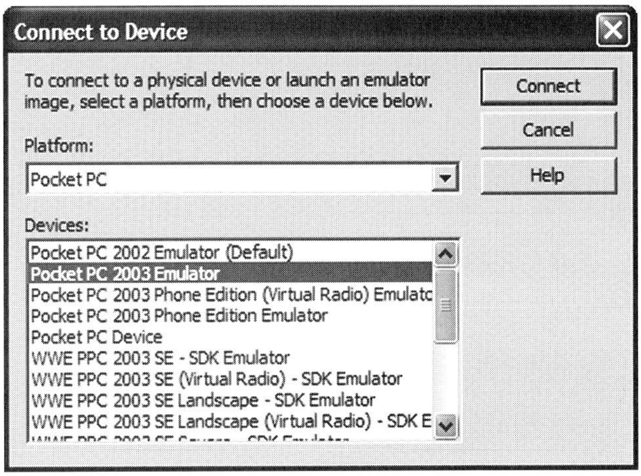

Figure 9–1. Connect to Device dialog box.

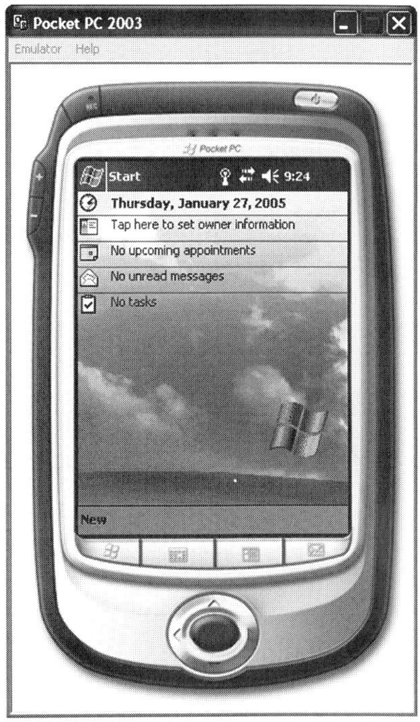

Figure 9–2. Pocket PC 2003 emulator.

5 The status bar in the IDE displays "Device Connected."

6 From the emulator, click on Start/Settings and the Connections tab, and then double click on Connections. When the Connections Settings window appears, click on the Advanced tab (Figure 9–3).

7 Click on the Select Networks button. Assuming your LAN is connected to the Internet directly, select My Work Network for both drop-down boxes (Figure 9–4), which indicates that it automatically connects to the Internet and private network using the My Work Network option. If your LAN is connected to the Internet through a proxy, you need to click on the Edit button to configure the proxy settings.

8 Click on OK two times and close out Settings until you see Start, as shown in Figure 9–2.

9 Click on Start/Internet Explorer and enter the following URL: *http://mobile.msn.com*.

The Pocket Internet Explorer (PDA version of Internet Explorer) appears and displays the home page of *mobile.msn.com* (Figure 9–5).

✔ **NOTE:** *You cannot use "localhost" to view any web page on your local machine using the Pocket PC 2003 emulator. You need to use the name or IP of your local machine instead.*

Now you can enter any other URL in the Address Bar to surf the Internet. Using the emulator, developing wireless applications is similar to developing web applications and testing them with web browsers on your PC.

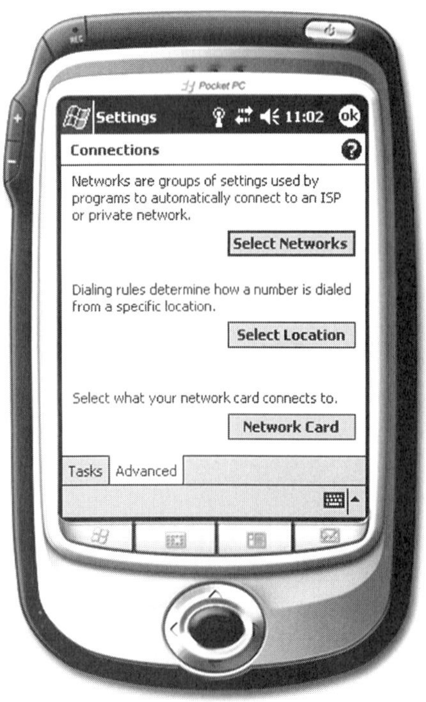

Figure 9–3. The advanced Connections setting interface for the Pocket PC 2003 emulator.

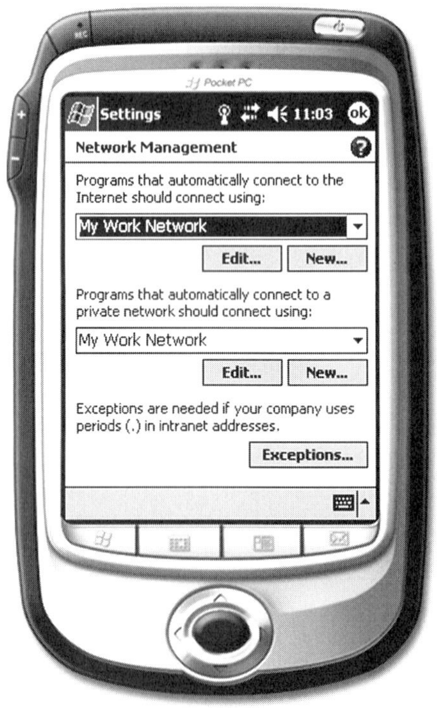

Figure 9–4. Network Management setting interface for the Pocket PC 2003 emulator.

Figure 9–5. MSN Mobile home page shown in Pocket Internet Explorer on the Pocket PC 2003 emulator.

Installing BlackBerry Simulator

The BlackBerry Simulator can be downloaded from the BlackBerry web site for developers. You need to download the BlackBerry SDK. BlackBerry uses its own proprietary technology, BlackBerry Java Development Environment (JDE), which is based on Java instead of Windows Mobile. The BlackBerry Simulator is included in its JDE. After you download it, you need to install the SDK.

✔ **NOTE:** *At the time of writing, the latest version of BlackBerry JDE is 4.0, which is available from its download web site at the following address.*

- *BlackBerry JDE:* http://www.blackberry.com/developers/na/java/tools/index.shtml

You need to provide your contact information and e-mail address on a form before downloading.

To install the simulator, you must log on your PC with administrator privileges so that you can install or remove the simulator. The BlackBerry JDE is a self-extraction file. Click on the file to start the installation, accept the license agreement, and then follow the instructions on screen (take all default settings and conduct a complete installation). If the installation is successful, click on Start/Programs/Research In Motion and you will see "BlackBerry Java Development Environment 4.0." To start the BlackBerry Simulator, you need to do the following:

1 Click on Start/Programs/Research In Motion/BlackBerry Java Development Environment 4.0/MDS Simulator. A Windows command prompt pops up (Figure 9–6).

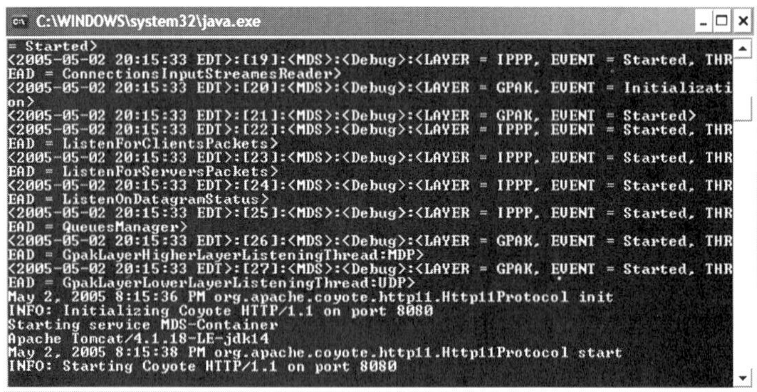

Figure 9–6. The BlackBerry MDS simulator (a Java program that runs inside a Windows command prompt).

2 Click on Start/Programs/Research In Motion/BlackBerry Java Development Environment 4.0/Device Simulator. A Device Simulator Windows command prompt pops up and then the BlackBerry Simulator follows (Figure 9–7) with the BlackBerry Radio Simulation control panel displayed.

3 Use the Up and Down arrow keys to highlight an application/option. Highlight the Browser and press Enter to see the Bookmarks page.

4 Highlight Home Page and press Enter. The BlackBerry home page displays (Figure 9–8). If you would like to enter your own favorite URL, follow steps 5 through 7. Otherwise, go to step 9.

5 Click on the Trackwheel on the simulator or press the Right arrow key to bring up a menu (Figure 9–9).

6 Use the Down arrow to highlight the Go To option and then press Enter. A URL dialog box pops up (Figure 9–10). The BlackBerry browser does not have the location window nor address bar.

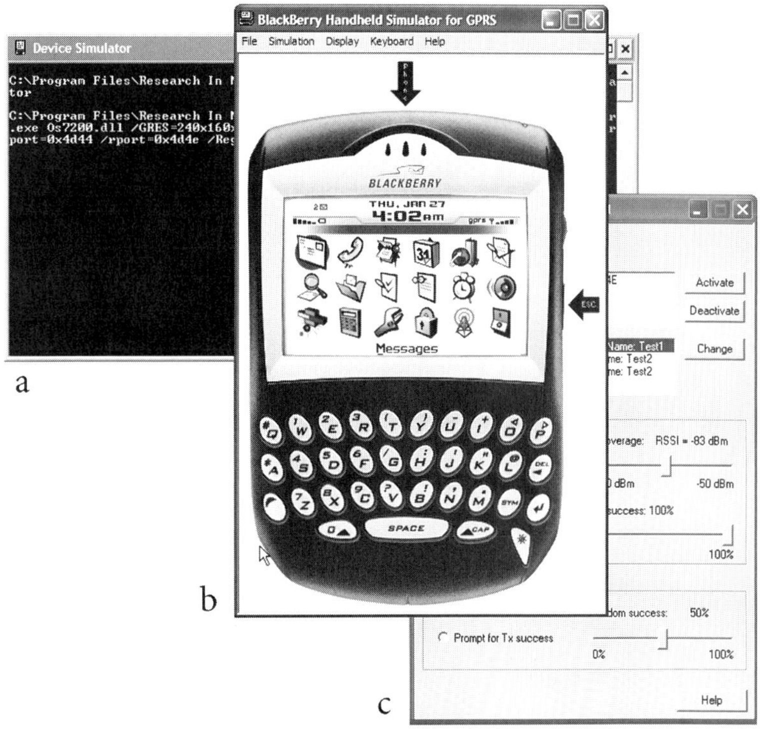

Figure 9–7. (a) The BlackBerry Device Simulator Windows command prompt (back), (b) the BlackBerry simulator (front), and (c) the BlackBerry Radio Simulation control panel (right).

Figure 9–8. BlackBerry home page shown in the BlackBerry browser on the BlackBerry simulator.

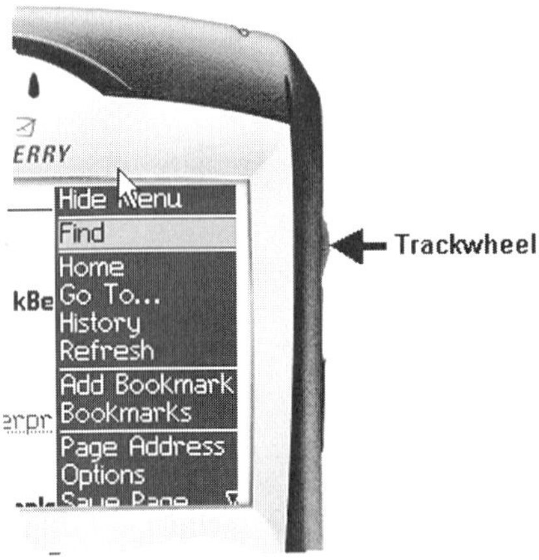

Figure 9–9. Web operation menu of the BlackBerry browser.

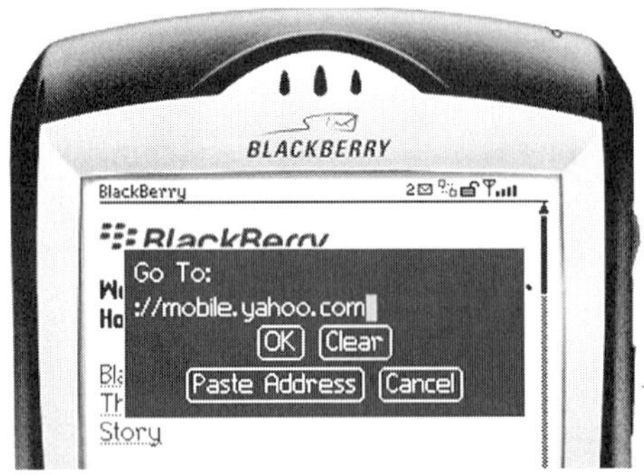

Figure 9–10. URL dialog box for the BlackBerry simulator.

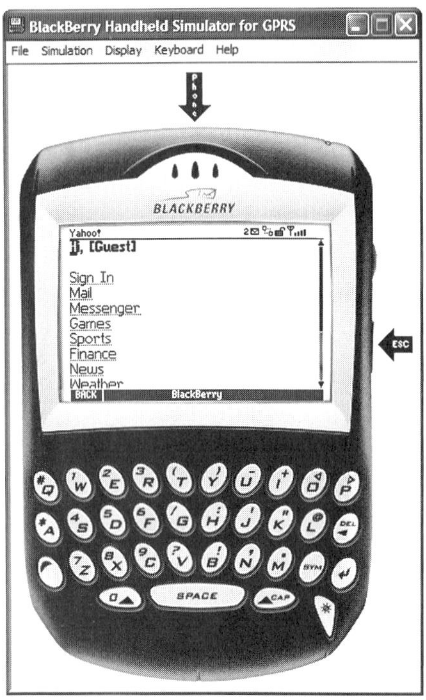

Figure 9–11. Yahoo Mobile home page shown in the BlackBerry browser on the BlackBerry simulator.

7 Enter your URL (e.g., *mobile.yahoo.com*). Use the Down arrow to highlight the OK button and then press Enter. Yahoo's Mobile home page displays in the BlackBerry browser (Figure 9–11).

8 Click on the Windows Close button on each of the simulators to quit the simulator.

✔ **NOTE:** *You must first start with the MDS simulator and then the BlackBerry simulator. MDS stands for Mobile Data Service. It is a feature of the BlackBerry Enterprise Server that enables always-on/push-based access to enterprise applications and information using the BlackBerry device, browser, and software development tools. MDS also provides HTTP/HTTPS connectivity as an open, extensible, and secure interface for extending corporate applications and the corporate intranet based on existing web standards.*

Well, you may ask why BlackBerry cannot lump those two simulators as a single click instead of two clicks. We posed this question to Blackberry. Basically, because all BlackBerry services go through the BlackBerry Enterprise Server, there is no exception for the BlackBerry simulator. Starting MDS first simulates starting the BlackBerry Enterprise Server, from which you fire up the BlackBerry simulator.

Installing the Openwave Phone Simulator

The Openwave phone simulator can be downloaded from the Openwave Developer Network web site. You only need to download the Openwave phone simulator if you use it as a standalone simulator. If you want to use it in conjunction with Visual Studio .NET so that you can test and preview your applications in the simulator without having to leave your IDE, you would need to download the Openwave Simulator VS Integration add-in. You must register before downloading the software if you do not have an active account with Openwave.

✔ **NOTE:** *At the time of writing, the latest version of the Openwave phone simulator is 7.0 Preview, which is available from its download web site at the following address.*

- http://developer.openwave.com/dvl/tools_and_sdk/openwave_ mobile_sdk/phone_simulator/index.htm

The latest version of the Openwave Simulator VS Integration add-in is available at:

- http://developer.openwave.com/dvl/tools_and_sdk/ide_integration/microsoft/visualstudio-addin.htm

To install the simulator, you must log onto your PC with administrator privileges so that you can install or remove the simulator. The Openwave phone simulator is a self-extraction file. Click on the file to start the installation, accept the license agreement, and then follow the instructions on the screen (take all default settings and conduct a complete installation). If the installation is successful, click on Start/Programs/Openwave V7 Simulator Preview/Openwave V7 Simulator Preview to start the simulator. According to Openwave, the Openwave Phone Simulator 7.0 Preview is connected via WAP protocol by default via the Openwave Developer Mobile Access Gateway running on *devgate2.openwave.com*. You need to add an HTTP configuration to browse your applications using the simulator. Perform the following steps to configure the simulator for a *localhost* HTTP.

Figure 9–12. Openwave simulator console (back) and Openwave simulator 7.0 (front).

1 Click on Start/Programs/Openwave V7 Simulator Preview/Openwave V7 Simulator Preview to start the simulator. The Openwave Simulator Console window pops up, followed by the Openwave Simulator 7.0 (Figure 9–12). When the Openwave simulators first launch, they may appear in the background, so you may have to bring them to the foreground.

2 From the simulator, enter the following URL:

http://localhost/MobileWebApplication1/MobileWebForm1.aspx.

At this time, the simulator screen displays "HTTP Error :404 Not Found" (Figure 9–13). This is caused by the default WAP protocol, as mentioned previously.

3 From the Openwave Simulator 7.0 Settings tool, click on Server Profiles. The Server Configuration Profiles dialog box pops up with the Openwave profile highlighted and checked (Figure 9–14).

4 Highlight "http-localhost" and click on the Activate button.

The "http-localhost" profile becomes checked (Figure 9–15), indicating that you are going to use your local web server via HTTP instead of WAP. You can also define a new profile for a different web server other than the *localhost* by clicking on the New button.

5 Click on the Close button to save the configuration.

6 Reenter the URL of step 2. The simulator screen will show a greeting message of "Hello, mGIS!", as shown in Figure 9–16. Hallelujah! You are in business.

✔ **NOTE:** *According to Openwave, the following are the requirements for running the Openwave simulator inside Visual Studio .NET.*

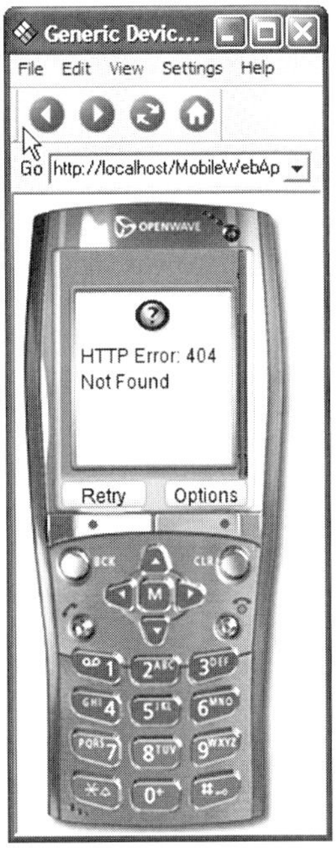

Figure 9–13. An HTTP 404 error shown in the Openwave simulator 7.0.

Figure 9–14. Server Configuration Profiles dialog box for the Openwave simulator 7.0 with the Openwave WAP protocol checked by default.

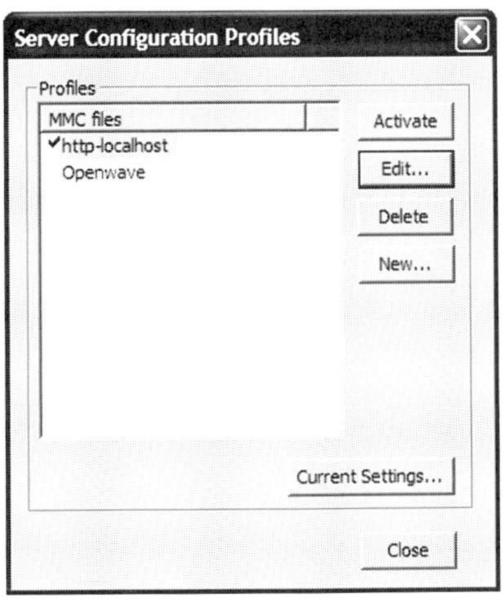

Figure 9–15. Server Configuration Profiles dialog box showing the status after activating the "http-localhost" profile.

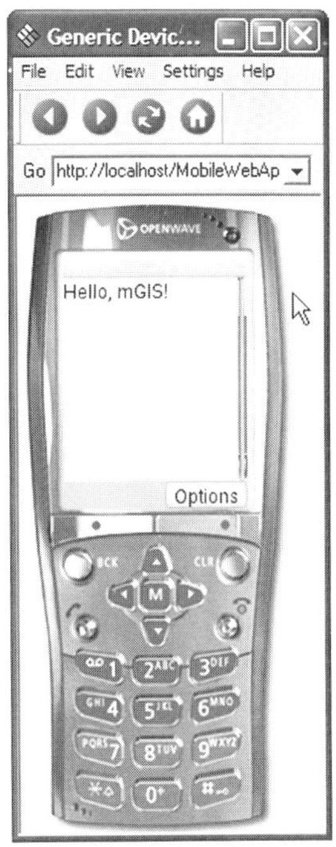

Figure 9–16. The "Hello, mGIS!" message is showing via the Openwave simulator 7.0.

- *Visual Studio .NET*
- *Openwave Simulator VS Integration add-in* (OpwvSimVSSetup.msi)
- *Openwave SDK 4.1.1, SDK 6.1, SDK 6.2, and/or SDK 7.0*

To install the Openwave Simulator VS Integration add-in for Visual Studio .NET, you need to do the following.

1 If Visual Studio .NET is running, close it.

2 Double click on the Openwave Simulator VS Integration setup file (*OpwvSimVSSetup.msi*) to start the installation.

3 Follow the setup prompts to complete the installation.

4 Restart Visual Studio .NET.

After the installation, you need to configure the Openwave Simulator VS Integration add-in prior to its first use. You do so as follows.

1 From the Visual Studio .NET Tools menu, select Openwave Simulator/Options.

2 The Options dialog box appears (Figure 9–17), with tabs corresponding to the Openwave SDK 4.1.1, SDK 6.1, SDK 6.2, and SDK 7.0.

3 Click on the SDK 7.0 tab. The correct path to the Openwave SDK 7.0 executable (C:*\Program Files\Openwave\V7 Simulator Preview\bin\phone.exe*) should have been set already.

4 Leave the Debugging Integration setting at Use Project Default, unless you really want to use Simulator 7.0 to debug your codes. Debugging inside Visual Studio .NET would make your coding time more efficient than debugging the code using a simulator/emulator. You can also have this setting applied to general web projects and not just mobile web projects, by selecting the checkbox on the bottom.

5 Click on OK to save your settings and close the dialog.

Now you can view content with the Openwave Simulator VS add-in. However, you can only do it in design mode. You first open an ASP.NET Mobile project, then from the Tools menu select Openwave Simulator > Openwave Simulator > View in 7 to view the current content in the Openwave Simulator 7.0. Make sure you have configured the simulator via HTTP before viewing the content inside Visual Studio .NET. If you have not done that, see the HTTP configuration described previously.

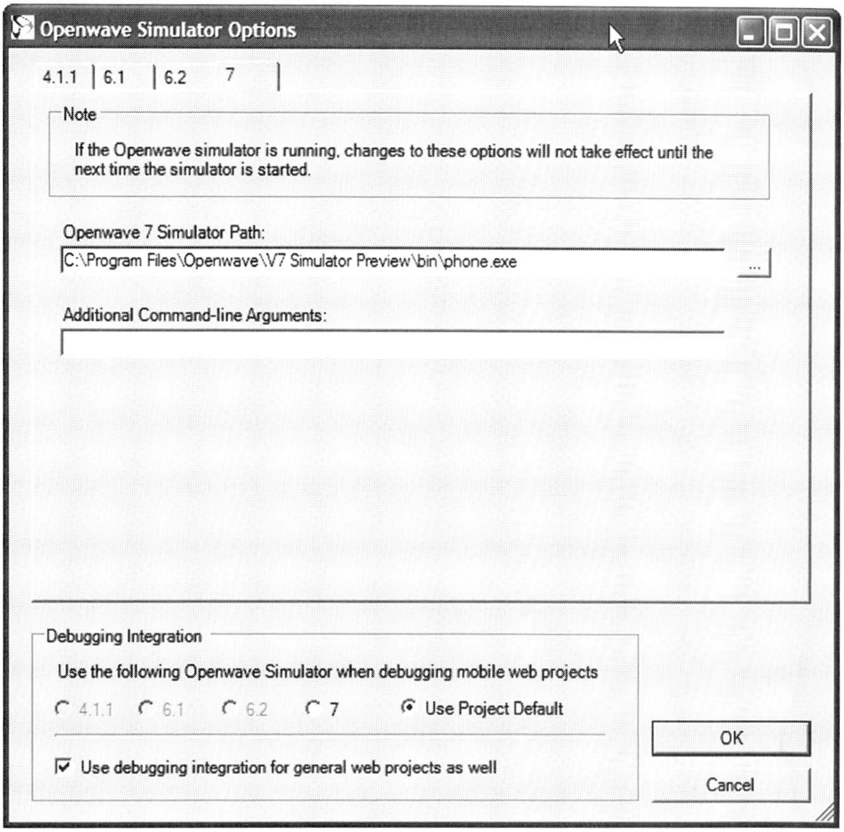

Figure 9–17. Openwave simulator Options dialog box for Visual Studio .NET.

✔ **NOTE:** *We suggest that you download Openwave Simulator 4.1.1 and install it for Visual Studio .NET as well. Openwave Simulator 4.1.1 is good at testing WML pages with black-and-white images.*

Mobile Controls

In Chapter 2, we created our first ASP.NET project *WebApplication1*. It had a web form and a *Label* control on the form. A greeting message of "Hello ArcGIS Server" was then set to the *Text* property of the *Label* control. We will follow the same logic to create an ASP.NET Mobile project called *MobileWebApplication1* to display a greeting message of "Hello mGIS!". After we finish the project, you will find out how similar *MobileWebApplication1* is to *WebApplication1*.

Creating First ASP.NET Mobile Web Project

We will use the ASP.NET Mobile Web template and a *Mobile Label* control. The following are the steps for creating a new project.

To create a Mobile web form:

1. Launch Visual Studio .NET and click on the New Project button. The New Project dialog box appears.
2. In the Project Type pane, click on Visual Basic Projects, and in the Templates pane select ASP.NET Mobile Web Application.
3. Name your application *MobileWebApplication1* in the Location box (Figure 9–18).
4. Click on OK.

The *MobileWebApplication1* project is created with the default Mobile web form called *Form1* displayed (Figure 9–19). This is different from *WebApplication1*, which has only one form. You can have many mobile forms with ASP.NET Mobile. We discuss that subject in the "Mobile Web Forms" section.

To add the Mobile web controls:

5. Drag a *Label* control on the Mobile Web Forms tab of the toolbox onto the Mobile web form *Form1* and set the *Text* property to "Hello mGIS!".
6. Click on Debug/Start or on the Start Button on the toolbar to run the first mobile web application. Because we set Use Project Default for debugging, which uses Internet Explorer by default, an IE window pops up with the greeting message (Figure 9–20).
7. Write down the URL and then close IE.
8. Click on the Save All button or File/Save All to save the project.
9. Close Visual Studio .NET.
10. Take the URL from IE and then enter it in the Pocket IE, BlackBerry browser, and Openwave browser (Figure 9–21). You should see the greeting message displayed in all emulators.

Now you can see that creating ASP.NET Mobile web projects is similar to building ASP.NET web projects except that the ASP.NET Mobile web application works in IE, the Pocket IE, the BlackBerry browser, and the

Openwave browser. The process is much more efficient. You build it once for all devices.

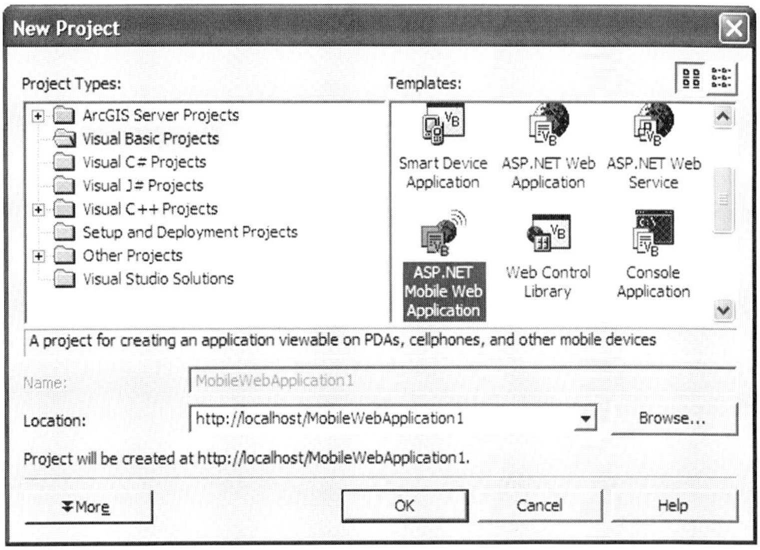

Figure 9–18. New Project dialog box showing the available templates, with the ASP.NET Mobile Web Application template highlighted.

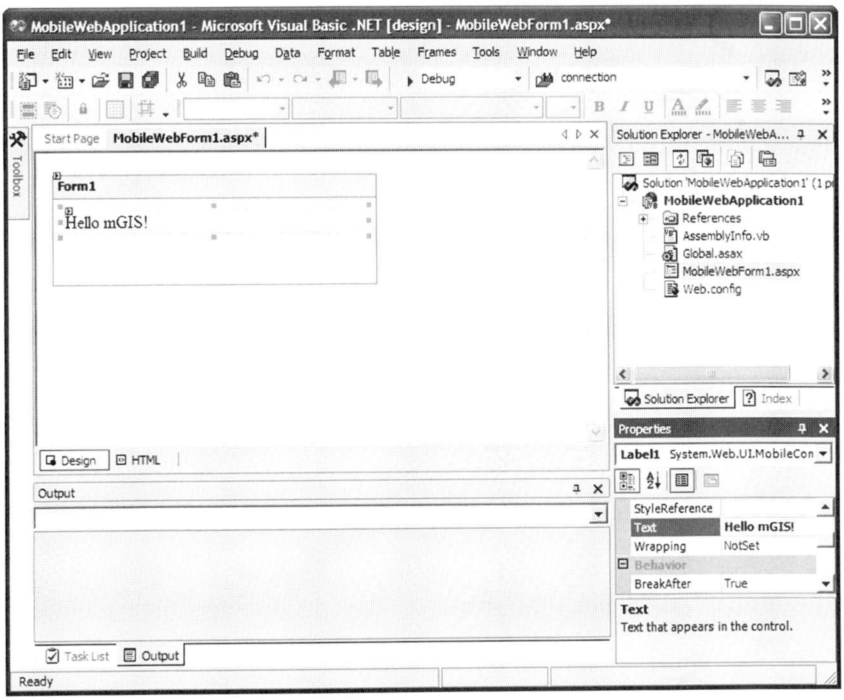

Figure 9–19. The Visual Studio .NET interface shows a design-mode ASP.NET Mobile web application.

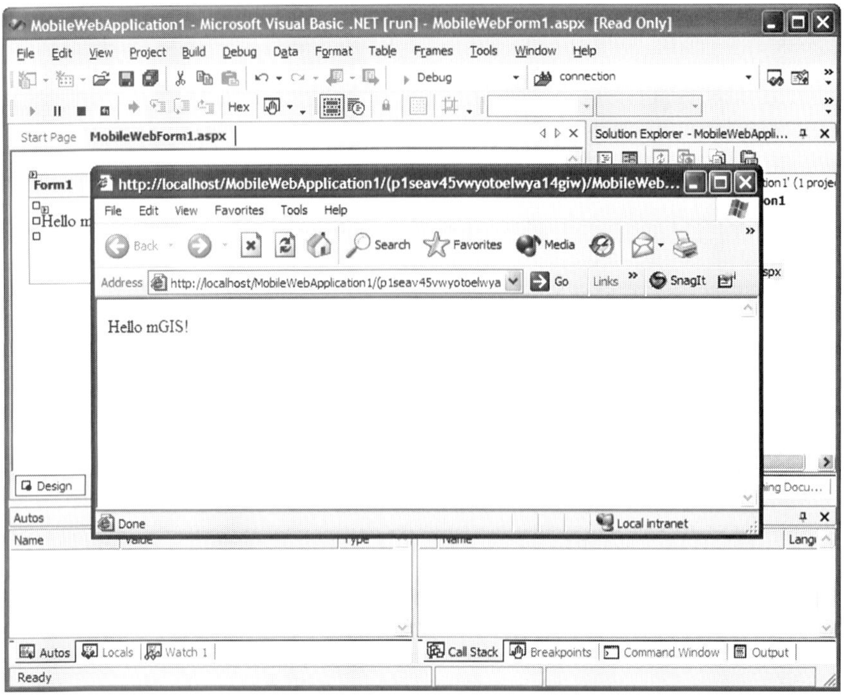

Figure 9–20. The IE browser shows the "Hello, mGIS!" greeting message with the Visual Studio project in the background.

Figure 9–21. Run MobileWebApplication1 *in the Pocket PC 2003 emulator (left), the BlackBerry simulator (middle), and the Openwave simulator 7.0 (right).*

Adding an Event Handler

If you recall, we added an interaction to *WebApplication1* to create *WebApplication2* in Chapter 2. Basically, we added a *Button* control to *WebApplication1* and a clicking handle so that the greeting message only shows up when the button is clicked. We will now follow that example using the ASP.NET Mobile web project *MobileWebApplication2*. The following are the steps for creating a new project.

To create a Mobile web form:

1 Launch Visual Studio .NET and then click on the New Project button.

2 In the Project Type pane, click on Visual Basic Projects, and in the Templates pane select ASP.NET Mobile Web Application.

3 Name your application *MobileWebApplication2* in the Location box.

4 Click on OK.

To add the Mobile web controls:

5 Drag a *Label* control on the Mobile Web Forms tab of the toolbox onto the Mobile web form *Form1* and set the *Text* property to blank.

6 Drag a *Command* control on the Mobile Web Forms tab of the toolbox onto the Mobile web form *Form1* (Figure 9–22). The *Mobile Command* control is the same as the *Button* web control.

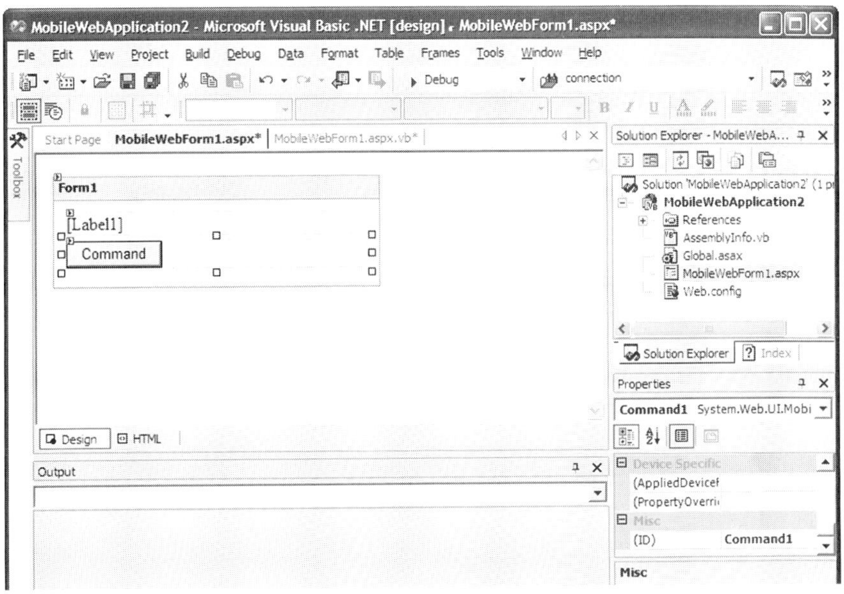

Figure 9–22. Command control shown at design mode in Visual Studio .NET.

7 Double click on the *Command* control and add the following code to the `Command1_Click` event.

```
'Declare a string variable for the greetings
Dim myGreetings As String = "Hello mGIS!"
Label1.Text = myGreetings
```

8 Click on the Start button or press F5 to run the application. A web page with the button displays (Figure 9–23).

9 Click on the button on the web page. The web page will refresh and show the "Hello mGIS!" greeting message (Figure 9–24).

Figure 9–23. Example of using the Command control.

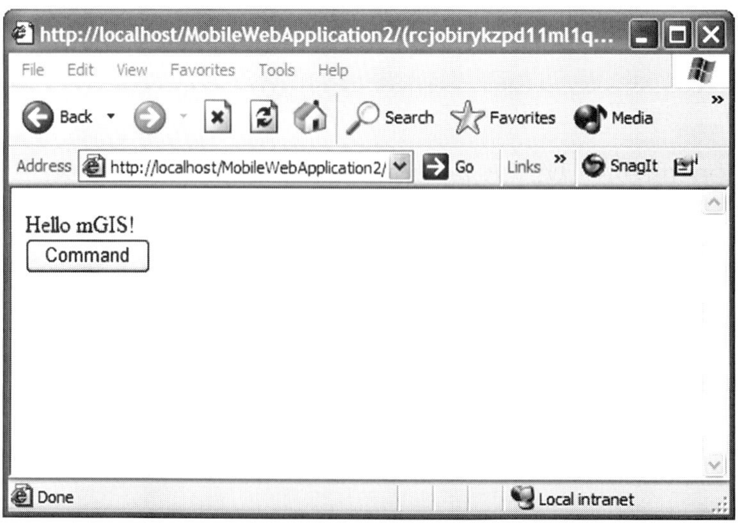

Figure 9–24. Example of using the Command control to invoke a greeting message "Hello, mGIS!".

10 Write down the URL and close out the web browser.

11 Take the URL from IE and enter it in the Pocket IE, BlackBerry browser, and Openwave browser (Figure 9–25).

Again, you can see that adding event handlers for mobile applications is similar to that for ASP.NET Web projects. The ASP.NET Mobile web application works in IE, Pocket IE, the BlackBerry browser, and the Openwave browser. You build it once for all of these devices.

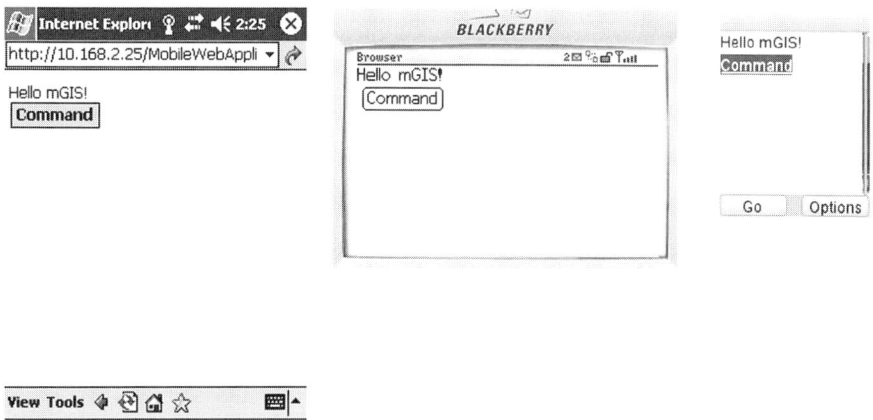

Figure 9–25. Run MobileWebApplication2 *to show the "Hello, mGIS!" greeting message in the Pocket IE (left), BlackBerry simulator (middle), and Openwave simulator (right).*

Container Controls

In addition to the *Panel* control, which is included in web controls, ASP.NET mobile controls include the *Form* control. Both the *Panel* control and *Form* control can contain a number of other controls. The *Form* control differs from the *Panel* control in that the former supports posting of data, whereas the latter is only used to group related controls. Also, forms cannot be nested but panels can.

Although the *Panel* control is used to group related controls, different devices render this control differently. Thus, do not spend too much time on designing the layout unless you know the target device. Even so, you should test your application thoroughly before deploying it.

The *Form* control is the outermost container of controls within a mobile web page. A mobile web page can have multiple *Form* controls, but only one form is rendered to the browser at a time. Let's modify

MobileWbApplication2 by using two *Form* controls. The first contains a *Command* control, and the second holds a *Label* control. When the Command button is clicked on the first form, the control is passed on to the second form and the *Text* property of the *Label* control is set to "Hello mGIS!". We name this project *MobileWebApplication3*. Listing 9-1 is the *.aspx* file with two *Form* controls.

Listing 9–1. *.aspx* file with two *Form* controls.

```
<%@ Register TagPrefix="mobile"
    Namespace="System.Web.UI.MobileControls"
    Assembly="System.Web.Mobile" %>
<%@ Page Language="vb" AutoEventWireup="false"
    Codebehind="MobileWebForm1.aspx.vb"
    Inherits="MobileWebApplication3.MobileWebForm1" %>
<HEAD>
   <meta name="GENERATOR" content="Microsoft Visual Studio .NET 7.1">
   <meta name="CODE_LANGUAGE" content="Visual Basic .NET 7.1">
   <meta name="vs_targetSchema"
      content="http://schemas.microsoft.com/Mobile/Page">
</HEAD>
<body Xmlns:mobile="http://schemas.microsoft.com/Mobile/WebForm">
<mobile:Form id="Form1" runat="server">
   <mobile:Command id="Command1" runat="server">Command</mobile:Command>
</mobile:Form>
<mobile:Form id="Form2" runat="server">
   <mobile:Label id="Label1" runat="server"></mobile:Label>
</mobile:Form>
</body>
```

Similarly we add the following code to the `Command1_Click` event of the first form.

```
'Declare a string variable for the greetings
Dim myGreetings As String = "Hello mGIS!"
'switch to Form2
ActiveForm = Form2
Label1.Text = myGreetings
```

We set `ActiveForm` to *Form2*. The *ActiveForm* is a property of the *Form* control. In this project, we set the property in the `Command1_Click` event in *Form1*. Run the project in the Openwave browser and you will see *Form1* with only a Command button. After you click on the button, you will see *Form2* with the greeting (Figure 9-26). Because wireless devices have limited real estate, we spread out the logic onto multiple forms, which also makes application programming much easier to control and debug.

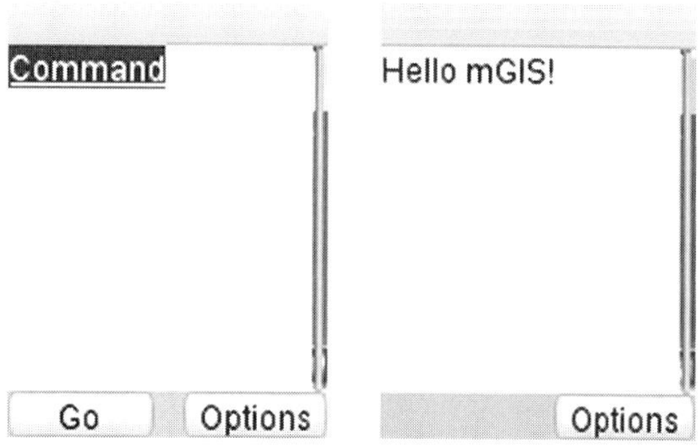

Figure 9–26. The two-form mobile web application was built to show the "Hello, mGIS!" greeting message.

Link Control

The *Link* control serves as a hyperlink. The hyperlink can also be used to navigate between forms. You specify the link's text through the *Text* property. For output devices such as cellular phones that support soft keys, you can define the soft key's caption through the *SoftkeyLabel* property. A soft key is a programmable button available on mobile phones. You specify the link's destination URL through the *NavigateURL* property. Let's add a *Link* control to *Form2* in *MobileWebApplication3* for the navigation back to *Form1*. The following are the steps for adding a *Link* control.

1 Drag a *Link* control from the Mobile Web Forms tab of the toolbox onto *Form2*, and set the *Text* property to "Go Back" and the *Softkey-Label* property to "Back". The default value for the *SoftkeyLabel* property is "Link".

2 Set the *NavigateUrl* property to "#Form1". You can set *NavigateUrl* in one of the following ways.

- Directly enter a URL into the *NavigateURL* property.
- In the Properties window, click on the *NavigateURL* property and then a drop-down list appears. Choose Select URL and the Select URL dialog box appears. Browse to the location you want to link to and click on OK.
- In the Properties window, all available forms are shown in the drop-down list next to the *NavigateURL* property (both *#form1* and *#form2* are listed in *MobileWebApplication3*). Select one form from the list.

After the addition of the *Link* control, the *.aspx* file looks as follows.

```
<%@ Register TagPrefix="mobile"
  Namespace="System.Web.UI.MobileControls"
  Assembly="System.Web.Mobile" %>
<%@ Page Language="vb" AutoEventWireup="false"
  Codebehind="MobileWebForm1.aspx.vb"
  Inherits="MobileWebApplication3.MobileWebForm1" %>
<HEAD>
<meta name="GENERATOR" content="Microsoft Visual Studio .NET 7.1">
<meta name="CODE_LANGUAGE" content="Visual Basic .NET 7.1">
<meta name="vs_targetSchema" content="http://schemas.microsoft.com/Mobile/Page">
</HEAD>
<body Xmlns:mobile="http://schemas.microsoft.com/Mobile/WebForm">
 <mobile:Form id="Form1" runat="server">
 <mobile:Command id="Command1" runat="server">Command</mobile:Command>
 </mobile:Form>

 <mobile:Form id="Form2" runat="server">
  <mobile:Label id="Label1" runat="server"></mobile:Label>
  <mobile:Link id="Link1" runat="server" NavigateUrl="#Form1"
        SoftkeyLabel="Back">Go Back</mobile:Link>
 </mobile:Form>
</body>
```

The *<Mobile:Link>* tag is inserted after the *<Mobile:Label>* tag. No Visual Basic .NET code addition is needed. Now you can launch the Openwave browser and enter the URL for *MobileWebApplication3* in the Address Bar to run the application (Figure 9-27). *Form1* still shows the Command button, and *Form2* now displays the greeting and the *Link* control with the "Go Back" test. The soft key is now displayed as "Back". After clicking on the soft key, the application goes back to *Form1* (as shown in Figure 9-27).

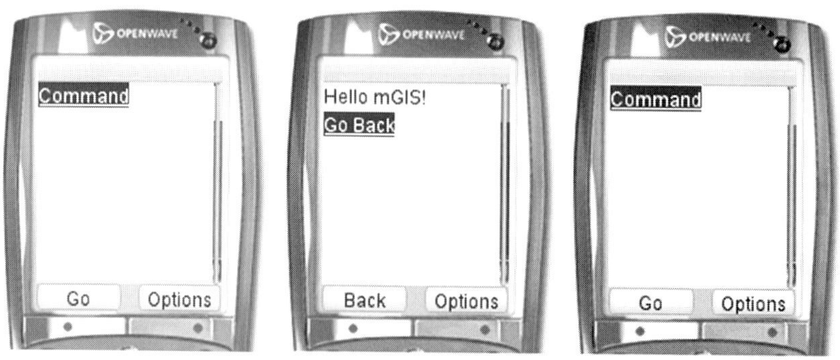

Figure 9-27. The entry screen of the "Hello, mGIS!" example (left), the greeting message screen with a Link control and a soft key of "Back" (middle), and back to the entry screen (right).

Image Control

The ability to publish images to wireless devices is critically important in building wireless web GIS applications. Images are still an effective way of delivering maps over the wireless Web, but they pose a unique challenge for authoring mobile web applications because their appearance is so dependent on the characteristics of the device they will be displayed on. ASP.NET Mobile provides an *Image* control that allows different versions of an image to be rendered, based on device characteristics.

Because of the nature of wireless devices, a single image format will not be appropriate for all devices. Some browsers are incapable of displaying certain image formats, and not all image sizes and color content are appropriate for all devices. For example, an image formatted as a *.gif* file will be displayed on cHTML or HTML capable browsers, but will not be displayed on early versions of WML-capable browsers. For WML-capable browsers, it is best to use an image formatted as a *.wbmp* file, which stands for Wireless BitMap. Unfortunately, a *.wbmp* file is black and white, which is not suitable for a multiple-feature GIS map. Fortunately, WML 1.1 started supporting the PNG format, and WML 2.0 now includes GIF, JPEG, and other image formats.

With the *Image* control, you can specify multiple types of image files: *.gif, .jpg, .jpeg, .bmp, .wmf, .png*, and *.wbmp*. Thus, you can create the same image in multiple file formats. The control chooses the correct image file based on device characteristics. The *Image* control itself is independent of the type of image file you use. You can use any type of image file that is compatible with the targeted browser.

ASP.NET Mobile allows you to specify device-specific behavior for mobile devices. When an *Image* control is rendered on a mobile device, the control evaluates the capabilities of the device. Based on this evaluation, one of the following can occur.

- The text in the *AlternateText* property is rendered as a *Label* control if the device is not capable of displaying images.
- For image-capable devices, the control checks for a URI to a registered image specifically for the incoming device. To find a registered image, the control checks for a *<DeviceSpecific>* tag inside the *Image* control. If one exists, the control evaluates and extracts a value from a matching *<Choice>* clause. This value is treated as the URI of an image file and is used to render the file on the device.

The following example presents four alternate images to be chosen by the control, based on device characteristics. The control chooses alternate text where no images are appropriate.

```
<mobile:Image id="Image1" runat="server" AlternateText="smile"
    ImageUrl="images/color/happy.bmp">
 <DeviceSpecific>
  <Choice Filter="isUP4x" ImageUrl="images/bw/smile.bmp"></Choice>
  <Choice Filter="isWML11" ImageUrl="images/color/happy.bmp">
  </Choice>
  <Choice Filter="isCHTML10" AlternateText="GN Logo"
   ImageUrl="images/color/gn_logo.gif"></Choice>
  <Choice Filter="isHTML32" AlternateText="ESRI Logo"
   ImageUrl="images/color/esri_logo.jpg"></Choice>
 </DeviceSpecific>
</mobile:Image>
```

To add an Image *control to an ASP.NET Mobile web form:*

1 Drag an *Image* control from the Mobile Web Forms tab of the toolbox onto *Form1* and set the *AlternateText* property to "Smile" and the *ImageUrl* property to "image/color/happy.bmp".

If a device is not capable of displaying images, it will display the text "Smile" instead.

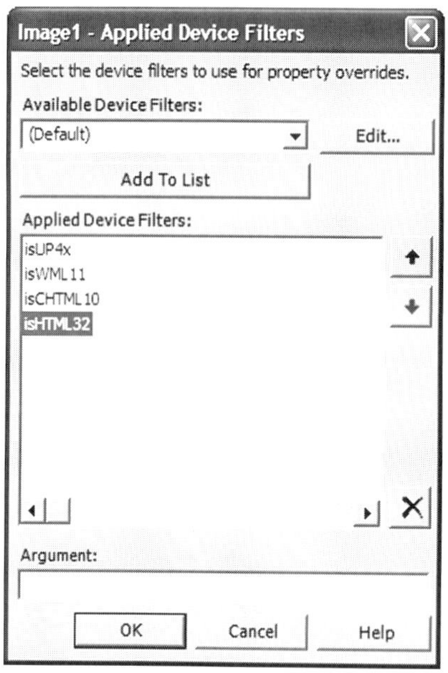

Figure 9–28. Applied Device Filters dialog box.

2 From the Properties window, click on the *AppliedDeviceFilters* property and an Ellipsis button appears. Click on the button and the Applied Device Filters dialog box appears (Figure 9–28).

3 Use the *AvailableDeviceFilters* drop-down list to select an available device filter. After you select a filter, click on Add to List to add the selection. Add four filters as specified in Table 9–2.

Table 9–2: Device Filters Used in Image Control Example

Device Filter	Description	AlternateText	ImageUrl
isUP4x	For mobile browsers that can only display black-and-white images.	Smile	images/bw/smile.bmp
isWML11	For mobile browsers that can display color .bmp files.	Happy	images/color/ happy.bmp
IsCHTML10	For browsers that only support cHTML 1.0.	GN logo	images/color/ gn_logo.gif
isHTML32	For browsers that support HTML 3.2 or above.	ESRI logo	images/color/ esri_logo.jpg

4 From the Properties window, click on the *PropertyOverrides* property and an Ellipsis button appears. Click on the button and the Properties Overrides dialog box appears (Figure 9–29).

5 Use the *AvailableDeviceFilters* drop-down to set the *AlternateText* and *ImageUrl* properties for each filter. Only those four filters chosen in step 4 are listed. The alternate text and image URL for each filter are shown in Table 9–2.

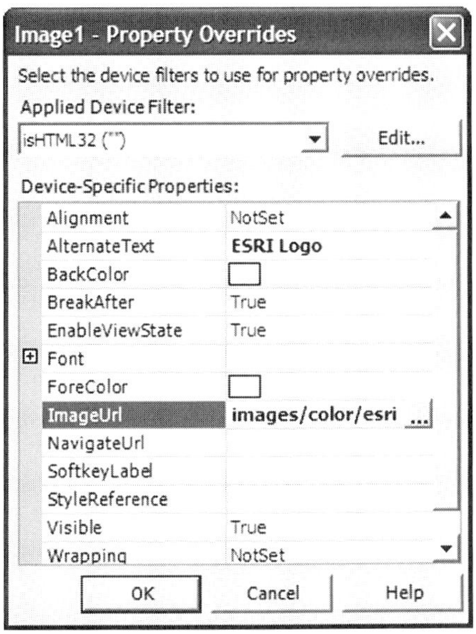

Figure 9–29. Property Overrides dialog box for the device filters.

6 Click on Tools/Connect to Device to select a PDA Pocket PC 2003 simulator to run the application, and click on Tools/Openwave Simulator/View in 7 and Tools/Openwave Simulator/View in 4.1.1 to run the same application.

The appropriate image displays for the specific browser (Figure 9–30). In Figure 9–30, the *Image* control example was viewed in (left to right) Openwave Phone Simulator 4.1.1, Openwave Phone Simulator 7.1, and Pocket PC 2003 Simulator. Openwave Phone Simulator 4.1.1 is only capable of displaying a black-and-white bitmap, so the BMP file of *smile.bmp* was displayed. Openwave Phone Simulator 7.1 is able to show color bitmaps, so the color BMP file of *happy.bmp* was shown on the device screen. The Pocket PC 2003 simulator supports HTML 3.2, so the JPEG file of *esri_logo.jpg* was rendered.

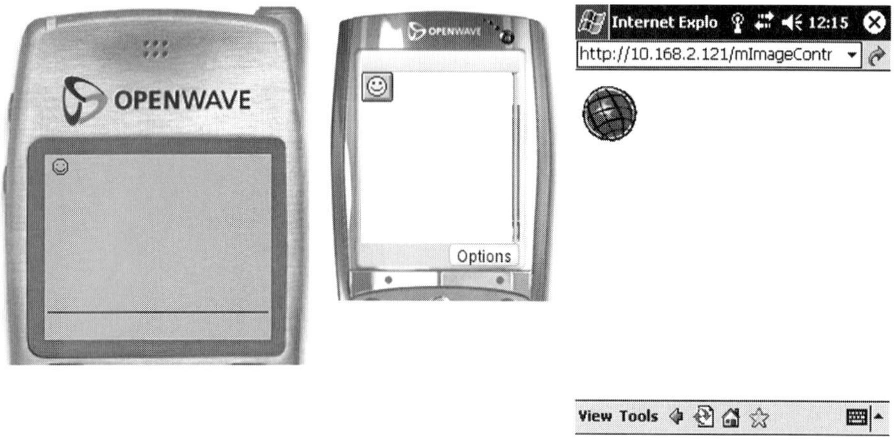

Figure 9–30. An image is displayed via an Image control in Openwave Simulator 4.x (left), Openwave Simulator 7.0 (middle), and in Pocket IE (right).

There is an alternative method for displaying the appropriate image format. You can write an image DLL that converts the original image (let's say in GIF format) to the proper format for a specific device based on the `<DeviceSpecific> tag`. This DLL can be enhanced to convert the original image to a size that fits the device screen.

List Controls

There are three types of list controls: *List*, *SelectionList*, and *ObjectList*. The *List* control is used to display a static list of items. To allow user interaction with the list, you can use the *SelectionList* control. To display database information in tabular format, use the *ObjectList* control. Note that you must use data binding to fill an object list with data for the *ObjectList* control, whereas you can either enter or program a list of static items for the other two list controls. A *List* control must be placed within a *Form* control or *Panel* control.

In the following example, we will create a project called *mListControls* that contains examples of the three list controls, each with its own Mobile web form. A *List* control example is contained in *mList.aspx*, a *SelectList* control example in *mSelectionList.aspx*, and an *ObjectList* control example in *mObjectList.aspx*. To run each example, simply make the appropriate Mobile web form as a start page. In Visual Studio .NET, highlight the target web form in the Solution Explorer window, right-mouse click on the *.aspx* file, and select Set As Start Page. You need to perform this step whenever you want to establish a different web form as the start page.

To add a List control to an ASP.NET Mobile web form:

1 Drag a *Label* control from the Mobile Web Forms tab of the toolbox onto *Form1* and set the *Text* property to "Selections" and the *Font/Bold* property to *True.*

2 Drag a *List* control from the Mobile Web Forms tab of the toolbox onto *Form1*. A *List* control can be placed on a *Form* or *Panel* control on a Mobile web form page. Set the *ID* property to "frm1LstOrg".

3 From the Properties window, click on the *Items* property and an Ellipsis button appears. Click on the button and the List Properties dialog box appears (Figure 9–31).

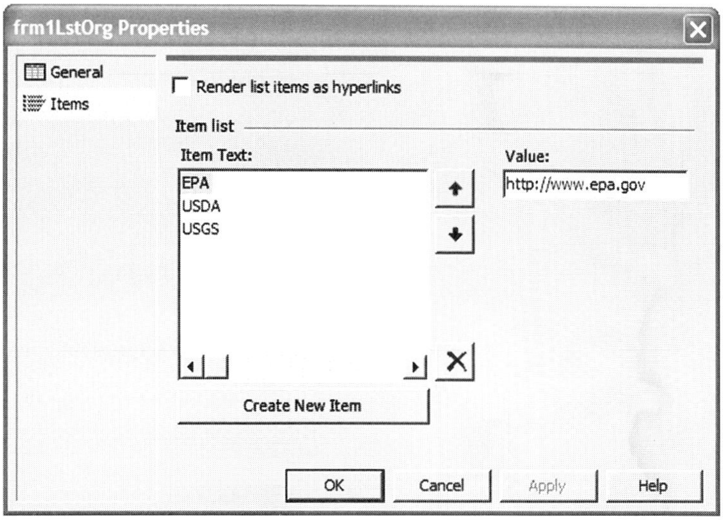

Figure 9–31. List Properties dialog box for specifying list items for the List *control.*

4 Enter the items you want to appear in the list.

- Click on the Create New Item button.
- In the "Item Text" text box, a new item appears showing "text". Type *EPA* as the text for that item.
- Click on the Value text box, and the text you just entered will appear. Type *http://www.epa.gov* for the value.
- Repeat the above process to enter two items of *USDA* and *USGS* with the URLs of *http://www.usda.gov* and *http://www.usgs.gov*, respectively.

✔ **NOTE:** *You can select Render list items as hyperlinks to make the items as hyperlinks. If you choose this option, you need to specify a* *navigation link for each item value. Or you can set the ItemAsLinks property to True to achieve the same goal. With this configuration, a click action forwards to the web site specified by the hyperlink for the chosen organization.*

5 Click on OK.

6 Set the *Decoration* property to itemize the list with *bullets.*

7 Drag another *Form* control from the Mobile Web Forms tab of the toolbox and place it in the design area beneath *Form1*.

8 Drag four *Label* controls from the Mobile Web Forms tab of the toolbox onto *Form2*. Set the properties specified in Table 9–3 for the *Label* controls.

Table 9–3: Property Setting Values for the Four *Label* Controls

ID	Text	Font Formatting
Label2	Organization:	Bold: True
LblOrg	Label (default)	
Label4	URL:	Bold: True
LblUrl	Label (default)	

9 Add the following code to `frm1LstOrg_ItemCommand`.

```
lblOrg.Text = e.ListItem.Text
lblUrl.Text = e.ListItem.Value
ActiveForm = Form2
```

The first line assigns the clicked item text of the *List* control to the *Text* property of the *Organization* label, the second line assigns the clicked item value of the *List* control to the *Text* property of the *URL* label, and the third line makes *Form2* active. In other words, when a user clicks on an item of the *List* control on *Form1* both the text and value of the clicked item are captured and displayed on *Form2*.

10 Click on Tools/Connect to Device to select a PDA Pocket PC 2003 simulator to run the application (Figure 9–32).

The left-hand image in Figure 9–32 shows the list of the government organizations. When a user clicks on "USDA", for example, the next page pops up and displays the selected organization name and URL (as shown in the right-hand image).

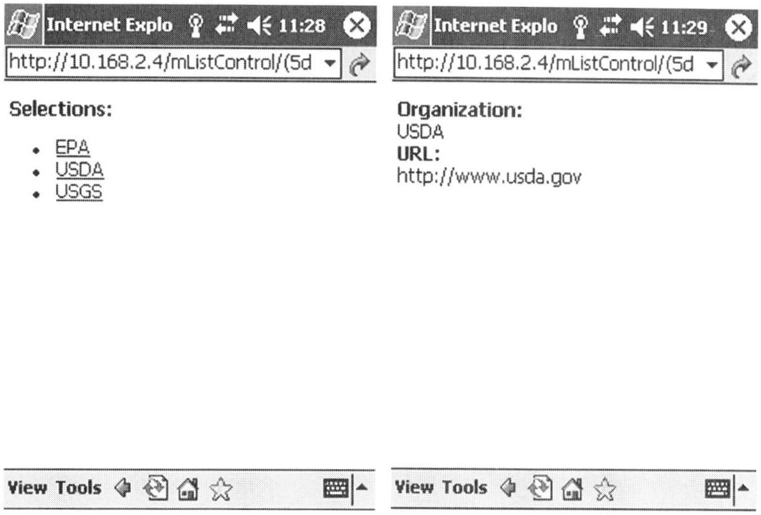

Figure 9–32. Example of using the List *control to list three governmental organizations (left) and to show the organization URL when the ItemCommand is triggered (right).*

Listing 9–2 is the *.aspx* file for the Mobile web form with a *List* control.

Listing 9–2. *.aspx* file for the Mobile web form with a *List* control.

```
<mobile:Form id="Form1" runat="server">
   <mobile:Label id="Label1" runat="server">Select One:</mobile:Label>
   <mobile:List id="frm1LstOrg" runat="server"
      Decoration="Bulleted">
      <Item Value="http://www.epa.gov" Text="EPA"></Item>
      <Item Value="http://www.usda.gov" Text="USDA"></Item>
      <Item Value="http://www.usgs.gov" Text="USGS"></Item>
   </mobile:List>
</mobile:Form>
<mobile:Form id="Form2" runat="server">
   <mobile:Label id="Label2" runat="server"
      Font-Bold="True">Organization:</mobile:Label>
   <mobile:Label id="lblOrg" runat="server">Label</mobile:Label>
   <mobile:Label id="Label4" runat="server"
      Font-Bold="True">URL:</mobile:Label>
   <mobile:Label id="lblUrl" runat="server">Label</mobile:Label>
</mobile:Form>
```

This example lists three government organizations (EPA, USDA, and USGS), with hyperlinks to the individual organization web sites. The list was created as bulleted items as specified by the *Decoration* attribute of the *<mobile:List>* tag.

To change an item on the *List* control, simply go back to the List Properties dialog box. Double click on the item you want to change, or click on the item and then press F2, which makes the item editable. Type the new text for that item and press Enter. Click on the OK button to close out the List Properties dialog box. We can also dynamically add an item to the *List* control. For example, you could add the following code to `Page_Load`.

```
Dim myItem As MobileListItem = New MobileListItem("FEMA", _
   "http://www.fema.gov")
frm1LstOrg.Items.Add(myItem)
```

These two lines of code add "FEMA" as the fourth item to the list when the page is loaded. The last item shown in Figure 9-33 was dynamically added using the code.

To add a SelectionList control to an ASP.NET Mobile web form:

1 Drag a *Label* control from the Mobile Web Forms tab of the toolbox onto *Form1* and set the *Text* property to "Selections" and the *Font/Bold* property to *True*.

2 Drag a *SelectionList* control from the Mobile Web Forms tab of the toolbox onto *Form2*. A *SelectList* control can be placed on a *Form* or *Panel* control.

3 From the Properties window, use the drop-down list next to the *SelectType* property to choose the *Radio* type of *SelectionList* to use. You have five options to choose from the drop-down list: *DropDown* (default), *ListBox*, *Radio*, *MultiSelectListBox*, and *CheckBox*.

4 From the Properties window, click on the *Items* property and an Ellipsis button displays. Click on the button and the SelectionList Properties dialog box appears (Figure 9-34).

5 Enter the items you want to appear in the list.

- Click on the Create New Item button.
- In the Item Text textbox, a new item appears showing "text". Type *EPA* as the text for that item.
- Click on the Value text box and the text you just entered will appear. Type *"http://www.epa.gov"* for the value.

- Click on the Selected checkbox to set the EPA as the selected item by default. We highly recommend that you select this checkbox.
- Repeat the above process to enter “USDA” and “USGS” with the URLs of “http://www.usda.gov” and “http://www.usgs.gov”, respectively. Because the Radio option was chosen, you do not need to click on the Selected checkbox for these two items. If you did, only the first checked item would be set as the selected item by default, which is “EPA” in this case.

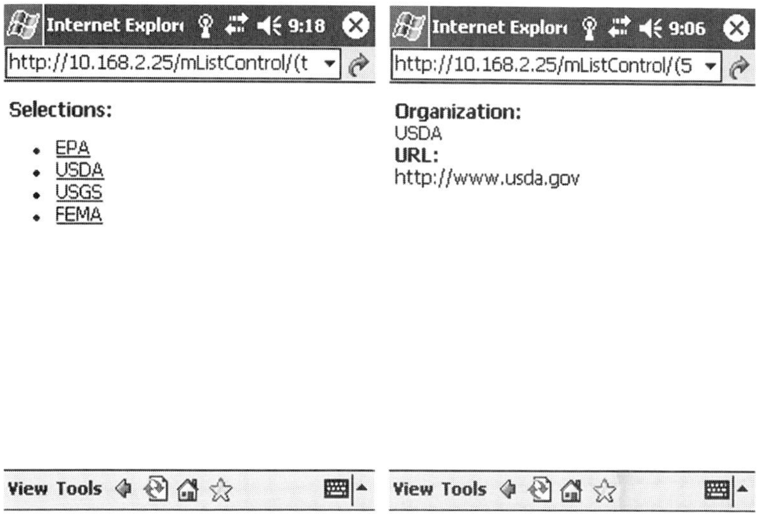

Figure 9–33. Three predefined items and one dynamically added item were listed in a List *control (left), and the web page is refreshed with the URL for USDA after USDA is selected (right).*

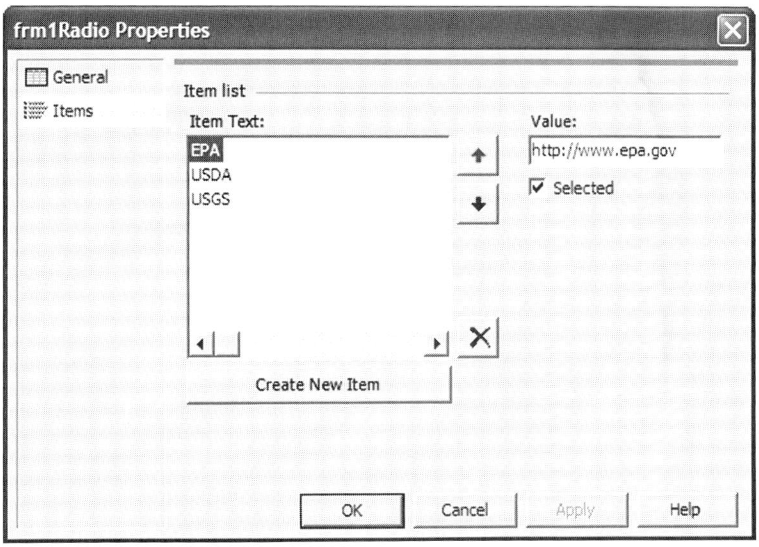

Figure 9–34. SelectionList *Properties dialog box for specifying list items for the* SelectionList *control.*

6 Click on OK.

7 Drag a *Command* control from the Mobile Web Forms tab of the toolbox onto *Form1* and set the *Text* property to "Submit" and the *ID* property to "btnSumit".

8 Drag a *Form* control from the Mobile Web Forms tab of the toolbox onto the Visual Studio .NET design area beneath *Form1*.

9 Drag four *Label* controls from the Mobile Web Forms tab of the toolbox onto *Form2*. Set their properties as specified in Table 9–2.

10 Add the following code to `btnSumit_Click`.

```
lblOrg.Text = frm1Radio.Selection.Text
lblUrl.Text = frm1Radio.Selection.Value
ActiveForm = Form2
```

The first line assigns the checked item text of the *SelectionList* control to the *Text* property of the *Organization* label, the second line assigns the checked item value of the *SelectionList* control to the *Text* property of the *URL* label, and the third line makes *Form2* active. In other words, when a user checks an item of the *SelectionList* control on *Form1*, both the text and value of the checked item are captured and displayed on *Form2*.

11 Click on Tools/Connect to Device to select a PDA Pocket PC 2003 simulator to run the application (Figure 9–35).

The left-hand image in Figure 9–35 shows the list of the government organizations with EPA selected by default. When a user clicks on the

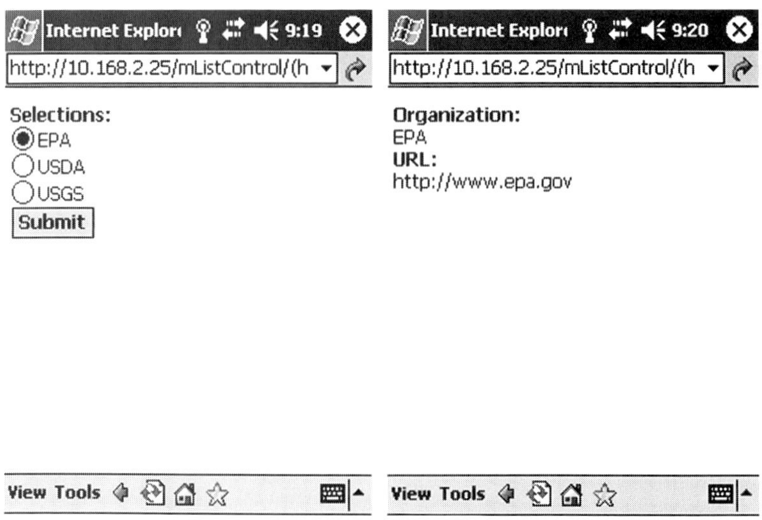

Figure 9–35. An example of using the SelectionList control to list three governmental organizations (left) and the result after the button click event is triggered (right).

Submit button, for example, the next page pops up and displays the organization name and URL for EPA (shown in the right-hand image).

Listing 9-3 is the *.aspx* file for the Mobile web form with two Mobile web forms and a *SelectionList* control.

Listing 9–3. *.aspx* file for the Mobile web form with two Mobile web forms and a *SelectionList* control.

```
<mobile:Form id="Form1" runat="server">
   <mobile:Label id="Label1" runat="server"
      Font-Bold="True">Selections:</mobile:Label>
   <mobile:SelectionList id="frm1Radio" runat="server"
      SelectType="Radio">
      <Item Value="http://www.epa.gov" Text="EPA"
         Selected="True"></Item>
      <Item Value="http://www.usda.gov" Text="USDA"></Item>
      <Item Value="http://www.usgs.gov" Text="USGS"></Item>
   </mobile:SelectionList>
   <mobile:Command id="btnSubmit" runat="server">Submit</mobile:Command>
</mobile:Form>
<mobile:Form id="Form2" runat="server">
   <mobile:Label id="Label2" runat="server"
      Font-Bold="True">Organization:</mobile:Label>
   <mobile:Label id="lblOrg" runat="server">Label</mobile:Label>
   <mobile:Label id="Label4" runat="server"
      Font-Bold="True">URL:</mobile:Label>
   <mobile:Label id="lblUrl" runat="server">Label</mobile:Label>
</mobile:Form>
```

This time we list three government organizations (EPA, USDA, and USGS) in radio buttons. Thus, the *SelectType* attribute of the *<mobile:SelectList>* tag was set to “Radio”. Similarly, you can use each of four types of *SelectList* control with the *SelectType* attribute set to *CheckBox*, *ListBox*, *DropDown*, or *MultiSelectListBox*. Both *MultiSelectListBox* and *CheckBox* support multiple selections.

To change an item on the *SelectList* control, simply go back to the SelectList Properties dialog box. Double click on the item you want to change, or click on the item and then press F2, which makes the item editable. Type the new text for that item and press Enter. Click on the OK button to close out the SelectList Properties dialog box.

To use the *ObjectList* control, you must have a bindable data source. In most cases, such a bindable data source is set via code. Let’s create an application to show population by region, as shown in Table 9-4.

Table 9–4: Population by Region (Source: 2000 Census)

ID	Region	Population	Male	Female
1	Northeast	53594378	25897327	27697051
2	Midwest	64392776	31555438	32837338
2	South	100236820	49057320	51179500
3	West	63197932	31543478	31654454

To add an ObjectList control to an ASP.NET Mobile web form:

1 Drag an *ObjectList* control from the Mobile Web Forms tab of the toolbox onto *Form1*. An *ObjectList* control can be placed on a form or inside a *Panel* control.

2 Add the following code to the top of *mObjectList.aspx.vb* before the class.

```
Imports System.Data.OleDb
```

3 Add the following code to the `Page_Load` event.

```
If Not IsPostBack Then
 Dim dbConn As OleDbConnection = New OleDbConnection( _
  "Provider=Microsoft.Jet.OLEDB.4.0;data source=" & _
  Server.MapPath("mdb/chapter9.mdb"))
 Dim sql As String
 Dim dbComm As OleDbCommand = dbConn.CreateCommand()
 dbComm.CommandText = "SELECT * FROM Region"
 dbconn.Open()
 Dim dbRead As OleDbDataReader = dbComm.ExecuteReader()
 ObjectList1.DataSource = dbRead
 ObjectList1.DataBind()
 dbRead.Close()
 dbConn.Close()
End If
```

4 Set the *AutoGenerateFields* property to *False*. Otherwise, each field will be displayed twice.

5 From the Properties window, click on the *Fields* property and an Ellipsis button displays. Click on the button and the ObjectList Properties dialog box appears (Figure 9–36).

6 Enter the items you want to appear in the list.

- Leave *Create fields automatically at run time* unchecked. If you check this option, you do not need to specify anything else on the dialog box. But you need to go back and set *AutoGenerateFields* to *True*.

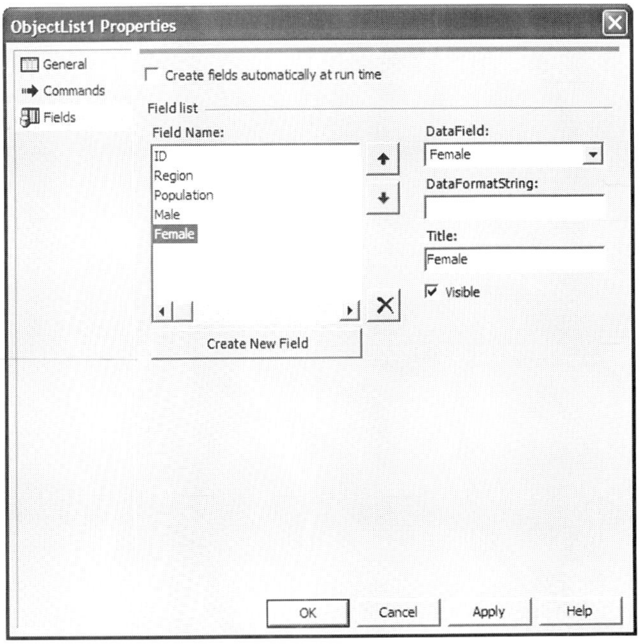

Figure 9–36. ObjectList Properties dialog box for specifying list items for the ObjectList *control.*

- Click on the Create New Field button to specify a field you want to display in the *ObjectList* control.
- In the Field Name text box, a new field name appears showing "Field1". Type the text for that field name.
- Enter the same text for the DataField text box and the Title text box. You can enter different text for the Title, but the DataField needs to be one of existing field names in the data source. In other words, you could set the DataField to *Region* for the Field Name of *ID*.
- Leave *DataFormatString* blank. You can specify a Microsoft .NET Framework format string in the box if you want to format a field.
- Leave Visible selected for each field name. If you uncheck it, that field will not be displayed at runtime.

7 Set the *LabelField* property to *Region* (Figure 9–37) so that the *ObjectList* item will be itemized by region name. The default is the first column in the data source.

8 Click on Tools/Connect to Device to select a PDA Pocket PC 2003 simulator to run the application.

A web page with a list of regions displays (Figure 9–38). Click on any of regions to display the detailed population information on the region.

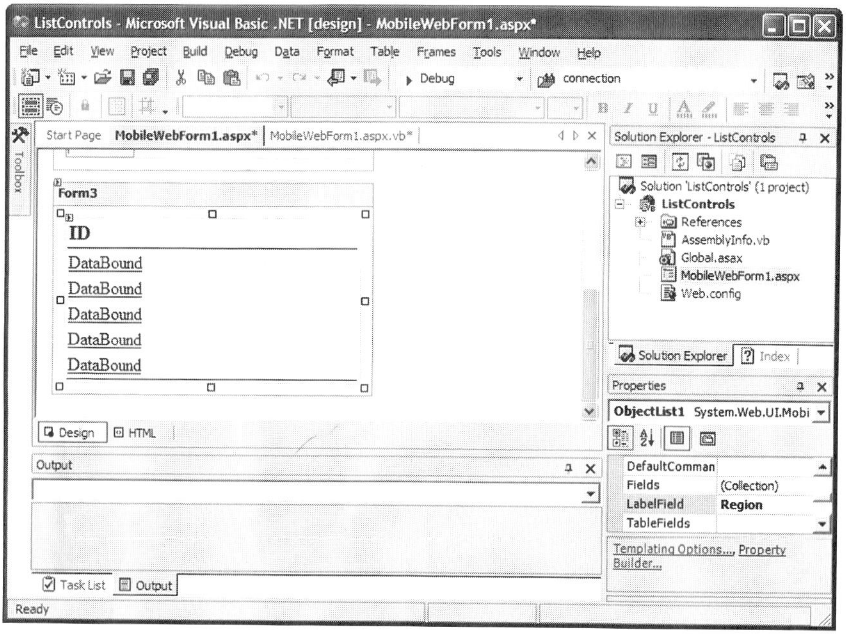

Figure 9–37. Setting the LabelField *property in Visual Studio .NET.*

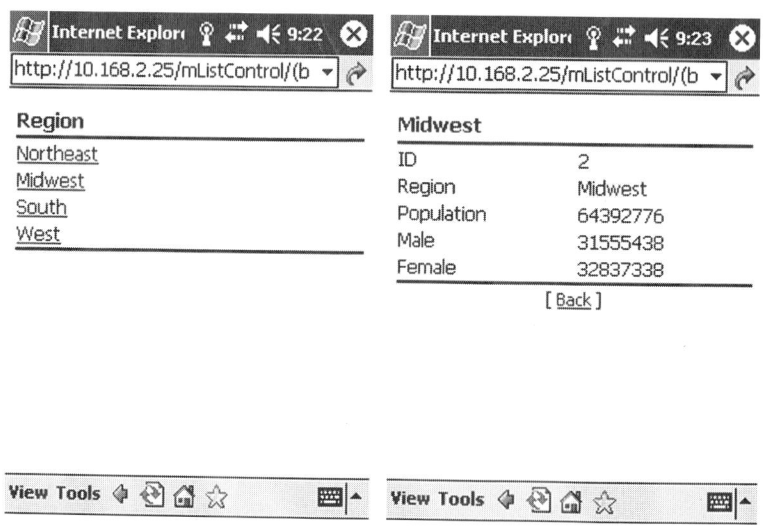

Figure 9–38. An example of using the ObjectList control to list regions (left) and show the regional population information in a new screen (right) when a region is selected.

Listing 9–4 is the *.aspx* file for the Mobile web form and an *ObjectList* control.

Listing 9–4. *.aspx* file for the Mobile web form and an *ObjectList* control.

```
<mobile:Form id="Form1" runat="server">
   <mobile:ObjectList id="ObjectList1" runat="server"
      CommandStyle-StyleReference="subcommand"
      LabelStyle-StyleReference="title"
      AutoGenerateFields="False" LabelField="Region">
      <Field Title="ID" Name="ID" DataField="ID"></Field>
      <Field Title="Region" Name="Region" DataField="Region"></Field>
      <Field Title="Population" Name="Population"
         DataField="Population"></Field>
      <Field Title="Male" Name="Male" DataField="Male"></Field>
      <Field Title="Female" Name="Female" DataField="Female"></Field>
   </mobile:ObjectList>
</mobile:Form>
```

We can also modify the application to show all regional population information all at once. All we need to do is set the *TableFields* property to *Region;Population;Male;Female* for the *ObjectList* control. Run the application again and a list of population for all regions will be displayed (Figure 9–39).

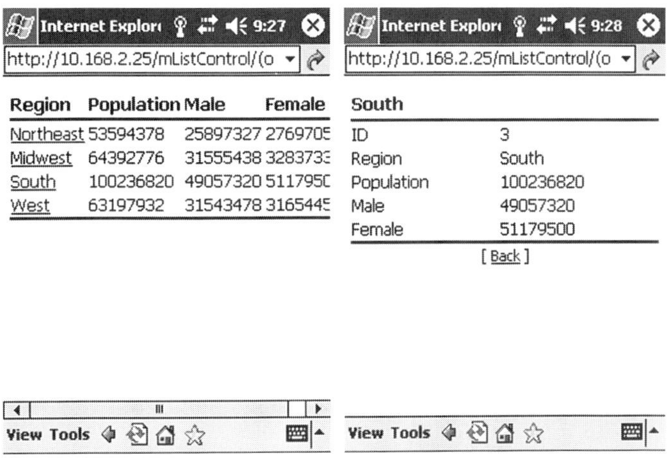

Figure 9–39. Example of using the ObjectList *control to list regional population by predefining the* TableFields *property (left) and show the regional population information in a new screen (right) when a region is selected.*

The corresponding HTML content shows an additional attribute of *TableFields* for the `<mobile:ObjectList>` tag, as highlighted in Listing 9–5.

Listing 9–5. Additional attribute of *TableFields* for the `<mobile:ObjectList>` tag.

```
<mobile:Form id="Form1" runat="server">
   <mobile:ObjectList id="ObjectList1" runat="server"
      CommandStyle-StyleReference="subcommand"
      LabelStyle-StyleReference="title"
      AutoGenerateFields="False" LabelField="Region"
      TableFields="Region;Population;Male;Female">
      <Field Title="ID" Name="ID" DataField="ID"></Field>
      <Field Title="Region" Name="Region" DataField="Region"></Field>
      <Field Title="Population" Name="Population"
         DataField="Population"></Field>
      <Field Title="Male" Name="Male" DataField="Male"></Field>
      <Field Title="Female" Name="Female" DataField="Female"></Field>
   </mobile:ObjectList>
</mobile:Form>
```

Text Controls

There are two types of *Text* controls: *TextBox* and *TextView*. The *TextBox* control is used for a single-line text input, whereas the *TextView* control is used for a multiline text input. Text inputs such as ZIP code, city name, or county name can use the *TextBox* control. If you would like to display a page of formatted text, you need to use the *TextView* control.

The *TextBox* control can be used to obtain numeric, string, or password input from the user. It must be placed within a *Form* control or *Panel* control. A *TextBox* control for passwords hides a user's entry as it is typed, if the device supports that capability. To add a *TextBox* control, simply drag a *TextBox* control from the Mobile Web Forms tab of the toolbox onto a *Form* or *Panel* control on a Mobile web form page. If you have some initial text, set the *Text* property to the initial value. If your text includes a numeric or password input, set the *Numeric* property or the *Password* property to *True*. You can control the number of characters a user can input by setting up the *Size* and *MaxLength* properties. The following shows three types of *TextBox* inputs.

```
<mobile:TextBox id="TextBox1" runat="server" Size="10" MaxLength="10">
    </mobile:TextBox>
<mobile:TextBox id="TextBox2" runat="server" Numeric="True">
    </mobile:TextBox>
<mobile:TextBox id="TextBox3" runat="server" Password="True">
    </mobile:TextBox>
```

The *TextView* control is used to display multiline static text on the mobile screen. In this control, you can specify style text to appear with normal, bold, or italic formatting and to make use of line breaks, paragraph markers, and hyperlinks. A *TextView* control must be placed within a *Form* control or *Panel* control. You can align the text with the *Alignment* property by setting its value to *NotSet*, *Left*, *Centered*, or *Right*, depending on the mobile devices (the setting values vary among devices). You can alter the default appearance of text with the *ForeColor*, *Font*, and *StyleReference* properties. You enter text for the *TextView* control through the *Text* property. You have limited ways of formatting the text using the markup tags outlined in Table 9–5. Listing 9–6 shows an example of the *TextView* control. Figure 9–40 shows what the *TextBox* control and *TextView* control look like on the Openwave Phone Simulator 7.0.

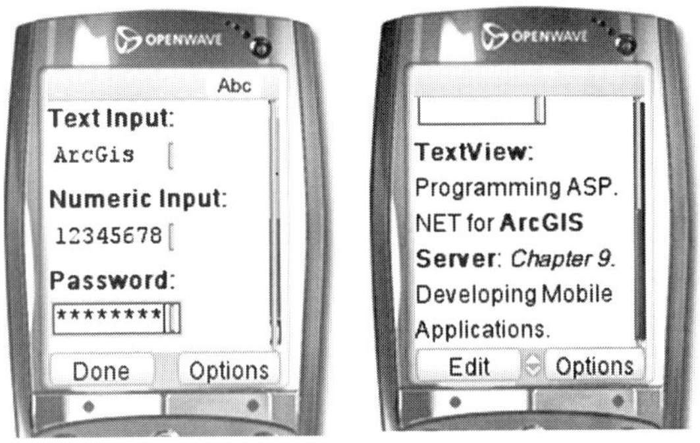

Figure 9–40. Example of using the TextBox *control to show text, number, and password (left), and another example of using the* TextView *control to show a paragraph of text (right).*

Table 9–5: Markup Tags for Formatting Text in the *TextView* Control

Markup Tag	Description
<a href = "URL">hyperlink</a>	Provides hyperlink capability. URL represents the navigation URL and hyperlink represents the link text
<b> </b>	Makes all text between these two tags bold
 	Inserts a line break
<i> </i>	Displays all text between these two tags in italics
<p> </p>	Contains a paragraph

Listing 9–6. Example of the *TextView* control.

```
<mobile:TextView id="TextView1" runat="server" Wrapping="Wrap"
   BackColor="Red" ForeColor="White"
   Font-Size="Small" Font-Name="Arial Narrow">
   Programming ASP.NET for <b>ArcGIS Server:</b> <i>Chapter 9.</i>
   Developing Mobile Applications.
</mobile:TextView>
```

Validation Controls

ASP.NET mobile controls provide a group of validation controls. These mobile validation controls are based on the web validation controls listed in Table 2–5. Validation controls generally do not appear on a form at runtime. Instead, they validate the data in *SelectionList* and *TextBox* controls on the same form. Validation occurs when the form is posted to the server. The validation controls test the user's input, and if the data fails any of the validations the server sends the page back to the client device with predefined error messages.

All validation controls except the *ValidationSummary* control must reference a *SelectionList* or *TextBox* control. The reference is set in the *ControlToValidate* property. Because the *ValidationSummary* control does not reference other controls, it does not contain a *ControlToValidate* property.

Like the web validation controls, the mobile validation controls support an *IsValid* property. This property stores the results of the validation but it cannot be set equal to a value because it is not a declarative property. Therefore, the *IsValid* property is not shown in the Properties window. However, you can write code to test the *IsValid* property. To test the *IsValid* property, loop through the page's *Validators* collection, which contains references to all of the validation controls. Then you can test the *IsValid* property for each validation control. For example, you can add the following code to *Page_Load* to achieve this goal.

```
If (Me.IsPostBack) Then
  Me.Validate()
  If (Not Me.IsValid) Then
   Dim msg As String
   ' Loop through all validation controls to see
   ' the error(s) if any.
   Dim aValidator As IValidator
   For Each aValidator In Validators
     If aValidator.IsValid = False Then
       msg = msg & "<BR>" & aValidator.ErrorMessage
```

```
      End If
    Next
    Label1.Text = msg
  End If
End If
```

Let's create a project called *MobileValidationRegex* to validate a ZIP code text box. Of course, we launch Visual Studio .NET and create the project. Then we drag and drop a form, a label, and a text box. Now we add a *RegularExpressionValidator* to the project.

To add a RegularExpressionValidator *control to an ASP.NET Mobile web form:*

1 Drag a *Label* control from the Mobile Web Forms tab of the toolbox onto *Form1* and set the *Text* property to "ZIP Code: (22202-3013)", the *Font/Bold* property to *True*, and the *Font/Name* to "Arial".

2 Drag a *TextBox* control from the Mobile Web Forms tab of the toolbox onto *Form1* and set the *Size* property to "10" and the *MaxLength* property to "10".

3 Drag a *RegularExpressionValidator* control from the Mobile Web Forms tab of the toolbox onto *Form1*. A *RegularExpressionValidator* control can be placed on a *Form* or *Panel* control on a Mobile web form page.

4 Click on the arrow next to the *ControlToValidate* property in the Properties window to select the ID of the control you want to validate. Because the *RegularExpressionValidator* only validates *SelectList* and *TextBox*, only *TextBox1* is available in the list for selection.

5 Take the regular expression for ZIP code validation from Table 3–10 and then enter a regular expression into the *ValidationExpression* property. The regular expression validates 5 plus 4 ZIP code format. Alternatively, click on the Ellipsis button next to the *ValidationExpression* property to bring up the Regular Expression Editor dialog box. You can also select from a set of standard (predefined) regular expressions or type an expression.

6 Enter "invalid ZIP code" into the *ErrorMessage* property. The *RegularExpressionValidator* control displays this text when the control being validated fails the test.

If you enter a string into the *Text* property, the *RegularExpressionValidator* control displays that text when an error occurs. If the *Text* property is left empty, the *RegularExpressionValidator* control displays the string in the *ErrorMessage* property.

7 Click on Tools/Connect to Device to select a PDA Pocket PC 2003 simulator to run the application.

Figure 9–41. Example of using the RegularExpressionValidator *control to validate the ZIP code information.*

Because the regular expression rule says that a ZIP code needs to be entered in 5 plus 4 format with a hyphen between the first five digits and last four digits, the *RegularExpressionValidator* control will display "invalid ZIP code" if you enter any ZIP code that does not meet the rule (Figure 9–41).

For mobile devices, validation controls are more important than for web applications because you cannot use client-side validation for most of the devices. Server-side mobile validation controls are the only type of validation for some mobile devices. Using server-side mobile validation controls increases the volume of traffic between the wireless gateway and the device on the wireless Internet, which cannot even begin to compare to the wired Internet in terms of speed and available bandwidth. Keep this limitation in mind when you design a wireless Internet application. Also, you need to keep *ErrorMessage* as short as possible.

Mobile Web Forms

An ASP.NET web page can contain only one form per page. However, an ASP.NET Mobile web page allows several forms to be defined on the same page. This multi-form mobile web page design was specifically implemented to handle mobile devices via WAP. Those devices use WML. A WML page is called a deck, consisting of multiple cards, each of which is equivalent to a form in ASP.NET Mobile. Also, because mobile devices typically have smaller screens, applications require multiple cards/forms. Multi-form implementation provides several benefits for wireless devices. For example, if it still only allows one form per web page, a large number of small pages would have to be created for a web site. With multi-form implementation, it helps to manage the complexity of a web site by avoiding a large number of small pages. You can organize your mobile web application in the same way as your desktop web application, because you can still have a single mobile web page with multiple forms, with each for a simple task. In addition, it reduces the traffic between server and client because multiple screens are sent in a single response.

When a client accesses a mobile web page for the first time, the first form is shown by default. You can then use the *ActiveForm* property of the mobile web page to programmatically navigate to other forms, or you can define a *Link* control to allow a user to navigate between two forms, as shown in *MobileWebApplication3*.

When you create an ASP.NET Mobile web project using Visual Studio .NET, it comes with *Form1* by default. As we introduced in Chapter 2, the *Form* control is one of the mobile web controls, as shown in Table 2-6. To add form control, simply drag-and-drop a *Form* control from the Mobile Web Forms tab of the toolbox to the mobile web page. Then drag and drop other mobile controls necessary onto the new form. Make sure to set the *ActiveForm* property or a *Link* control between the two connected forms.

ASP.NET mobile controls also provide a mechanism for automatically separating the large content within a form into smaller groups of rendered content. This option is called *pagination*. When you use pagination, grouped content is automatically formatted to fit the target device. The *Form* control also renders other mobile controls you can use to browse to other pages.

By default, pagination is set to *False* in forms. To activate pagination, set the *Paginate* property of the *Form* control to *True*. The *Form* control provides other properties (such as *PageCount*, *CurrentPage*, and *PagerStyle*), which allow you to control pagination behavior. Keep in mind that when you develop applications for wireless devices you should keep the display simple and compact. You can design creative pagination that runs well on emulators, but it may not work at all on a real device because of memory limitations.

Developing Mobile GIS Applications

Developing mobile GIS applications is not trivial because the ArcGIS Server web controls do not work on the ASP.NET Mobile web forms. This means that you cannot simply drag and drop an ArcGIS Server *Web* control onto the mobile web form. You need to be creative. As discussed earlier in this chapter, you can use the *Image* control. If you recall, the *Mobile Image* control has a property called *ImageURL*, which allows you to set a URL image for the control. In other words, you can set that property to a remote URL of a web form application developed based on ArcGIS Server. That web application performs all GIS functionality and streams images in GIF, JPG, and BMP. We can create that web application using an ArcGIS Server web control, as we discussed in earlier chapters, because that is not using mobile web forms.

In this section, we are going to build a mobile web application to find a city or a place and display the location on the map. This wireless Internet application consists of two components: mobile web forms for the input and matching the input with candidates and a regular web form that produces images. We will use an ESRI ArcWeb service for the matching of the input text strings with candidates. The service is called *PlaceFinderSample* and is available from the ArcWeb web site at *http://arcweb.esri.com*, which is a worldwide gazetteer service. Because the ArcWeb service is a SOAP/WSDL web service, we can use it easily inside Visual Studio .NET by adding that service as a web reference (see Chapter 8 for the details on consuming web services).

✔ **NOTE:** *The ESRI ArcWeb Services are subscription-based commercial web services. They offer both GIS data content and capabilities via SOAP/WSDL web services on the Web. With a subscription, you can access the GDT Streets map service using ASP.NET, for example. You can literally build a web-based mapping application without owning any data content. ESRI offers a 30-day evaluation trial (*http://www.esri.com/software/arcwebservices/evaluate.html*).*

The PlaceFinderSample service is free and does not require any authentication to use it. It is available at http://arcweb.esri.com/services/v2/PlaceFinderSample.wsdl

Building a Web Form for Image Streaming

In the following, we will build a web form application with the ArcGIS Server web controls that publish the JPEG images for a given pair of latitude and longitude coordinates. The *Map* and *Impersonation* controls will be used in the web application. We will use the "gnet" map server object created in Chapter 4, which shows an image map service from the Geography Network web site. For simplicity of discussion, we will predefine the image size as 200 x 200 pixels.

To add a Map control to an ASP.NET web form:

1. In Visual Studio .NET, create a new Visual Basic .NET project using the ASP.NET Web Application template and name it "mapImage".
2. Drag and drop an *Impersonation* control from the ESRI tab of the toolbox onto the web form.
3. From the Properties window, click on the *Identity* property and an Ellipsis button displays. Click on the button and the Impersonation1 Identity dialog box appears. Enter a valid user and password and click on OK. You can leave the "Domain or Machine Name" box blank.
4. Drag a *Map* control from the ESRI tab of the toolbox onto the web form and set the *ID* property to "map1", the *Height* property to "200px", the *Width* property to "200px, and the *Host* property to "local host".
5. Click on the *ServerObject* property and a drop-down list displays. Select the "gnet" server for the *ServerObject* property.
6. Expand the *CopyrightText* folder and set the *Text* property to "ASP.NET for ArcGIS Server", the *Halo* property to *True*, and the *Font/Size* property to 8.
7. Rename "WebForm1.aspx" to "*getImage.aspx*".
8. Click on the Start button or press F5 to test the application. If the "gnet" map server object displays, go to the next step. Otherwise, check the map server object by repeating steps 1 through 7.
9. Stop the application. Set the *Visible* property to *False* for the *Map* control, and save the project.

Now we are going to add some code to display a pair of latitude and longitude coordinates for a selected location on the map. In Chapter 6 we developed a geocoding application. When a match was found for an address, a pair of geocoded latitude and longitude coordinates was obtained and displayed on a map. That logic applies to what we are trying to do right now—displaying a pair of given latitude and longitude coordinates on a map. Thus, we will reuse the code from that geocoding application. We need to make a change to the server context. In the geocoding application, the server context was obtained from the geocode server object. Now we need to retrieve the server context from the "gnet" map server object. After we make that change, we can use the remaining code.

To add code to display a point on the map:

10 Switch to the code view of the project and start to *edit* `Page_Load`.

11 Add the following *Imports* statements before the class.

```
Imports ESRI.ArcGIS.Server
Imports ESRI.ArcGIS.Server.WebControls
Imports ESRI.ArcGIS.esriSystem
Imports ESRI.ArcGIS.Geometry
Imports ESRI.ArcGIS.Carto
Imports ESRI.ArcGIS.Display
```

12 Declare a pair of real variables for the latitude and longitude coordinates passed in from the Web. If both X and Y are valid, convert them to real numbers (because they are passed in as *String*).

```
Dim x, y As Double
If (Request("x") <> Nothing And Request("y") <> Nothing) Then
 x = Convert.ToDouble(Request("x"))
 y = Convert.ToDouble(Request("y"))
'
'to be completed
'
End If
```

13 Declare a `webMap` object based on the current *Map* control and retrieve the server context.

```
Dim webMap As WebMap = Map1.CreateWebMap()
Dim context As IServerContext = webMap.ServerContext
```

14 Create a point object for the map server and set the X and Y values based on the longitude and latitude.

```
Dim pt As IPoint = context.CreateObject("esriGeometry.Point")
pt.X = x
pt.Y = y
```

15 Draw the point on the *Map* control, zoom to a proper extent, and refresh the map.

```
DrawAddress(webMap, pt) 'sub from Chapter 6's geocoding app
webMap.Zoom(0.0005, pt)
webMap.Refresh()
```

16 Retrieve the map image URL and redirect it.

```
Response.Redirect(Map1.ImageUrl)
```

17 Build the application and test it.

You can test it using a web browser and enter the following URL:

http;//localhost/chapter9/mapImage/getImage.aspx?x = -77.1234&y = 38.89923

The select location is Arlington, Virginia, as shown in Figure 9–42. Now we have a web application that can stream an image for a pair of given latitude and longitude coordinates. We will proceed to the next step—building a Mobile web form application.

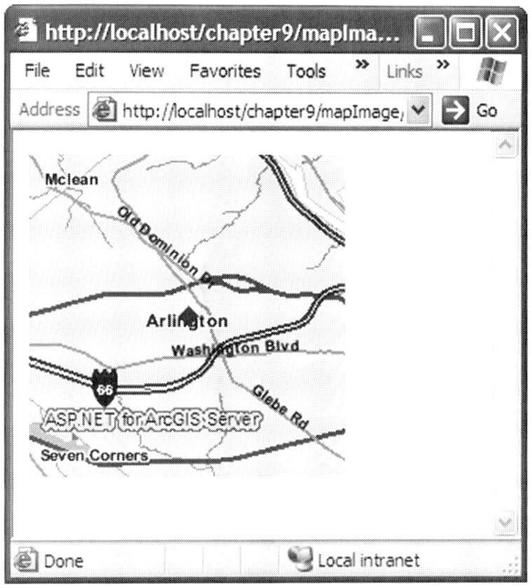

Figure 9–42. The image output of the "MapImage" example showing Arlington, Virginia.

Building Mobile Web Forms

We are going to build a mobile web application with three mobile web forms. The first mobile web form is for the entry page, the second is used to show candidate results, and the third is used to display the user-confirmed place on a map. In some cases, places share the same name but are located in different geographic areas. Thus, we need a list control to hold all candidates and let the user to confirm the selection. We cannot use the mobile *ListBox* control because it does not have any interaction capabilities. Instead, we will use the *SelectionList* control to list all candidates and at the same time allow the user to choose the desired city or place. The *Label, TextBox, Image,* and *Link* controls will be also used to develop the application. We will apply most of ASP.NET Mobile programming skills discussed so far in this chapter to the development of this project.

To build an ASP.NET Mobile application using Visual Studio .NET:

1. In Visual Studio .NET, create a new Visual Basic .NET project using the ASP.NET Mobile Web Application template and name it "PlaceFinder".
2. Drag a *Label* control from the Mobile Web Forms tab of the toolbox onto *Form1*, and set the *Text* property to "Place Name", the *Font/Bold* property to *True*, and the *Font/Name* property to *Arial*.
3. Drag a *TextBox* control from the Mobile Web Forms tab of the toolbox onto *Form1* and set the *ID* property to "txtPlace".
4. Drag a *Command* control from the Mobile Web Forms tab of the toolbox onto *Form1* and set the *Text* property to "Next" and the *ID* property to "btnNext".
5. Drag another *Form* control from the Mobile Web Forms tab of the toolbox onto the Visual Studio .NET design area beneath *Form1*. It will have *Form2* as its ID by default.
6. Drag a *Label* control from the Mobile Web Forms tab of the toolbox onto *Form2* and set the *ID* property to "lblPlaces", the *Font/Bold* property to *True*, and the *Font/Name* property to *Arial*.
7. Drag a *SelectionList* control from the Mobile Web Forms tab of the toolbox onto *Form2* and set the *SelectType* property to *ListBox* and the *Wrapping* property to *Wrap*.
8. Drag a *Command* control from the Mobile Web Forms tab of the toolbox onto *Form2* and set the *Text* property to "Submit" and the *ID* property to "btnSubmit".
9. Drag another *Form* control from the Mobile Web Forms tab of the toolbox onto the Visual Studio .NET design area beneath *Form2*. It will have *Form3* as its ID by default.
10. Drag a *Label* control from the Mobile Web Forms tab of the toolbox onto *Form3* and set the *ID* property to "lblDesiredArea", the *Font/Bold* property to *True*, and the *Font/Name* property to *Arial*.
11. Drag an *Image* control from the Mobile Web Forms tab of the toolbox onto *Form3* and set the *AlternateText* property to "Desired Area".
12. Drag a *Link* control from the Mobile Web Forms tab of the toolbox onto *Form3*.

13 In the Properties window, set the *Text* property from "Link" to "New Place" and the *SoftkeyLabel* property to "New".

14 Set the *NavigateUrl* property to "#Form1".

We have completed the GUI design for the mobile web application. We need to add some events to find the candidates and list them. Once the user makes a selection, the selected city or place needs to be displayed on the map. We cannot have any map control because that control does not work inside ASP.NET Mobile, but we can use an *Image* control. We will set its *ImageUrl* property to the URL of the "mapImage" web application.

15 Add the following *Imports* statement before the class.

```
Imports System.Web.UI.MobileControls
```

16 Before we can consume the ArcWeb *PlaceFinderSample* service, we have to add a web reference to it. The following are the steps for adding a web reference.

- Highlight Reference of the *PlaceFinder* project inside Solution Explorer.
- Right-mouse click on Reference and select Add Web Reference.
- The Add Web Reference dialog box appears (Figure 9–43). Enter *http://arcweb.esri.com/services/v2/PlaceFinderSample.wsdl* for the URL box, and click on the Go button.

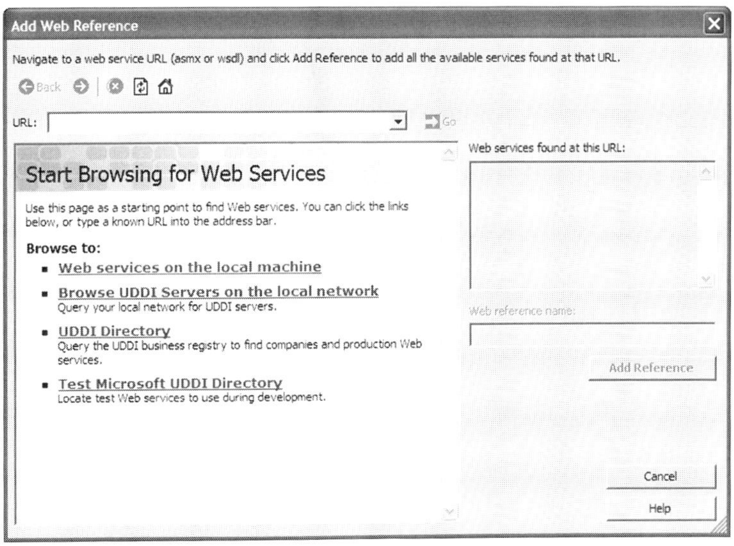

Figure 9–43. Add Web Reference dialog box of Visual Studio .NET.

- The *PlaceFinderSample* service info displays as shown in Figure 9–44. After briefly examining the service, click on Add Reference.

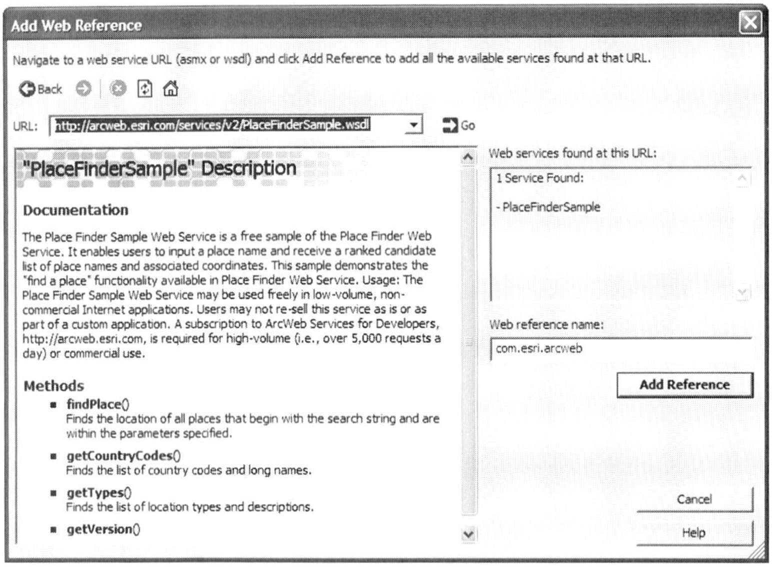

Figure 9–44. The "PlaceFinderSample" web service is shown in the Add Web Reference dialog box of Visual Studio .NET.

- The *PlaceFinderSample* web service will be added to the project as one of the Web References and will be named "com.esri.arcweb". Rename it "arcweb" for simplicity.

17 Add the following code to the `btnNext_Click` event.

```
Dim pl As arcweb.PlaceFinderSample = New arcweb.PlaceFinderSample
Dim myOpts As arcweb.PlaceFinderOptions = New arcweb.PlaceFinderOptions
myOpts.dataSource = "ESRI.Gazetteer.World"
Dim locInfo As arcweb.LocationInfo = _
 pl.findPlace(txtPlace.Text, myOpts)
Dim loc As arcweb.Location
Dim i As Integer
For i = 0 To locInfo.candidates.Length - 1
 lblPlaces.Text = locInfo.candidates.Length.ToString &" places found:"
 lblPlaces.Font.Bold = BooleanOption.True
 loc = locInfo.candidates(i)
 Dim lValue As String = "x=" & loc.point.x.ToString & _
           "&" & "y=" & loc.point.y.ToString
```

```
 Dim lText As String = loc.description1
 Dim myItem As MobileListItem = New MobileListItem(lText, lValue)
 sLstPlaces.Items.Add(myItem)
Next
txtPlace.Text = ""
ActiveForm = Form2
```

We first instantiate a local copy of *PlaceFinderSample* and retrieve the candidate name description *Description1* and the longitude and latitude coordinates, *X* and *Y*. Then we create a mobile list item with the text of the candidate name description and the value of the longitude and latitude coordinates. Before we do that, we concatenate the longitude and latitude as a URL phrase (e.g., *x=-77.1234&y=38.8923*). The concatenated URL phrase will be set as the value for a mobile list item. When the item is selected from the list, the URL phrase can be directly attached to the URL of the image-streaming application. We clear the user entry before we go to the next mobile web form.

18 Add the following code to the *btnSubmit_Click* event.

```
lblDesiredArea.Text = sLstPlaces.Selection.Text
imgDesiredArea.ImageUrl = _
 "http://10.168.2.4/chapter9/mapImage/getImage.aspx?" & _
 sLstPlaces.Selection.Value.ToString
sLstPlaces.Items.Clear()
ActiveForm = Form3
```

19 When a candidate is chosen, retrieve the text of the selected item and assign it to the label (`ID=lblDesiredArea`) on *Form3*. The image URL for the *Image* control is set to the URL of the image-streaming application plus the value of the selected item. As mentioned previously, the item value is a concatenated longitude and latitude coordinate. We clear the candidate list before we move to the next mobile web form.

✔ **NOTE:** *You cannot use "localhost" if you are planning to run this application using the Pocket PC emulator.*

20 Save the project and click on the Start button or press F5 to run the application. When the web browser appears, enter "Falls Church, VA" for the place name and then click on Next (Figure 9–45a). A list of all candidates appears in a new screen (Figure 9–45b). Select an item from the list and click on Submit. A map of Falls Church, Virginia, appears in the next screen (Figure 9–45c). You can run the same application with the Pocket PC emulator and you will see the same result as shown in Figure 9–46.

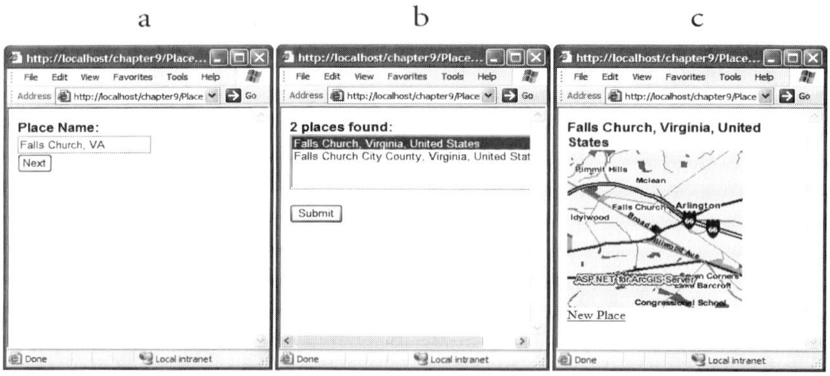

Figure 9–45. A mobile GIS application is viewed in IE. (a) The entry page, (b) the candidate list, and (c) a map of Falls Church, Virginia, with the "New Place" link.

Figure 9–46. The same mobile GIS application shown in Figure 9–45 as viewed in Pocket IE. (a) The entry page, (b) the candidate list, and (c) a map of Falls Church, Virginia, with the "New Place" link.

Summary

In this chapter, we have introduced the wireless Internet and the devices that can access the wireless Internet, including cellular phones, PDAs, and BlackBerry devices. Every device has its own web browser that can be used to browse the wireless Internet. Most of these browsers can communicate via both WAP and HTTP. WML, cHTML, and HTML are three types of markup languages that can be used to publish a web page via the wireless Web.

Due to the compactness of mobile devices, it would be very difficult to port a regular web page to a mobile web page. We need to use Mobile web forms instead of web forms. ASP.NET Mobile offers mobile web form and mobile web controls for mobile devices. You can develop ASP.NET Mobile web applications using Visual Studio .NET. You drag and drop mobile web controls onto mobile web forms in the same way as you do with regular web controls. Multiple mobile forms can be used in the same application. You can even have specific device profiles for individual mobile controls.

The mobile web applications developed in ASP.NET Mobile can be viewed with Internet Explorer or Netscape. There are wireless simulators offered by individual vendors. Three mobile simulators are introduced in this chapter: the Microsoft PDA emulator, the BlackBerry simulator, and the Openwave phone simulator. Microsoft and Openwave provide add-ins so that their simulators can work within Visual Studio .NET. BlackBerry does not have one yet. With the add-ins, you can view a wireless web application without leaving the IDE.

Mobile web controls are the LEGOs for building mobile GIS applications. You can use them in the same way as you do with regular web controls. Some of the mobile controls have been enhanced over the counterpart web controls. For example, the *Mobile TextBox* control differentiates text, number, and password modes. You do not have to write a number validation control anymore. The *Mobile Image* control is the only control that can be used to display map images on mobile devices. The ArcGIS Server web controls do not work with ASP.NET Mobile, but you can develop ASP.NET applications or web services that publish map images. Then you use the *Mobile Image* control in your wireless GIS applications to consume those images.

You need to download the simulators before you start developing any mobile GIS application to ensure that you have the latest version of the simulator. Unlike other software, we have found that simulators generally do not outpace the real devices. On the contrary, it is common for a device to be released and it is nearly impossible to find a current simulator, which makes it very difficult to program for that device. Simulators are made to facilitate programming and application development. It is also beneficial to have some insight about the target devices for your mobile GIS application. Mobile application development will benefit mobile GIS development activities.

Mobile devices are compact and the bandwidth of the wireless Internet is still very limited. Therefore, you need to make each single mobile web page as compact as possible. Because you can have multiple mobile web forms within the same application, you should spread out controls across forms. You need to create each mobile web form page as simple as possible so that it loads in a reasonable amount of time. You can still build a sophisticated wireless web application by composing simple and compact individual mobile web form pages.

APPENDIX

GIS Web Service Sources and Further Information

THERE ARE MANY GOVERNMENT AND COMMERCIAL GIS WEB SERVICE sites available on the Web. Some of them offer images and others offer stream features. Most of these sites provide web services as ArcIMS images and features and as WMS, WFS, and SOAP/WSDL formats.

GIS Web Service Sources

The sections that follow describe web services (arranged alphabetically) for GIS applications available publicly and commercially via the Internet.

ArcWeb Services (www.esri.com/software/arcwebservices)

ESRI's ArcWeb Services are subscription-based commercial web services. Both GIS data content and capabilities are delivered via SOAP/WSDL with the ArcWeb Services. With a subscription, for example, you can access the GDT US Streets map service using ASP.NET. You can literally build a web-based mapping application without owning any data. ESRI offers a 30-day evaluation trial *(http://www.esri.com/software/arcwebservices/evaluate.html)*. In addition to the GDT US Streets map service, other well-known services include Address Finder and current and forecast weather services.

EPA EnviroMapper Services (maps.epa.gov)

The Environmental Protection Agency (EPA) first brought GIS to the Web via EnviroMapper for the public in 1998, and has provided that service ever since. Their EnviroMapper services incorporate commercial and governmental GIS web services. One of the best EnviroMapper services is called Window to My Environment (WME). WME is a community-based portal that provides environmental information on air, water, and land for a given geographic area. All nonspatial information is spatially indexed, which means that when you zoom to a specific area WME only shows the environmental information directly related to that specific area. The EnviroMapper web services include the ArcIMS image service, ArcXML, GML, and generic XML on the EPA's federally regulated facilities.

In addition, EnviroMapper has a public open access functionality called OpenLink. It was made available since the first release of EnviroMapper. Given a geographical area, you can embed a dynamically generated map image from EnviroMapper in your web page or directly link to EnviroMapper for that area via an HTTP request. You can also visit the EPA geospatial one-stop node at *http://geodata.epa.gov* to check out available services in ArcIMS image, ArcIMS feature, WMS, and WFS formats.

ESRI Geography Network (www.geographynetwork.com)

The ESRI Geography Network is a free resource that offers ArcIMS image and feature services and Open GIS Consortium WMS image services. The ArcIMS image and feature services are retrieved and published using ESRI's Arc Extensible Markup Language (ArcXML). The WMS image service is retrieved as GIF or JPEG images via an HTTP `Get` call. You retrieve the WMS service information in XML by sending a *GetCapabilities* request via HTTP. You can also view ArcIMS image and feature services and WMS services using ArcGIS Desktop 9 or ArcExplorer 9. Well-known services include the USGS Shade Relief data set, FEMA Flood Plain data set, and Census 2000 TIGER/Line data.

There is a governmental version of the Geography Network available at *http://www.geocommunicator.gov.* This is a portal of land management and land records maintained by the Bureau of Land Management and the U.S. Forest Service.

Geospatial One-Stop Portal (www.geodata.gov)

The Geospatial One-Stop Portal is an intergovernmental project managed by the U.S. Department of the Interior in support of the President's Initiative for E-government. You can access geospatial information from federal, state, local, tribal, and private agencies through the comprehensive portal. For the first time, the portal has integrated metadata with geospatial data. You can search metadata first and then locate the geospatial data related to the metadata (or vice versa). Once you find the geospatial data, you can view it using the map viewer provided by the portal. You can also conduct a Google search for metadata and geospatial data records using the portal.

Microsoft TerraServer USA and MapPoint Web Service

The Microsoft TerraServer USA web site (*terraservice.net*) offers free web services of USGS digital ortho quadrangle (DOQ) aerial imagery data and digital raster graphic (DRG) topographical map data. The latest addition includes the 1-foot resolution natural color data set known as "UrbanArea data" for 133 major U.S. cities (when completed). Well-known services include aerial imagery and topographical map data in WMS.

Microsoft also offers the MapPoint *(www.microsoft.com/mappoint)* commercial web service, which provides map rendering, geocoding and reverse geocoding, proximity, and routing. The MapPoint web service is a SOAP/WSDL web service. Once you subscribe to the web service, you can simply add the MapPoint web service as a web reference to your Visual Studio .NET project and start using it.

National Map (nationalmap.gov)

The National Map service provides geospatial data for the entire United States and is a USGS-led collaborative effort of federal, state, and local partners. It offers both ArcIMS image and WMS services. You can even download the data from the web site. A unique feature of the National Map is that local geospatial data (where available) can be displayed and retrieved when you zoom to a specific area. Well-known services include National Land Cover data, LANDSAT 7 satellite imagery, the National Hydrography Dataset, and the Geographic Names Information System data set.

There is also another USGS web site called *nationalatlas.gov*. You can find a lot of data and metadata at this web site. Their map services are available from the ESRI Geography Network (see previous entry). Additional data and web services can be found at *gisdata.usgs.net*.

Web Search Engines

Geospatial technology has been integrated with web search engines including MSN Search and Google. MSN Virtual Earth is a new Microsoft geospatial web site *(virtualearth.msn.com)*, whereas Google Maps is another geospatial web site *(maps.google.com)*. They both provide geographical context search that combines maps with business locations, contact information, and driving directions. The resultant maps can be street maps, aerial or satellite images, or the hybrid of both. In addition, both search engines offer an EnviroMapper OpenLink type of functionality so that you can geospatially link your location to the search result page with a street map, aerial or satellite image, or the hybrid of both. Google Maps also offers Google Maps API at *http://www.google.com/apis/maps/*.

In addition, Google provides a new online tool called Google Earth *(earth.google.com)* that combines Google Maps and 3-D flying around the globe. But it is required to download the Google Earth client (free) to your desktop.

Further Information

The following references, mentioned in the text, are leading sources of information in their respective categories.

Ayala, D., C. Browne, V. Chopra, P. Sarang, K. Apshankar, and T. McAllister (2002). *Professional Open Source Web Services*. Birmingham, UK: Wrox Press.

Bustos, J., and K. Watson (2002). *NET Web Services Using C#*. Birmingham, UK: Wrox Press.

Esposito, D. (2003). *Programming Microsoft ASP.NET.* Redmond, Washington: Microsoft Press.

ESRI. (2004). *ArcGIS 9: ArcGIS Server Administrator and Developer Guide.* Redlands, California: Environmental Systems Research Institute.

MacDonald, M. (2003). *Microsoft Visual Basic .NET Programmer's Cookbook.* Redmond, Washington: Microsoft Press.

Index

B

C